GRASS PRODUCTIVITY

CONSERVATION CLASSICS

Nancy P. Pittman, Series Editor

With the Conservation Classics, ISLAND PRESS inaugurates a new series to again make available books that helped launch the conservation movement in America. When first published, these books offered provocative alternatives which challenged established methods and patterns of development.

Today, they offer practical solutions to contemporary challenges in such areas as multiple-use forestry, desertification and soil erosion, and sustainable agriculture. These new editions include valuable introductions from the leaders of today's conservation movement.

The inaugural titles in the series are:

BREAKING NEW GROUND
by Gifford Pinchot
Introduction by George T. Frampton, Jr.

PLOWMAN'S FOLLY
and
A SECOND LOOK
by Edward H. Faulkner
Introduction by Paul B. Sears

TREE CROPS
A Permanent Agriculture
by J. Russell Smith
Introduction by Wendell Berry

GRASS PRODUCTIVITY
by André Voisin
Introduction by Allan Savory

DESERTS ON THE MARCH
by Paul B. Sears
Introduction by Gus Speth

FROM *THE LAND*
edited by Nancy P. Pittman
Introduction by Wes Jackson

ISLAND PRESS
WASHINGTON, D.C.   COVELO, CALIFORNIA

GRASS
PRODUCTIVITY

By

ANDRÉ VOISIN

Membre de l'Académie d'Agriculture de France
Chargé d'Enseignement à l'École Nationale
Vétérinaire d'Alfort (Paris)

Translated from the French by
CATHERINE T. M. HERRIOT

Conversion of tables by
M. M. SANDILANDS, B.A. (LOND.)
(University of Nottingham)

ISLAND PRESS

Washington, D.C. □ Covelo, California

Cover design: studio grafik.

ABOUT ISLAND PRESS

Island Press, a nonprofit organization, publishes, markets, and distributes the most advanced thinking on the conservation of our natural resources—books about soil, land, water, forests, wildlife, and hazardous and toxic wastes. These books are practical tools used by public officials, business and industry leaders, natural resource managers, and concerned citizens working to solve both local and global resource problems.

Founded in 1978, Island Press reorganized in 1984 to meet the increasing demand for substantive books on all resource-related issues. Island Press publishes and distributes under its own imprint and offers these services to other nonprofit organizations.

Funding to support Island Press is provided by The Mary Reynolds Babcock Foundation, The Ford Foundation, The George Gund Foundation, The William and Flora Hewlett Foundation, The Joyce Foundation, The J. M. Kaplan Fund, The John D. and Catherine T. MacArthur Foundation, The Andrew W. Mellon Foundation, Northwest Area Foundation, The Jessie Smith Noyes Foundation, The J. N. Pew, Jr. Charitable Trust, The Rockefeller Brothers Fund, and The Tides Foundation.

For additional information about Island Press publishing services and a catalog of current and forthcoming titles, contact Island Press, P.O. Box 7, Covelo, California 95428.

LIBRARY OF CONGRESS
Library of Congress Cataloging-in-Publication Data

Voisin, André, 1903–1964.

[Productivité de l'herbe. English]
Grass productivity / by André Voisin : translated from the French by Catherine T. M. Herriot.
(Conservation classics)
Translation of: Productivité de l'herbe.
Reprint. Originally published: New York: Philosophical Library, 1959.
Bibliography: p.
Includes index.
ISBN 0–933280–63–7   ISBN 0–933280–64–5 (pbk.)
1. Pastures—Management. 2. Grazing—Management. 3. Dairy cattle—Feeding and feeds.
4. Grasses. 5. Agricultural productivity. I. Title. II. Series.
SB199.V813   1988
636.2—dc19   88–18748
                                                                            CIP

Manufactured in the United States of America
10 9 8 7 6

FOREWORD

by M. McG. COOPER

B.Agr.Sc.(N.Z.), B.Litt.(Oxon), Dip.Rur.Econ.(Oxon), F.R.S.E.
Dean of Agriculture, University of Durham

ABOUT ten years ago I upset a lot of people by stating that British farming was still only at half-cock. This criticism was not applied to tillage farming which generally is of a very high standard in this country, but to the quality of our grassland farming which was, and still is, far from satisfactory. It has been estimated that the effective production of starch equivalent from our pastures is less than fifteen hundredweights per acre, and this is appreciably less than the yields we obtain from the normal run of cereal crops. If our grasslands were properly managed the offtake of nutrients would be well in excess of that from cereal crops and would compare favourably with the yield from such expensively produced crops as swedes, mangolds and fodder beet. But our grasslands are not well managed. Many, and especially the permanent pastures, are undernourished and the majority of them suffer from one or other of those twin evils of grassland mismanagement, under- or over-grazing. Too often farmers accept pasture for what it is and not for what it could be if they put into its management the same level of technical knowledge as they apply to their tillage crops.

You will understand the force of my arguments when you have read this book, for André Voisin has a story to tell which has a foundation of great achievement. He once belonged to that school of thoughtlessness which pays little or no attention to vital principles of pasture management. In those days, when he utilised his grassland by continuous grazing, he produced 1800–2000 lb. of starch equivalent per acre. Even by current standards this is a fairly respectable level of production, but it was well below the potential of his swards, for when he applied what he aptly calls *rational management* their yields were trebled and starch equivalent production exceeded that of any arable crop grown under similar conditions.

I am not going to steal M. Voisin's thunder and explain what is meant by rational management, for the pages that follow will give you this information. But I must comment on how refreshing it is to find someone who combines the viewpoint and understanding of the plant physiologist and of the animal physiologist in his endeavours to realise the potential of his grassland. Too often there has been a dichotomy of interests where the plant and the animal are regarded as isolates rather than as integrates. For instance, we have had plant breeders producing new varieties of herbage plants without consulting the animal, and ending up with something the animal does not like. S. 143

cocksfoot is a good example of what I mean. It has excellent agronomic characteristics, but it is really not very good for milk or meat production. On the other hand, we have graziers who do not recognise that grass and clover leaves have the function of feeding their parent plants as well as the stock that graze them. M. Voisin's system of grazing recognises this duality of function and the importance of understanding the interaction of plant and animal in the attainment of maximum profitability from grassland.

I must warn you before you start reading that you may not always agree with André Voisin. For my part I have a number of issues to raise with him when next we meet, though I have a sneaking suspicion that sometimes he is trailing his coat. He is like that in the flesh, for he is one of those stimulating people who have the gift of making people think. Certainly he has made me very thoughtful through the level of performance of his grassland, for I know of no farmer who has achieved an output of over 6000 lb. of starch equivalent per acre from his pastures. The careful reasoning of a man who can obtain an output of this magnitude cannot but make interesting and profitable reading.

One feature that interests me particularly is M. Voisin's regard for permanent grass. He does not obtain this high level of production from young leys, for his youngest pastures were established just after the war and in his estimation they have not yet reached their productive prime. We in Britain have had an emphasis on ley farming not just for the sake of increasing tillage crops but also with the object of increasing pasture productivity. Though I do not question the value of ley farming under appropriate conditions, where cash cropping is the most profitable way of using land, M. Voisin's experience increases my doubts of the wisdom of ploughing pasture merely to re-establish it, especially when there is so much that can be done by surface improvement. We have twelve million acres of permanent grass in this country which have resisted all attempts to get them ploughed, and to my mind the greatest single problem in our farming is to make this permanent grass more productive. M. Voisin has given us some valuable guidance to this end.

M. McG. COOPER

University of Durham,
March, 1959

CONTENTS

Part Six: COMMON ERRORS IN SUPPOSEDLY RATIONAL SYSTEMS OF GRAZING

Part Seven: TETHERING AND RATIONED GRAZING, SPECIAL SYSTEMS OF RATIONAL GRAZING

Part Eight: DIVISION OF PASTURES

Part Nine: RATIONAL GRAZING TRANSFORMS THE FLORA

Part Ten: SALIENT POINTS OF RATIONAL GRAZING

Part Eleven: WEALTH OF OUR PASTURES

Part Twelve: DIFFICULTIES OF YESTERDAY AND TO-MORROW

Conclusions: GREEN PASTURES

LIST OF TABLES

LIST OF FIGURES

LIST OF PHOTOS

NOTE
Conversion from Metric to British Units

ALL figures are shown both in the metric system of the original French and in the British system of weights and measures since some countries use one system and some the other.

Throughout the text, metric figures, mainly kilograms/hectare (*kg./ha.*), have been converted to British figures, pounds/acre (lb./acre), the metric figures being shown *italicised* and in brackets. The British equivalents do not always correspond exactly to the metric figures of the French edition. As it is often a matter of approximate or indicative figures it was decided to convert into round numbers (in most cases) of the British system in order to achieve clarity and simplification.

It should be noted that, where decimals appear, the decimal point of the British system has been used, even where the units are metric.

Attention is also drawn to the fact that " STARCH EQUIVALENT " (S.E.) in the British system is expressed in pounds avoirdupois, but in kilograms in the metric system.

INTRODUCTION TO THE ISLAND PRESS EDITION

by ALLAN SAVORY

Executive Director
Center for Holistic Resource Management, Albuquerque, New Mexico

I first discovered *Grass Productivity* some 25 years ago. I was a young biologist at the time, struggling to save the wildlife in Africa that was rapidly disappearing as the land deteriorated. Although I had never heard of André Voisin, the book's title suggested that it might hold some answers and I quickly bought it. After flipping through its pages, however, I put it away to gather dust on my bookshelves, greatly disappointed.

In my ignorance, I could see no connection between Voisin's work with cattle on lush green pastures in France and my work with buffalo, elephant, and many other big game species in the vastness of Africa's often arid veld. Apart from that, as a conservationist I was antagonistic to cows, which were overgrazing my beloved Africa to death.

In the years that followed, my field observations led me to believe that there was a connection between the hooves of herding game and healthy grasslands. Although the teaching of the day was that animal hooves damaged land, I found example after example where the opposite was clearly the case. In seeking to resolve this paradox, I stumbled on the notion that *time* might be a factor, but I could not yet say how.

At this point some Rhodesian cattle ranchers who loved their land and were distressed at its continued decline turned to me for help. Although my antagonism to cattle was well known, what I was saying publicly at the time made sense to them. They had religiously practiced the advice provided by our government research stations and extension service, but their once magnificent ranches had continued to deteriorate.

In tackling this new challenge to manage cattle and wildlife together while improving the land, I researched all of the literature I could find. In the process I dusted off my copy of *Grass Productivity* and was astounded to find that Voisin had already solved the riddle of *time*. He had proven that overgrazing had little relationship to the number of animals but rather to the *time* plants were exposed to the animals, the time of exposure being determined by the growth rate of the plants. If animals remained in any one place for too long, or returned to it too soon, they overgrazed certain plants. Suddenly I could see how trampling also could be either good or bad. Time determined that too. The disturbance needed for the health of the soil became an evil if prolonged too much or repeated too soon.

Voisin was a biochemist by training, but a farmer by inclination. Although extremely widely-read as a scientist, his greatest learnings appear

xv

to have come from his field observations, particularly during the long hours spent watching cows graze on his farm in Normandy. He realized more than most that the unknowns in science are far greater than the knowns and that simple observation of the cow at grass could teach us more about ecological relationships than the most sophisticated research yet developed.

Initially, in the belief that Voisin had solved our problems, we applied his "rational grazing" to a number of farms and ranches in Africa. While the plants and livestock flourished on the planted, fertilized, and heavily watered pastures, we ran into problems on the rangelands. Here the growing conditions were erratic and we had a tremendous variety of grasses, forbs, brush and trees — all growing at different rates. We also had wildlife running with the stock, seriously affecting our time calculations.

We would not unravel the whole mystery for a long time, but we knew that Voisin had set us on the right path. In setting out to improve pasture productivity, he had made brief references to arid rangeland problems, believing, correctly as it turned out, that the same thinking would apply. But for rangelands we had to develop a sophisticated planning process to deal with the many variables "rational grazing" did not address. Voisin's discoveries and interpretations of research data provided a valued anchor in our quest and the simple techniques he devised for measuring forage consumption and quality are an integral part of the planning process we now use.

In discovering that *time* was key in grazing management, Voisin contributed more to science than he realized, for that discovery has helped us understand the causes of desertification — one of the greatest problems mankind now faces — and given us vital clues about how to combat it.

Like so many people who have made major breakthroughs in thinking, Voisin did not receive the recognition he deserved. In 1978 I was on a lecture tour of eight American universities and to my amazement found that not a single range scientist had read *Grass Productivity*, even though it had been published in five languages, including English. Although rotational grazing was applied in many countries, including America, often under the guise of "rational grazing," the instigators ignored the importance of the time factor. Voisin pointed this out to them time and again, but tragically, it is as true today as it was 20 years ago.

Voisin's work has been kept alive in Brazil, where *Grass Productivity* survives in a Portuguese edition. In the early 1980s, Bill Murphy, an American agronomist working in Brazil, read Voisin's work and brought it back with him to Vermont. With the publication of Murphy's book *Greener Pastures on Your Side of the Fence: Better Grazing with Voisin Grazing Management*, there has been a resurgence of interest in "rational grazing" in the New England states. However, it is in New Zealand that people took Voisin's work most to heart, and that fact underlies New Zealand's unequaled success in pasture management today.

In the field of pure pasture management I have never been able to better

Voisin's work and have encouraged those managing pastures to read the original, which now is once more readily available.

While *Grass Productivity* is still the best basic pasture management book available, I must caution those involved in handling livestock on rangeland watersheds and catchments that much ground has been covered in developing Voisin's and others' ideas since *Grass Productivity* was first published. While the new knowledge is vital to the better management of such lands it does not detract in any way from this original work.

Water is the Achilles' heel of modern industrial and post-industrial civilizations, and its quantity and quality are determined by the state of the land on which it falls. These water "catchments" are to a large degree grasslands. In managing them as Voisin would have us, we ensure our own survival. When the world awakens to that fact, the debt of gratitude owed André Voisin by billions of people in all walks of life can be paid.

If you are engaged in any way with the management of livestock, I sincerely hope that this book will help you as much as it has helped me and the many people whom I have in turn been able to help. It will always be a deep regret of mine that I was unable to meet Voisin before his death, as there are few scientists whose work I have admired for as many years. I am grateful indeed to be able to express my admiration publicly for so brilliant a farmer and so humble a scientist.

Allan Savory
Executive Director
Center for Holistic Resource Management
Albuquerque, New Mexico

January 1988

Introduction

THE MEETING OF COW AND GRASS

What is grazing?

SIMPLE questions often help us to understand problems better; and I think it indispensable, at the beginning of this work, to ask a question which appears simple in the extreme:

"What is grazing?"

The answer is generally as follows:

"Causing grass to be eaten by an animal."

That is correct! But here is another answer which, to my mind, is more realistic:

"Causing the grass and the animal to meet."

Since this book is almost exclusively concerned with grazing by cattle, I propose the following definition to the reader, requesting him to allow it to become well impressed upon his mind:

Grazing is the meeting of cow and grass.

The study of pasture plants

Pasture studies have been particularly concerned with the plants of which the pastures are composed. These plants have been selected from the botanical point of view to produce a higher yield, better resistance to pests and to diseases. The influence on these factors of fertilisers, methods of soil cultivation, time of sowing, etc., has been studied.

In experimental and research centres throughout the world there are millions of little plots sown for the botanical study of grasses and legumes.

True, it has not been forgotten that these grasses provide feed for cattle, and a multitude of chemical analyses have been carried out on them. But unfortunately these analyses in actual fact provide us with only a very approximate idea of the actual value of the plant to the animal. Will the chemical analysis of a plant give us the slightest idea of its taste? A plant found to be admirable in the laboratory will not always be eaten with the same admiration by the cow.

Chemical analysis has not yet been able to reveal the elements which give rise to bloat. Now there have been, and still are, catastrophes with a certain variety of white clover which gave better yields than our old, ordinary, white

1

clover but showed a great tendency to cause bloat. One of my neighbours sowed a pasture with a mixture containing the white clover variety in question. When he returned from the market one evening he found a dozen bloated, dead cattle in his field!

Recently, in a tour of the *Départements Finistère* and *Côtes du Nord* (*Brittany*) I was able to confirm the ravages that can be caused by bloat on pastures reseeded with new varieties of white clover.

We must therefore never forget the animal when we are studying the grass.

The cow influences the pasture

Moreover, if the grass is there to be eaten by the cow we must remember that the cow has a profound effect on the pasture that it eats. I will only call to mind the flora of weeds, so different on mown meadow and grazed sward, a well-known example which suffices to illustrate the enormous influence of the cow on the pasture.

But here is another very characteristic example:

At an American experimental station they were studying different types of white clover from the botanical point of view on small plots. The young professor accompanying us said: "Strain A gives higher yields than strain B, but it is of no interest, because at the beginning of summer it is attacked and destroyed by the Potato Leafhopper (Empoasca Fabæ). Variety B, on the other hand, is not attacked."

We went on to another American station which was likewise experimenting with the two strains A and B of white clover. This time, however, it was not a case of botanical experiments on small plots, but an actual grazing trial with cows. The professor explained to us that strain B was non-existent by comparison with strain A, which gave vastly superior milk yields. "But," we said, "have you no potato leafhopper in this region?"

"We are infested with it," was the reply. And the professor, guessing our thoughts, added with a smile: "Potato leafhopper attacks Variety A *when it is* NOT *grazed*. But reproduction of the leafhopper in a grazed sward is hindered by the hoof and tooth of the grazing animal."

One can therefore understand the errors which might arise from a botanical study in itself, forgetting the relations between plant and animal.

Feeding the cow in the stall

All our studies and tables on feeding of cows are concerned with the cow in the stall. When one wanted to investigate the feeding value of some green fodder one was content with bringing it to the cow's feeding-trough after it had been cut.

Take any treatise on animal nutrition or any work on grass and see how many pages are devoted to the behaviour of the animal as it grazes.

D. E. Tribe (111) considers H. I. Moore's *Grassland Husbandry* (76) as one of the best works on grasses. But he adds: "Of the 126 pages in this work, there are hardly 6 concerned with what may be called *animal aspects of grass.*"

And when D. E. Tribe himself studies the behaviour of the animal at grass he devotes more than half of his article to the tastes observed in rats in the laboratory. The remainder deals with sheep, horses, etc., mainly, moreover, in order to note what one does not know rather than what one knows.

Botanists and animal experts should get together

We can therefore say that the botanists have studied the plants in themselves while the animal experts have studied the cow in this closed receptacle known as the stall or the respiration calorimeter.

There is grass in itself, and the cow in itself; **but above all, there is the cow that grazes the grass, and for eight months in the year that is just what it does do.**

It is essential therefore that botanists and animal experts meet and fill in the gap separating their two sciences.

The demands of the grass and of the cow

It is a meeting of this nature, or at least the first steps towards such an end, that I want to attempt in this book.

We will not study the grass and the cow separately. We will always consider them simultaneously and together, in such a manner as best to satisfy the demands of each.

When we think of the cow, we will not forget the demands of the grass. When we examine the grass, we will always bear in mind the demands of the cow.

It is by satisfying as far as possible the demands of both parties that we will arrive at a rational grazing, which will provide us with maximum productivity on the part of the grass while at the same time allowing the cow to give optimum performance.

PART ONE
THE GRASS

Chapter 1

WHAT IS A HERBAGE PLANT?

Cutting and successive re-growth

A pasture plant must be capable of growing *again* after it has been cut either by the tooth of the animal or by the blade of the mower.

When this plant is cut it retains very little, and sometimes indeed hardly any of the green aerial part capable, by photosynthesis, of creating the elements necessary for the formation of new plant cells: that is, for the initial re-growth of the plant.

It is therefore indispensable that the plant, at the moment when it is cut, should have, in its roots or at the foot of its stalks, sufficient reserves to allow the formation of a certain green portion which, by photosynthesis, will then permit the normal growth of the plant.

Every new growth, that is to say every re-growth of our herbage plants, takes place at the expense of the organic substances elaborated previously (before cutting) in excess of what was necessary for the maintenance and growth of the plant. These substances have been stocked in the roots and lower aerial portions. If one cuts the plant before the roots and the part not cut have stored up sufficient reserves, re-growth will be difficult and may even not take place at all.

There is a period in which wheat can be grazed without being destroyed

This evolution of reserves in our herbage and forage plants is a question which, unfortunately, has been very insufficiently studied by plant physiologists until now. We know very well that there is a moment in the course of a plant's development when the reserves in the roots are at their maximum and when, in consequence, the conditions for re-growth are optimum. Take our old graminaceous friend, wheat. Grazing wheat as it emerges from the soil destroys it. At harvest-time, when we cut the wheat with its grain formed and ripe, the stubbles of our fields do not produce re-growth. On the other hand, *between these two extremes* there is a period in which it is possible to graze the wheat and yet allow it to grow again and thus produce a reasonable harvest.

7

Definition of a herbage plant

We will therefore answer the question asked at the beginning of this chapter by stating that: **A herbage plant is a plant which is capable, several times in the course of a year, of accumulating in its roots (and at the foot of its stalks) sufficient reserves to allow it to grow again after every cut.**

Let us look quickly at a few points concerning the evolution and nature of these reserve substances which are indispensable to the re-growth of the grass, after cutting with the blade of the mower or shearing with the teeth of the animal.

Evolution of quantities of reserves in the plant

As Professor Klapp tells us (70, p. 350), the production of green matter by our herbage plants is not a continuous process throughout the period of vegetation; but accumulation and expenditure of substance alternate with each other. At the end of the summer and in the autumn the accumulation of reserve substances (as a result of the production of assimilation products by the leaves) permits re-growth in the ensuing spring, followed eventually by development up to flowering and the formation of seeds. An analogous phenomenon takes place after every cut, if the latter does not kill the plant.

Different plants differ enormously in the time and also in the speed of this assimilation and in the storing up in reserve of the substances assimilated.

Alternating rhythm of accumulation and exhaustion of the reserves

The Polish research worker Osieczanski (82, p. 65) has very clearly summarised this alternating rhythm of exhaustion and accumulation of reserves:

"Part of the products of photosynthesis is immediately utilised for the construction of the cells of those organs of the plant situated above and below the soil. Another part of these products of photosynthesis is used to satisfy the physiological requirements (respiration, metabolism). The remainder of these products is put into reserve for a time when there is no synthesis, or at least when the products of this synthesis are completely utilised to satisfy the needs of the plant organs. These reserves allow the plant to survive critical periods, such as, for example, the winter period, during which the balance of the phenomena of assimilation is negative.

"The reserve substances of grass are utilised for respiration, formation of stalks, leaves, seed, roots etc. and in particular for the respiratory processes at low temperatures (below 32° F. [$0°$ C.]) and at high temperatures (above 85°–95° F. [$30°–35°$ C.]); temperatures at which respiration uses up more energy than is supplied by the processes of assimilation. These reserves will also be utilised during periods when the plant is growing strongly as, for example, during tillering or the formation of seed. **This will be the case in** particular after cutting or grazing when the grass will have to re-create green surfaces supplying the products of assimilation. . . ."

Nature of the reserve substances

Under identical conditions as regards the quantities, or proportion, of reserve substances remaining after cutting, the re-growth of the same plant can vary greatly, concomitant with such other factors as day-length, soil moisture, amount of assimilable fertiliser elements present in the soil, rainfall, etc.

It would therefore be particularly desirable if we had better knowledge of the way in which the reserves are accumulated in our herbage plants: this would help us to use them more profitably.

At present, however, no firm conclusion has been reached even concerning the nature of the reserve substances. Sullivan and Sprague (102) have published a detailed review of the different theories put forward regarding these reserves. We refer the reader to these authors for this bibliographical review, and also for their study of the reserve carbohydrates of rye-grass (*vide* also Weinmann, 140).

Can grass build up a reserve of growth hormones?

In general, one considers as reserve substances all the fats and the nitrogen-free extract. As mentioned above, it is essential that the plant contains in its roots and the part not cut the maximum possible of these reserve substances. But these indispensable substances are probably not sufficient. Our herbage plants have also a stock of other substances which allow them to grow away again after being cut. Here, it is probably one or more hormones which allow the growth of the plant to be set in motion once more. R. O. Whyte (144), who is a plant physiologist, reminds us of this in well-chosen words:

> "The physiologist studying herbage plants cannot fail to wonder at the remarkably small effect which repeated removal of leaves and damage to tender growing points of the plant has on the physiological behaviour of the plant and on its development.
> "It does not appear out of place therefore to put forward a few hypotheses: is it not possible, when a plant goes to seed every year or every two years, that all (or almost all) the growth (or re-growth) hormone is removed in the seed? There would then be no more hormone left to revive the meristematic activity at the base and lead to formation of new tillers. Might it not be that in a herbage plant, only a proportion of the hormonal content is removed with the part cut off and enough remains at the base to meet the needs of the new tiller growth? The higher the concentration of hormone remaining, the more active the new tillering of the grass . . ." (*vide* also Söding, 98).

If these hypotheses are correct, it would obviously be of interest to know the fluctuations which take place in the reserves of this re-growth hormone in the parts not cut, and how we could augment these reserves by our different methods of cultivation (fertilisers, for example). Unfortunately we have no answers as yet to these important questions.

Comparison of the quantities of reserve substances and of their distribution in two gramineae

As McIntyre (73a) has reminded us, the recuperation of plants from defoliation is dependent on:

(*a*) the extent to which the photosynthetic surface has been eliminated;
(*b*) the extent of stored material which is accessible to the animal;
(*c*) the rapidity with which the plant can replace its reserves.

TABLE 1

Comparison of reserve materials in roots and leaf bases of two grasses according to the number of cuts taken per year

Number of cuts per year	Grass	Weight of reserve materials in roots and leaf base		Relative variation in reserve materials in roots and leaf base
		lb.	[grams]	
4 {	Cocksfoot	0·992	*[450]*	29
	Smooth-stalked meadow grass	1·109	*[503]*	40
3 {	Cocksfoot	1·367	*[620]*	39
	Smooth-stalked meadow grass	1·530	*[694]*	55
2 {	Cocksfoot	2·652	*[1203]*	76
	Smooth-stalked meadow grass	2·211	*[1003]*	79
1 {	Cocksfoot	3·481	*[1579]*	100
	Smooth-stalked meadow grass	2·778	*[1260]*	100

N.B. 1. Reserve materials represent the total of ether extract and nitrogen-free extractives.
2. The weights shown are those of 100 plants at the time of cutting (average).
(Calculated by the author from Klapp, 64)

Professor Klapp has studied the evolution of reserve substances in the course of the development of cocksfoot [1] and smooth-stalked meadow grass [2] with different numbers of cuts per annum. In addition, he determined the distribution of the reserve substances between the cut, green, aerial part and the roots and the base of the green part left by the cut.

The short summary of the data (Table 1) is taken from Klapp's table and clearly shows the difference in the behaviour of cocksfoot and smooth-stalked meadow grass in the face of frequent cropping close to the ground.

---

[1] American term: Orchard grass.
[2] American term: Kentucky blue grass.

We see that where three and four annual cuts are made, cocksfoot retains only 29 and 39% of the reserves in its possession when it is cut only once per annum. This proportion is reduced to only 40 and 55% respectively in the case of smooth-stalked meadow grass. It is understandable, therefore, that repeated increase in the frequency of cutting will weaken cocksfoot much more than smooth-stalked meadow grass. This corroborates Weinmann's judicious observation (140): "The effects of repeated defoliation are *cumulative*, and progressively deplete the reserves more and more . . ." (*vide* p. 24).

Chapter 2

THE CURVE OF GRASS GROWTH

Kinetics of plant growth

WHEN a plant emerges from its seed, it grows slowly to begin with and then accelerates its growth until it reaches the flowering stage, when the growth slows down again.

In their admirable work *Principles of Plant Physiology* (9, pp. 322–325)

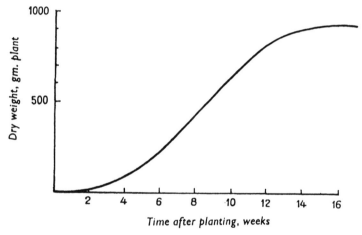

FIG. 1. Typical S-shaped growth curve of maize (corn).
From Bonner and Galston (9).

Bonner and Galston have provided us with an explanation of the kinetics of growth:

"Suppose we follow the growth of an intact plant through its life cycle by means of measurements of height or of total dry weight. We shall find, in general, that the dry weight of the seedling plant first tends to decrease slightly following germination, as the reserves of the seed are depleted.

"This is followed as photosynthesis becomes established in the new leaves, by a rapidly increasing growth rate, which finally becomes constant at some relatively high level (Fig. 1). The growth rate during this period is often

12

remarkably rapid. The bamboo stem may grow as much as 24 in. [*60* cm.] per day, and staminal filaments of certain grasses have been observed to elongate as much as 0·11 in. [*3* mm.] per minute over short periods of time. Growth continues at this rapid rate until the approach of maturity at which time its rate slowly declines and approaches zero. The dry weight of the plant may even decrease in the final stages of senescence.

"The 'S', or sigmoid, shape of the curve is typical of the growth of the plant as a whole, as well as of the growth of living organisms generally.

"The sigmoid growth curve of an entire organism is the resultant of the individual sigmoid curves of each of its component organs. For example, during the later phases of the growth of a plant, increase in dry weight may be largely manifested in the developing seeds and fruit, the vegetative organs contributing but little.

"In all of these instances we may distinguish three stages which together make up the so-called 'grand period of growth':

"1. An early period of slow growth.
"2. A central period of rapid growth.
"3. A final period of slow growth."

Let us now see how this universal, biological curve applies in the case of a grass growing up again after defoliation.

The curve of re-growth in grass

The curve of re-growth in grass is also sigmoid in shape, that is S-shaped, the characteristic and universal form of growth in all living organisms, as we have just seen (Fig. 1).

At first the grass, having only its reserves and an infinitesimal number of chlorophyll workshops at its disposal, grows slowly and with difficulty. Then it succeeds in creating a sufficiency of green cells, the photosynthesis of which will furnish building material for the rapid creation of other green cells, that is, of a large mass of grass per unit time. This is the *blaze of the grass's growth*. Towards the end of this period of rapid growth the grass renews its reserves and then slows down its synthesis of green cells in order to devote all its efforts to the production of flowers and seed.

This is what is shown in Fig. 2, where we have reproduced the typical sigmoid curve showing, in this instance, the quantity (in lb. or kg.) of green grass present per acre (or hectare) as influenced by the number of days which have passed since the grass was grazed, that is, since it was sheared with the animal's teeth.

In practice, the curve is much less regular. The increase in weight of the dry matter presents a serrated curve; but, on an average, this S-shaped curve is a good representation of the actual re-growth of the grass.

We have assumed two seasons in which the growth is different. For the sake of simplicity, the growth of grass in August–September is taken as being *twice* as slow as in May–June.

This relationship, of course, is theoretical: it varies with the region and the prevailing climatic conditions in any season. Nevertheless, one may say that it is more or less the average relationship in many regions of North-West Europe where grass growth is almost half as rapid in August as in May: this means that with well-conducted rational grazing the rest period for the

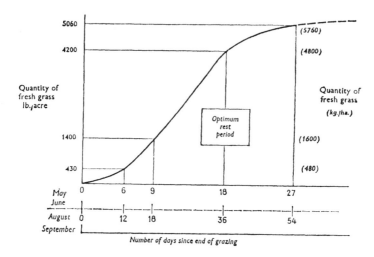

FIG. 2. Daily growth and total production of fresh grass, lb./acre (kg./ha.), at two different seasons.

grass between two successive rotations will have to be twice as long in August as in May (*vide* Voisin, 128 and 129).

The optimum times in this connection (subject to annual climatic variations) are, on the average, 18 days in May and 36 days in August (*vide* Voisin, 134).

We assume that during these optimum rest periods there has been a regrowth of 4200 lb. harvestable grass per acre [*4800 kg./ha.*].

We see then that:

(1) With a rest period of half the optimum time, production is reduced to a third 1400 lb./acre [*1600* kg./ha.] against 4200 lb./acre [*4800* kg./ha.].

(2) With a rest period equal to one third of the optimum, production is reduced to a tenth 430 lb./acre [*480* kg./ha.] against 4200 lb./acre [*4800* kg./ha.].

(3) With a rest period half as long again as the optimum, production is only increased by 20% 5060 lb./acre [*5760* kg./ha.] against 4200 lb./acre [*4800* kg./ha.].

Productivity curve of grass

What I shall arbitrarily describe as the "productivity of grass" is the daily quantity of grass re-growth per acre (or hectare), underlining the fact that this is a restricted conception of productivity.

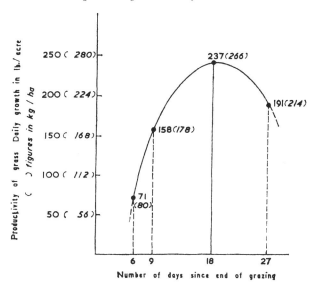

FIG. 3. Productivity curve in May and June.

We assume that during these optimum rest periods there has been a re-growth of 4200 lb. harvestable grass per acre [*4800* kg./ha.].

Two productivity curves will make quite clear to us the necessity for observing optimum rest periods so that the grass may be allowed to do its work with maximum productivity.

In Figs. 3 and 4 we have shown the lb./acre [kg./ha.] of grass produced daily as a function of the number of days of re-growth at the two periods of the year under consideration here, namely, May–June and August–September.

In actual fact, it is the same curve in both graphs but with different scales.

In each case a net maximum manifests itself corresponding with the maximum productivity of the grass, viz., 18 days in May–June (Fig. 3) and 36 days in August–September (Fig. 4).

These curves make it even more clear to us how low is the productivity of grass during short rest periods corresponding more or less to those pertaining between two bites where cattle grazing is continuous.

It is again emphasised that the rest period of 18 days, which corresponds with maximum productivity in May–June, corresponds only with low productivity in August–September. To obtain maximum productivity during

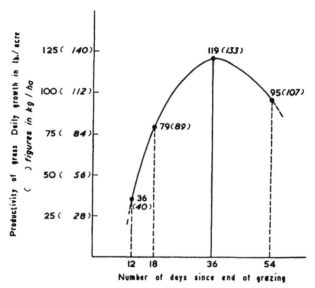

FIG. 4. Productivity curve in August–September.

this latter period one must double the resting time to 36 days. This demonstrates the necessity of **varying the rest period of grass according to the season** in order to obtain maximum productivity (subject to satisfying the demands of the cow).

As we shall see later in more detail, these prolonged rest periods make for a considerable increase in annual production of grass and of the nutritive elements per acre (per hectare).

When we choose an optimum rest period it will be necessary to go beyond, rather than to stay within, this optimum period for many grasses.

In actual fact, when, in May–June, we prolong the rest period by 9 days beyond the optimum, productivity falls to 191 lb./acre [214 kg./ha.], while if we reduce the optimum period by these same 9 days the productivity falls to 159 lb./acre [178 kg./ha.].

Of course, there is reason to take the nutritive value of the grass into account, and the next chapter will provide us with some data on this point.

The grass must be sheared at the appropriate time

Subject to the requirements of the cow, the grass must be sheared by the animal's teeth after the rest period (or number of days of re-growth) corresponding with the maximum point of the productivity curves in Figs. 3 and 4, that is, 18 days in May–June and 36 days in August–September.

In plain language: **there is a time when the grass is fit for being** *sheared* **with the teeth of the** *animal*, **just as there is a time when the grass is fit for** *cutting* **by the blade of the** *mower*.

A British observation on grass growth

In the course of observations (under controlled conditions which are not described to us), the Irish worker Linehan (73) noted that during the first 21 days of the re-growth period, there was a total grass dry matter production of 582 lb./acre [*630* kg./ha.], making a daily growth of 28 lb./acre [*30* kg./ha.] dry matter. Then in the 10 days following (from the 21st to the 30th day) there was a growth of 732 lb./acre [*820* kg./ha.] dry matter, making 73 lb./acre [*82* kg./ha.] per day.

In other words, the daily growth of grass during these 10 days was almost *three* times as strong as during the preceding 20 days.

We will suppose that the grass in question had a dry-matter content of 20–22%: this corresponds after 20 days to a green mass of 2587 lb./acre [*2900* kg./ha.] and after 30 days to 6227 lb./acre [*6900* kg./ha.].

Inserted in Fig. 5 are the two figures for green matter corresponding to 20 and 30 days of growth respectively. Then we have traced the S-curve which probably fits in with these two points. In addition, we have indicated a hypothetical point corresponding to 10 days of growth which lies on the level part at the base of the sigmoid. The figures we now have are more or less analogous to those of the sigmoid in Fig. 2.

With this curve in the case where (in August–September) one tripled the growth (or rest) period of the grass, increasing it from 12 to 36 days, one multiplied by 10 the quantity of green matter produced, thus making it from 430 to 4200 lb./acre [*480* to *4800* kg./ha.].

With the curve deduced from Linehan's experiences, in tripling the growth period (from 10 to 30 days), we also multiply the green matter by 10, making it 6227 lb./acre [*6980* kg./ha.] instead of 623 lb./acre [*698* kg./ha.].

In Linehan's observations the daily growth of green grass (lb./acre) [kg/ha] amounted to:

$$\frac{2587}{20} = 129 \text{ with a rest period of 20 days} \left[\frac{2900}{20} = 145\right]$$

$$\frac{6227}{30} = 207 \text{ with a rest period of 30 days} \left[\frac{6980}{30} = 232\right].$$

We find, therefore, that the growth figures lie between those noted on the hypothetical sigmoids for May–June and August–September.

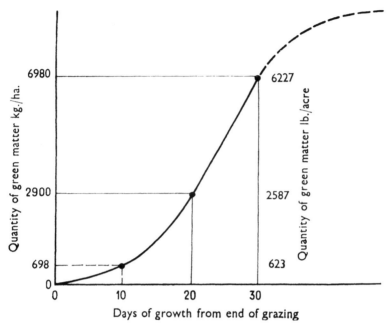

FIG. 5. Growth of grass according to Linehan's experiments.

Need for short periods of occupation [1]

The curves in Figs. 3 and 4 make us see the danger of too long periods of occupation in any "rational" grazing system, be it on permanent or temporary pasture. In fact, after 6 days in May–June and 12 days in August–September, a blade of grass will probably have re-grown sufficiently to be seized upon and cropped once more by the animal. In spite of the dividing up of pastures, or the use of the electric fence, we will then be working with productivities of 71 lb./acre [80 kg./ha.] per day (instead of the optimum 240 lb./acre [266 kg./ha.]) in May–June, and 36 lb./acre [40 kg./ha.] per day (instead of the optimum 119 lb./acre [133 kg./ha.]) in August–September.

Whatever our system of "rational" grazing, whether we call it rotational, rationed, etc., **the grass production will be low if the period of occupation is sufficiently long to allow the animal, within one rotation, to shear for a second time grass sheared during the initial days of occupation on the plot in question.**

The time limit obviously varies with the type of animal; a sheep being able to grasp and shear short grass which the cow is incapable of grazing.

[1] For the definition of "period of occupation" *vide* p. 143.

Prolongation of the period of occupation beyond the limit when grass can be sheared in the course of the same occupation of a plot (that is, in the course of the same rotation), leads to a considerable decline in production.

We will see later (Table 51, p. 236), in studying rationed grazing, the enormous falls in yield which can be brought about by repeated prolongation of the period of occupation on one section of the plot.

It is established in this table that repeated prolongation (in each rotation) of the period of occupation by 3 days (combined with a corresponding reduction in the rest period) causes the grass production to fall by almost a half. We will also see that the unfavourable effect of this protraction of the period of occupation is the more marked, the drier the season.

In almost all the so-called "rationed" systems of grazing I have observed in most countries, the allowing of the animals to go back for water (pp. 220–236) leads to a considerable increase in the time during which the sections of the plots near to the watering points are occupied.

Productivity of grass under continuous grazing

Without committing any great error, we can say that very short rest periods of 6 days in May–June and 12 days in August–September correspond more or less with what takes place in the case of "*continuous* grazing" with cattle.

In fact, with continuous grazing the grass which was previously sheared by the grazing animal will, after rest periods of this order (at these times of year), have grown again to a sufficient length to be seized upon and snapped up once more by the cow.

These are **hidden** rest periods, for one cannot observe them, since they do not appear in a visible form.

With these "hidden" rest periods in continuous grazing the grass is working with a very low productivity, viz.:

  1. In May–June, with a productivity of 71 lb./acre [*80* kg./ha.] per day against 237 lb./acre [*266* kg./ha.] with the optimum rest period of 18 days.

  2. In August–September, with a productivity of 36 lb./acre [*40* kg./ha.] per day against 119 lb./acre [*133* kg./ha.] with the optimum rest period of 36 days (exactly the same as with too long periods of occupation).

In other words, in continuous grazing we are probably working with a "productivity" equal to approximately one-third of that obtained with well-managed "rational" grazing.

The sigmoid curve of grass re-growth and the productivity curves we have deduced therefrom allow us to see immediately and clearly the reasons for the superiority of "rational" over continuous grazing.

Chapter 3

REST PERIOD AND ANNUAL PRODUCTION OF GRASS

Paucity of studies on the influence of the rest period on the herbage yield

STUDIES of the influence of rest periods on the annual production of grass are practically non-existent. The work which has been reported from the various research centres compares the yields of herbage plants (individually or in association) under different conditions (fertilisers, rainfall, etc.), the plants almost always, if not always, being cut **at equal intervals**: that is, every week, every two weeks, etc.

Moreover, variation in rest period according to the time of year must not be confused with the employment of different rest periods in different sets of experiments, the interval in the latter instance remaining constant throughout the season.

Let us take the example of a study entitled: "*The effect of varying the period of rest in rotational grazing*" (55). When we read the text we see that grazing was carried out:

> every 4 weeks under System A;
> every 2 weeks under System B;
> every 4 days under System C.

With a few very rare (and recent) exceptions, the same procedure has been adopted by all research workers.

When I read the reports of work of this nature, I want to ask: "Did you graze (or mow) every two weeks in winter also?" This may appear to be a stupid question; but, in my opinion, it is outrageous to apply the same rest periods in summer and winter as one employed in the spring.

In such studies the rest period must be varied according to the time of year, and these variable intervals must be applied under different systems.

The only study with which I am at present acquainted concerning the influence of variable and different rest periods on grass production taken as a whole throughout the year is that carried out by Professor Zürn (148) at the Admont Grassland Experimental Centre in Austria. To this study grassland science owes much.

Here are some of the figures obtained in the course of this experiment.

20

Study by Professor Zürn

Zürn's results are contained in Table 2, giving the yields of fresh grass in lb./acre [kg./ha.] per rotation and for the whole year (Zürn's results were given as dry matter).

Three systems of rest periods (variable according to the season) were used, being described as:

1. Short rest periods.
2. Medium rest periods.
3. Long rest periods.

TABLE 2

Influence of length of rest period on the yield of fresh grass, per rotation and for the whole year

Type of rest period	Short			Medium			Long		
Rotation No.	Days of rest between two rotations	Yield of fresh grass		Days of rest between two rotations	Yield of fresh grass		Days of rest between two rotations	Yield of fresh grass	
		lb./acre	[kg./ha.]		lb./acre	[kg./ha.]		lb./acre	[kg./ha.]
1	13	2,540	[2,840]	17	3,945	[4,420]	25	6,335	[7,100]
2	10	1,840	[2,060]	15	3,340	[3,740]	24	5,650	[6,330]
3	12	1,670	[1,875]	19	3,280	[3,680]	25	5,000	[5,600]
4	14	1,850	[2,070]	22	3,195	[3,580]	27	5,100	[5,720]
5	17	2,200	[2,470]	25	3,420	[3,840]	30	5,230	[5,860]
6	19	1,865	[2,090]	30	3,490	[3,910]	40	5,380	[6,030]
7	30	2,105	[2,360]	40	3,070	[3,440]	—	—	—
8	39	1,890	[2,120]	—	—	—	—	—	—
Total for year		15,960	[17,885]		23,740	[26,610]		32,695	[36,640]
Relative variation		100	[100]		149	[149]		205	[205]
Average per rotation		1,995	[2,240]		3,390	[3,810]		5,450	[6,100]

N.B. Calculations based on Zürn's results (148).

Taking the results as a whole, we see that the grass production over the year was twice as high with the long rest periods, although only six rotations were possible under this system, as compared with eight where short rest periods were applied.

Annual production of nutrients

The objection might be raised that this older grass has a lesser nutritive value. But Zürn has carefully analysed the grass produced with the three systems of resting.

Like the green matter, the production of starch equivalent is almost doubled, but the production of crude protein has only been increased by 41%: which in itself is an enormous increase in annual output per acre.

It should be observed here that it is not a case of protein but of Kjehldahl nitrogen multiplied by a factor akin to 6·20. Moreover, the biological value of the protein has not been considered, and I am not at all certain that by prolonging the rest period one does not get a grass with a nitrogen fraction of such a composition as to give it a higher biological value and, in particular, to render it less dangerous to the health of the animal; this question will be examined in detail in Chapter 9 of this present part and in Chapters 5 and 6 of Part Two.

TABLE 3

Total production of grass and nutrients during the season as a function of the length of the rest period

	Length of rest period					
	Short		Medium		Long	
Number of rotations in year .	8		7		6	
Fresh grass: Total production in year—						
lb./acre . . .	15,960		23,740		32,690	
[kg./ha.] . . .		[17,885]		[26,610]		[36,640]
Relative variation . .	100		149		205	
Crude Protein: Total production in season—						
lb./acre . . .	921		1,204		1,297	
[kg./ha] . . .		[1,032]		[1,351]		[1,454]
Relative variation . .	100		131		141	
Starch Equivalent: Total production in season—						
lb./acre . . .	1,760		2,730		3,945	
[kg./ha.] . . .		[1,975]		[3,060]		[4,420]
Relative variation . .	100		155		218	

Calculated from Zürn's results (148).

But leaving these considerations aside for the moment, we see finally that *prolongation* of the rest period, in allowing the grass to work with a high productivity, has led in the course of the year to an increased production of:

Green matter   .   .   . 105%
Crude protein   .   .   .   41%
Starch equivalent   .   . 118%

It should be noted that production was almost doubled where *long* rest periods replaced *short*. The latter, however, represented an enormous step forward over those existing (hidden) in the case of continuous grazing.

These results of Zürn's provide good confirmation of the theoretical considerations which we deduced from our study of the sigmoid curve of grass re-growth.

Daily growth of grass in Zürn's experiments

These considerations are equally well confirmed by Table 4, where we find the amounts of daily grass growth (fresh weight) for three types of rest period.

TABLE 4

Influence of the length of rest period on the daily growth of fresh grass

Rotation No.	Short rest period — Days of rest between two rotations	Short rest period — Daily growth of fresh grass (lb./acre)	([kg./ha.])	Medium rest period — Days of rest between two rotations	Medium rest period — Daily growth of fresh grass (lb./acre)	([kg./ha.])	Long rest period — Days of rest between two rotations	Long rest period — Daily growth of fresh grass (lb./acre)	([kg./ha.])
1	13	195	[218]	17	232	[261]	25	253	[284]
2	10	184	[206]	15	223	[249]	24	235	[264]
3	12	139	[156]	19	173	[193]	25	200	[224]
4	14	132	[148]	22	145	[163]	27	189	[212]
5	17	129	[145]	25	137	[154]	30	174	[195]
6	19	98	[110]	30	116	[130]	40	135	[151]
7	30	70	[79]	40	77	[86]	--	—	—
8	39	48	[54]	—	—	—	—	—	—
Average		124	[139]		158	[177]		198	[222]
Relative variation		100	[100]		127	[127]		160	[160]

N.B. Calculated from Table 2, p. 21.

For example, between the 3rd and 4th rotation the following amounts of daily growth, expressed as a function of the re-growth period, are:

Type of rest period	Days of re-growth	Daily growth of fresh grass (lb./acre)	[kg./ha.]
Short	14	132	[148]
Medium	22	145	[163]
Long	27	189	[212]

These figures are very similar to those deduced by us from the re-growth curve and also to Linehan's results (p. 17).

Cumulative effect of short rest periods on grass growth

If, in the course of the season, the grass is defoliated at too short intervals, that is after too short rest periods, the pasture reaches the end of the season worn out.

These over-short rest periods have exhausted the strength of the grass. Like a runner who has not made proper use of his resources, the grass cannot stay the course: that is to say, it cannot produce growth up to the normal end of the grazing season.

This is clearly shown by Table 5, also taken from Zürn, unfortunately it is based on *constant* rest periods (whatever the time of year):

of 4 weeks on the one hand;

of 2 weeks on the other hand.

TABLE 5

Daily growth of grass when it is cut or grazed every two or every four weeks

Time of year	Daily growth of grass when it is cut every			
	four weeks		two weeks	
	lb./acre	[kg./ha.]	lb./acre	[kg./ha.]
May	349	[380]	131	[147]
June	211	[237]	113	[127]
July	157	[176]	113	[127]
August	136	[152]	62	[70]
September	88	[99]	27	[30]

N.B. Calculations based on Zürn's results (149).

This table makes it clear that too short rest periods, of 2 weeks duration, fatigue the grass, which, in September, can only provide a daily re-growth of 27 lb./acre [30 kg./ha.] as opposed to 88 lb./acre [99 kg./ha.] in the case of grass which has only been asked to "work" every four weeks.

Now we understand even better the words of Weinmann (140) cited above (p. 11): "The effects of repeated defoliation are cumulative and progressively deplete the reserves . . ."

This cumulative effect of "fatigue" in grass which is defoliated too often is the more serious as to-day (as we shall see further on) we are in a position to sustain the vigour of our grass at the end of the season by feeding it with nitrogen. Thus we can prolong the grazing season and, at the same time, increase the yield in grass. But if, at the end of the season, the grass which receives the nitrogen is "fatigued" by too much and too frequent defoliation, in its state of exhaustion it will be incapable of utilising the fertiliser.

As we will often repeat in the course of this book, it is impossible, in the case of herbage, to separate the question of fertiliser usage from that of the method of utilisation. That was the first example: we will have occasion to see many others.

The "mown rotation" in the valley of Elorn

In the course of my peregrinations across the grasslands of the world I believe that I encountered the most refined system of grassland management in the valley of the Elorn, right at the tip of Europe, in the *Finistère Département*, in Brittany.

Each peasant in this valley owns a small acreage of grass which he irrigates regularly with water from the River Elorn; this is charged with the waste from the tanneries and the rich sewage from the town of Landivisiau, which dominates that part of the valley.

The grass is mown, to be carried in and fed to the animals in the stall (soilage). One might therefore describe this as a "mown rotation". In this way and with the very mild climate they succeed in obtaining eight or nine rotations. In winter they are aided by the fact that the temperature of the waters of the Elorn is 22° F. [*12*° C.] higher than that of neighbouring rivers.

The resting period and annual butter production

The peasants in this valley know exactly at what stage of growth the herbage must be cut if it is to provide an increased yield of green matter and lead to maximum production on the part of the cows fed with it in the stall (p. 98).

I had a long discussion with the Breton peasant seen mowing his grass on Photo No. 1 (facing p. 74). The grass seemed rather short to me, so I asked him if it was really a suitable height for cutting. He replied that he would be far better cutting longer grass, like some which he pointed out to me belonging to his neighbour (this can be clearly seen behind the peasant on the photo).

I asked him why and the answers he gave were evidence of his profound knowledge.

> "I am short of grass", he said, "and so I am forced to cut it a little too soon. That means I will lose some of my grass yield. . . . I am even going to lose milk for this grass contains too much water and is less nourishing. When I don't wait long enough to cut I know quite well that at the end of the year my wife will tell me that my grass in the valley didn't provide as much butter for sale as usual."

The peasant in the Elorn valley knows the grass productivity curves

Obviously a technical man would have said: "In view of the too short rest period I have given the grass it will not provide its maximum daily production, since I am not at the maximum of the productivity curve as shown in Figs. 3 and 4: the total output of grass will have diminished, as Zürn's work has shown (Table 2, p. 21)."

The Breton peasant did not need to see our curves: he, and his ancestors, were intuitively acquainted with them. They knew that by cutting their

grass after an insufficient period of rest their annual production would be
diminished.

The peasants' Departments of Chemical Analysis and Rural Economy

The peasant had also remarked that this grass, being too young, would
make his milk yields fall. He knew that rest periods of insufficient length
furnish a smaller quantity of grass which is less "nourishing". Obviously
this Breton peasant, learned without knowing it, did not speak of too
high a percentage of non-protein nitrogen in the grass or of the poor
nitrogen/carbohydrate balance; but the director of his Department of Rural
Economy, to wit, his wife, kept him regularly up to date with the statistics
of butter production per acre, that is to say, the amount of butter made in
relation to the acreage of grass in the valley.

The result of this technique and of these subtle statistics is that according
to the information collected (roughly estimated as it was) the peasants of the
Elorn valley harvest per annum more than 528 tons of grass per acre [120
tons/ha.]. The likeable young agriculture adviser for the Monts d'Arrée area
is going to undertake some more precise calculations at my request.

Observation of the necessary rest period is more important in dry than in humid regions

One often hears it said that rational grazing applies in relatively humid
regions but is of no interest in dry regions.

I sincerely believe that the contrary is true. Normandy has its dry years
too, and it is in these years that I can affirm the greatest difference between
rationally grazed pastures, which remain green, and those grazed all the
time, which are transformed into yellow straw matting.

The laws of grass growth and the necessity for optimum rest periods
between two successive grazings explain why the advantages of rational
grazing are most marked in dry weather.

The kinetic energy of very young grass, defoliated before it has accumulated sufficient reserves, will be much more affected by dry than by damp
weather. A weakened body easily succumbs to difficult environmental
conditions, although it has a chance of survival under favourable conditions.
It has been seen (Table 5, p. 24) that too short periods have a cumulative
effect on the strength of the grass.

Moreover, it will be shown (Table 51, p. 236) that a period of occupation
too long by 3 days, combined with a rest period reduced by 3 days, diminishes
grassland production to a much greater extent in dry than in damp weather.

The pasture must therefore be treated with special care when it is sickly
and delicate, as is the case in areas with dry summers. The principles of

rational grazing spare the grass and allow it to better withstand and to suffer less at the hands of unfavourable atmospheric conditions.

In winter, a time when grassland in all countries undergoes a hard test and is often very delicate, it will benefit in practice from an obligatory rest period of 100–150 days.

The animals do not return to grass until the spring, when the optimum rest period has allowed the herbage to attain the necessary degree of re-growth.

There are regions and seasons where the grass in summer is as weak and fragile as in winter. In these cases, in particular, the principles of rational grazing must be observed.

"Transhumance" is merely rotation on a grand scale

The herds of the hot dry regions of southern France have always practised, without knowing it, a form of rotation on a grand scale: "transhumance".

This is an enormous rotation in which, in the very hot weather, the grass fields of the plain (hot and dry) are abandoned in favour of the mountain pastures (fresh and humid): in this way the necessary rest periods for the plain grasslands are observed.

The absolute necessity of adhering to these rest periods during seasons of drought leads to the practice of a special rotation, known as transhumance, in spite of the considerable time and long journeys it involves.

There is no better illustration of the point to which the principles of rational grassland management must be observed in hot, dry regions.

The ravages of drought on the pastures of the ranches

Resting of grassland during periods of drought should be the more em-phasised where the climate is particularly hot and dry and where the pasture has not yet taken root. One often encounters extreme cases which provide a better illustration of the phenomena. Lady Eve Balfour (8), visiting the Texas Research Foundation at Renner, near Dallas (Texas), was able to see the efforts that had been made to re-establish permanent pastures on bare soils which at one time in the past had supported such grassland.

The Texas agronomists told Lady Balfour:

"Grass made our soils: and grass will bring them back."
[And they added:] **These young pastures can only take root and establish on condition that they are grazed when the grass has grown sufficiently and that the animals are subsequently removed to allow the pasture to rest.**

One might even wonder if this will not always have to be the practice if one does not wish to renew the destruction of grass and soil caused in the past by abusive grazing with enormous herds of ranch cattle.

If rest periods are not observed in dry regions, erosion ravages the soil

This necessity of granting the pasture the rest it requires to conserve its strength is particularly marked in extreme climates where torrents of rain follow abruptly on long periods of drought.

Non-observation of the rest period not only leads to degradation of the flora: it causes annihilation of the whole grass covering. The result is erosion and the dust bowls.

American farmers have exploited the soil like a mine and have forgotten to respect the demands of the grass. They have made it work like a slave: they have worn it out with work, neglecting the periods of rest it required. After the grass had died from overwork it was the soil itself that succumbed.

Washed away by erosion in the rivers, it went on to sleep its last sleep in the Gulf of Mexico.

Chapter 4

SEASONAL FLUCTUATIONS IN GRASS GROWTH

Variation in the growth fluctuation from year to year

THE vigour of the grass's re-growth varies with the season: it is generally non-existent in winter and sometimes also in summer.

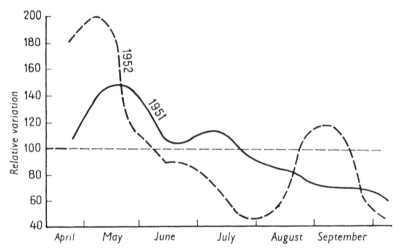

FIG. 6. Relative variations in daily growth of grass in two consecutive years (average of three pastures).
From Klapp (70) Fig. 134, p. 383.

Fig. 6, taken from Klapp, shows the relative variations in growth in the course of the years 1951 and 1952. The figures given are the average for three pastures.

The 1951 curve can be considered as fairly characteristic of North-West Europe. The 1952 curve corresponds with the extremely vigorous spring push and almost complete lack of growth at the end of July–beginning of August, such as one quite often encounters in the centre, and especially in the South of France (cf. the curves in Figs. 9 and 10, pp. 166 and 168).

29

Variation in grass growth according to region and country

Table 6 shows the variations in the growth of the grass in different regions of Europe. The figures indicate the lb./acre [kg./ha.] of fresh grass produced by daily growth.

We see that the maximum yields of 406 and 426 lb./acre [*455* and *477* kg./ha.] of fresh grass were observed in April–May in Berkshire (England) and in Switzerland. I myself have seen a growth of 400 lb./acre [*450* kg./ha.] fresh grass at the end of May in Normandy under very favourable conditions.

Reciprocal influence of rest period and daily growth

It must be emphasised that we do not know exactly under what conditions the figures included in Table 6 were calculated nor, in particular, what was the rest period, at various times of the season, which produced this daily grass growth.

If the rest period influences the average daily growth of the grass, it is this daily growth which determines the rest period to be observed between rotations. It is therefore a *reciprocal* function, and every mathematician knows how very complicated such functions can be.

TABLE 6

Average daily growth (in lb./acre or kg./ha. of fresh grass) during different grazing months in various parts of Europe

Period	Switzer-land (Kauter) (1)		Admond (Styria) (3)		Berkshire (Great Britain) (4)		Holland (T. Hart) (5)		Poppels-dorf (near Bonn) (2)		Holland (peaty soil)		Holland (clay soil) (6)	
	lb./ acre	[kg./ ha.]	lb./ acre	[kg./ ha.]	lb./ acre	[kg./ ha.]	lb./ acre	[kg./ ha.]	lb./ acre	[kg./ ha.]	lb./ acre	[kg./ ha.]	lb./ acre	[kg./ ha.]
Mar.–April .	288	[*323*]	—	[—]	49	[*55*]	40	[*45*]	—	[—]	—	[—]	—	[—]
April–May .	426	[*477*]	254	[*284*]	406	[*455*]	300	[*336*]	361	[*405*]	134	[*150*]	130	[*146*]
May–June .	272	[*305*]	236	[*264*]	219	[*246*]	264	[*296*]	207	[*232*]	312	[*350*]	159	[*178*]
June–July .	264	[*296*]	200	[*224*]	126	[*141*]	162	[*182*]	219	[*246*]	254	[*284*]	172	[*193*]
July–August	300	[*336*]	189	[*212*]	65	[*73*]	194	[*218*]	207	[*232*]	229	[*257*]	172	[*193*]
Aug.–Sept. .	247	[*277*]	174	[*195*]	236	[*264*]	150	[*168*]	228	[*255*]	236	[*264*]	104	[*116*]
Sept.–Oct. .	134	[*150*]	135	[*151*]	138	[*155*]	81	[*91*]	118	[*132*]	140	[*157*]	—	[—]
Oct.–Nov. .	—	[—]	—	[—]	49	[*55*]	40	[*45*]	—	[—]	—	[—]	—	[—]

N.B. Calculated by the author from:
    1. Ref. 62.
    2. Ref. 67, Klapp: the Poppelsdorf experiments (the growth has been considered to cover a rest period of 3 weeks).
    3. Ref. 149 (for long rest periods, see Table 4).
    4. Ref. 138.
    5. Ref. 70, Klapp: Fig. 151, p. 381, for the year 1950.
    6. Ref. 79.

Calculation of the rest period by successive approximations

Let us take Table 4 (p. 23) above, which I deduced from Zürn's results. There we find the amounts of fresh grass growth recorded in the course of the year, using three different rest periods: short, medium and long.

Let us suppose that a research station in Austria quotes the amount of growth given for the *short* rest period. Between the third and fourth rotations the daily growth of grass amounts to 132 lb./acre [*148* kg./ha.]. To get the 4200 lb./acre [*4800* kg./ha.] that should be got from the area in question the rest period will comprise

$$\frac{4200}{132} = 32 \text{ days} \qquad \left[\frac{4800}{148} = 32 \text{ days}\right]$$

But by prolonging the rest period we increase the daily growth of the grass (Fig. 3, p. 15) and it will be seen from Table 4 (p. 23) that with a rest period of 27 days the average daily growth (long rest period) has climbed to 189 lb./acre [*212* kg./ha.] instead of 132 lb./acre [*148* kg./ha.]; we then find that to obtain a yield of 4200 lb./acre [*4800* kg./ha.] of grass, a sufficient rest period will be:

$$\frac{4200}{189} = 23 \text{ days} \qquad \left[\frac{4800}{212} = 23 \text{ days}\right]$$

But then at 23 days the maximum of the productivity curves (Fig. 6, p. 29) would still not have been reached, and we see in fact from Table 4 (p. 23) that between the third and fourth rotations with a rest period of 22 days (average) (medium rest period) the daily re-growth is only 145 lb./acre [*163* kg./ha.], which means that, to attain a yield of 4200 lb./acre [*4800* kg./ha.] we must employ a rest period of:

$$\frac{4200}{145} = 29 \text{ days} \qquad \left[\frac{4800}{163} = 29 \text{ days}\right]$$

But with such a rest period (long rest period) we arrive at an amount of daily growth equal to 189 lb./acre [*212* kg./ha.].

We see, then, that as in the case of all *reciprocal* functions in mathematics, the problem can be solved only by successive approximations.

Agriculturalists have no need of such complicated calculations; but they do ask for indications to assist them in carrying out their policies of rational grazing.

Let us look to begin with at the variations in average optimum rest periods for climates favourable to grass growth.

Rest periods in Normandy and Austria

Tables 7 and 8 (p. 32) contain the rest periods (to be considered rather as the minima) which I have seen used in Normandy and those quoted by Professor Zürn in Austria.

My own table for Normandy (Table 7) contains an additional figure for a winter rest period: this is to indicate that the farmer, under compulsion, then gives his grass a necessary rest of 100–150 days.

An agricultural service for the future

One of the main tasks of regional Grassland Research Stations should be the study of the figures for grass growth as a function of soil and climate.

TABLE 7

Average rest periods for grass in Normandy

25–30 days in April
14–18 days in May
20–25 days in June and the beginning of July
28–35 days at the end of July, in August and at the beginning of September
40–60 days at the end of September, in October and November
100–150 days in winter

N.B. Compare Table 8 and Table 45, p. 162.

TABLE 8

Average rest periods for grass in Austria
(Professor Zürn)

Between 1st and 2nd grazings (May)	.	.	.	.	.	12–14 days
,, 2nd and 3rd ,, (May–June)	.	.	.	.	12–14 days
,, 3rd and 4th ,, (June–July)	.	.	.	.	16–18 days
,, 4th and 5th ,, (July–August) .	.	.	.	22–24 days
,, 5th and 6th ,, (August–September)	.	.	28–30 days
,, 6th and 7th ,, (September–October)	.	.	35–40 days

N.B. Compare Table 7 and Table 45.

But it must not be forgotten that these measurements, on a certain soil and under determined climatic conditions, will have to be made within the optimum systems of management for each region: that is, the amounts of growth will have to be measured for rest periods corresponding as nearly as possible to those employed in practice.

Moreover, it will not be sufficient to cut the grass with shears or a mowing machine; *it must be grazed*, which gives very different growth figures.

Priority must be given to the management employed

Scientific logic and common sense both demand that optimum methods of grassland management should be studied and clearly understood before studies which are a function of them are undertaken, whether they are concerned with amounts of grass growth, herbage yield or selection of a variety, etc.

It is clear, then, that overall investigations and measurements of herbage must be regulated by the method of management.

Alarm service provided by growth figures

Where regular and continuous measurement of the amounts of grass growth by our regional Grassland Research Stations is concerned, it can even be foreseen that in the more distant future when rational grazing is properly developed, these stations could issue warnings, analogous to those given at present regarding mildew and Colorado Beetle attacks, pointing out the variations in growth vigour as influenced by the prevailing climatic conditions. This service could warn grassland farmers, for example, that a failing grass growth should be sustained by a heavier dressing of nitrate of lime.

On the other hand, where growth was very vigorous, the agriculturalist would be advised that he could put the mower through some of his plots and reduce his dressings of nitrogen.

We will look further on (Part IV) at the different practical methods which allow us to compensate for seasonal fluctuations in the grass.

Influence of seasonal variations in climate on the rate of re-growth of individual plants

It is the environmental conditions, soil, climate and management, which determine the flora of the herbage. This subject will be studied in a work specially devoted to the "Dynamic ecology of herbage"; this differs quite clearly from the static ecology of herbage, which pays attention only to climate and soil in preparing its cartographic surveys and pays practically no attention at all to the effect of the methods of management.

Suffice it to recall here that the effect of seasonal variations in climatic conditions on the growth vigour of plants varies greatly from one plant to the other.

Precocity is a function of the growth vigour under the conditions pertaining at the beginning of the season, and a precocious plant, that is, a plant growing relatively vigorously under conditions of *reduced light and low temperature*, is not necessarily the plant which will grow most vigorously under the conditions of strong light and increased temperature in summer. In general, the contrary is even true.

The result is that if plots are sown with pure mixtures, with different precocities, the vigour of re-growth (and consequently the rest period) will not necessarily be the same in subsequent rotations. It might even happen that the *least precocious* plant at the beginning of the season will perhaps be the one with the *most rapid growth* in summer.

Finally, the order of grazing pastures sown with one species only will have to be modified (*vide* Part Nine, Chapter 2).

Chapter 5

INFLUENCE OF FERTILISERS ON THE VIGOUR OF GROWTH AND ON THE PRODUCTION OF GRASS

Influence of fertilisers on the daily re-growth of grass

TABLE 9, which has been extracted from an excellent study by Zürn, illustrates very well the influence exerted by phosphoric acid, potash and nitrogenous fertilisers on the vigour of grass growth and on the total annual yield of herbage.

The O columns contain the results where no fertiliser was applied. The PK columns show the results obtained with an annual dressing of 71 lb./acre [*80* kg./ha.] phosphoric acid and 107 lb./acre [*120* kg./ha.] potash (K_2O).

The NPK columns contain the results obtained with the same dressings of phosphoric acid and potash as given above (PK) and, in addition, an annual dressing of 76 lb./acre [*85* kg./ha.] of nitrogen apportioned as indicated in Table 9.

The nitrogenous fertiliser employed was nitrate of lime and ammonia (Kalkammonsalpeter).

The rest periods were varied, as in all well-conducted grazing management. The minimum was 19 days between rotations 1 and 2 and the maximum 41 days between rotations 5 and 6, that is, between the penultimate and final rotations.

Note that there is a rest period quoted before the first rotation; this corresponds to the number of days which had passed between the time when grazing was first started and the (theoretical) time when it was estimated that grass growth, for that year, had re-commenced.

We see, then, that, all other things being equal, the phospho-potassic fertiliser (PK) increased the yield by 33%, while the combined use of phospho-potassic and nitrogenous fertilisers brought about an increase of 70%. Without doubt, this increase in total yield merely reflected the increase in the vigour of grass growth as the result of a fertiliser dressing. The vigour of growth varies between 59 and 146 lb./acre [*66* and *164* kg./ha.] fresh grass per day where no fertiliser at all was applied. With the complete fertiliser the growth vigour reaches a maximum of 245 lb./acre [*275* kg./ha.] per day and falls to a minimum of only 87 lb./acre [*98* kg./ha.].

TABLE 9

Influence of artificial fertilisers on the daily growth of grass

Rotation No.	No. of days of rest	Nitrogen applied (lb./acre)	Nitrogen applied [kg./ha.]	Total yield of fresh grass — O (lb./acre)	O [kg./ha.]	PK (lb./acre)	PK [kg./ha.]	N (lb./acre)	N [kg./ha.]	Daily growth of fresh grass — O (lb./acre)	O [kg./ha.]	PK (lb./acre)	PK [kg./ha.]	NPK (lb./acre)	NPK [kg./ha.]
1	20	22	[25]	2,300	[2,580]	3,290	[3,690]	5,660	[6,350]	99	[111]	143	[160]	209	[243]
2	19	18	[20]	2,780	[3,120]	3,480	[3,900]	4,640	[5,200]	146	[164]	183	[205]	245	[275]
3	22	0	[0]	2,890	[3,240]	3,760	[4,220]	3,870	[4,340]	132	[148]	172	[193]	176	[197]
4	27	18	[20]	3,620	[4,060]	4,860	[5,450]	5,850	[6,560]	128	[143]	176	[197]	216	[242]
5	32	18	[20]	2,930	[3,280]	3,910	[4,380]	5,190	[5,820]	92	[103]	124	[139]	161	[180]
6	41	0	[0]	2,440	[2,740]	3,330	[3,730]	3,540	[3,970]	59	[66]	80	[90]	87	[98]
Total production				16,960	[19,020]	22,630	[25,370]	28,750	[32,240]	109	[122]	146	[164]	182	[204]
Relative variation				100	[100]	133	[133]	170	[170]	100	[100]	133	[133]	170	[170]

Total production / Relative variation. Average daily growth / Relative variation.

N.B. Calculated by the author from Zürn (149).

1. Average results for 10 consecutive years.

2. P = 71 lb./acre [80 kg./ha.] of phosphoric acid (P_2O_5), K = 107 lb./acre [120 kg./ha.] of potash (K_2O).

The action of phospho-potassic fertiliser is persistent

An even more important fact emerges from this study:

The phospho-potassic applied only once in the course of the winter exercised an effect throughout the six rotations, increasing the yields of fresh grass by:

1st Rotation	990 lb./acre [*1110* kg./ha.]
2nd Rotation	700 lb./acre [*780* kg./ha.]
3rd Rotation	870 lb./acre [*980* kg./ha.]
4th Rotation	1240 lb./acre [*1390* kg./ha.]
5th Rotation	980 lb./acre [*1100* kg./ha.]
6th Rotation	880 lb./acre [*990* kg./ha.]

respectively, over the production obtained with no fertiliser application.

Nitrogen acts immediately

By contrast, increase in yield and growth vigour due to dressings of nitrogen is obtained only in the rotation before which the nitrogenous fertiliser was applied.

In fact, the increases in the yield of green grass, by comparison with the production obtained where no nitrogen was applied (i.e., PK alone) are those contained in Table 10, below. It can be seen that the nitrogen acts immediately on the rotation for which it was applied, but its action is not persistent. We can say therefore: **nitrogen is rapidly consumed by growing grass and exercises very little influence, if any at all, on the grass,**

TABLE 10

Increase in production due to nitrogen in Zürn's experiments

Rotation No.	No. of days of rest	Nitrogen applied		Total increase in production of fresh grass due to nitrogen application		Increase in production of fresh grass per unit of nitrogen applied	
						Per rotation	Per day
		lb./acre	[kg./ha.]	lb./acre	[kg./ha.]		
1	20	22	[*25*]				5·30
	19	18	[*20*]	2380	[*2660*]	106	3·42
2	22	0	[*0*]	1160	[*1300*]	65	
3				107	[*120*]		
	27	18	[*20*]				2·03
4	32	18	[*20*]	990	[*1110*]	55	2·26
5	41	0	[*0*]	1290	[*1440*]	72	
6				210	[*240*]		
					Average	74	3·25

N.B. A unit of fresh grass means one pound or 1 kg. according to the measure system used.

which will grow up again following defoliation of the stand which consumed almost all the nitrogen applied in the preceding rotation.

In the present case, the result is the more remarkable in that it concerns ammonium nitrate, the action of which is much less rapid than that of other nitrogenous fertilisers, such as nitrate of lime.

This mode of action of nitrogen allows its use at the very time when one wants to obtain the greatest production. The result is that, by judicious distribution of nitrogenous dressings over the year, one can compensate in part for the seasonal fluctuations in the production of grass (*vide* pp. 165–171).

Production of grass by one unit of nitrogen

Table 10 provides us in addition with the valuable information that, under the conditions at Admont, one unit of nitrogen produced per rotation approximately 100 units of grass at the beginning of the season, the figure varying subsequently from 55 to 72, with an average of 74. The daily production of grass due to one unit of nitrogen also varies with the season, starting with 5·30 at the beginning of the season and falling to hardly more than 2 at the end of the season (the average being 3·25) (*vide* p. 168). These figures are extremely important where grazing is concerned. Unfortunately they are the only examples of their kind known to me. It would be interesting to study this question much more closely.

Increases in grass production with well-spaced dressings of nitrogen

By apportioning the nitrogen over the various rotations in this manner, one can use much greater quantities and obtain considerable increases in yield.

Table 11 (provided by the Ruhr Nitrogen Society) illustrates the production of starch equivalent which can be provided by rotational grassland on which the dressings of nitrogenous fertilisers are well apportioned.

TABLE 11

Increase in yield of rotational grassland by judicious application of nitrogenous fertiliser

Quantity of nitrogen applied		Production of starch equivalent from rotational grassland		Increase in production of starch equivalent due to application of nitrogen		Production of starch equivalent per unit of nitrogen
lb./acre	[kg./ha.]	lb./acre	[kg./ha.]	lb./acre	[kg./ha.]	
0	[0]	2281	[2557]	—	[—]	—
36	[40]	2759	[3092]	478	[535]	13
54	[60]	3140	[3515]	859	[962]	16
71	[80]	3283	[3680]	1002	[1123]	14
89	[100]	3569	[4000]	1288	[1443]	14·4
107	[120]	4680	[5245]	2399	[2688]	22·2

From *Ruhr Stickstoff Ges.* (86).

We see that with 107 lb./acre [*120* kg./ha.] of nitrogen, a production of 4680 lb./acre [*5245* kg./ha.] starch equivalent can be obtained: this corresponds to the possible maxima obtainable on arable land of the same quality.

Grass, with sugar beet, is the most efficient and economic factory for the conversion of very large quantities of nitrogenous fertiliser into protein without running the risk of lodging.

Phospho-potassic fertiliser must support nitrogen

As in the case of tillage, this use of large quantities of nitrogenous fertiliser is only possible if the dressings of phospho-potassic fertiliser correspond; these obviously vary with local conditions. In general, one may reckon that it is sufficient to apply annually 540 lb./acre [*606* kg./ha.] phosphoric acid, be it 360 lb./acre [*400* kg./ha.], 15% slag or 340 lb./acre [*380* kg./ha.] superphosphate.

In general usage slag is applied to herbage as the source of phosphoric acid, but it is possible (if not probable) that the exclusive use of slag, potassium chloride and nitrates, produces, in the long run, certain deficiencies of sulphur in the herbage. Also, I think it prudent to alternate slag and superphosphates according to the years. To simplify the work, it is even advisable to apply, before the beginning of the season, a complete fertiliser which will provide the first dressing of nitrogen, destined to activate the onset of vegetation and at the same time allow a grazing later on.

A reasonable dressing of potash seems to be in the region of 45 lb./acre [*50* kg./ha.] per annum, be it 90 lb./acre [*100* kg./ha.] chloride with 50% potash, or 108 lb./acre [*120* kg./ha.] of a sulphate with 40%. It may be wise to make the application in two dressings.

What one cannot repeat too often is that these quantities are merely indicative. On certain soils one would have to increase the phosphoric acid and reduce the potash: on other soils, the opposite would be necessary.

The trace-element question must not be neglected. It is very wise to add 4·5 lb./acre [*5* kg./ha.] of sulphate of copper every year when very large quantities of nitrogen are used.

Skill in the distribution of the fertiliser

The soils of every part of a pasture are not all strictly alike. In addition, the method of grazing can sometimes reveal differences in the long run. To cite my own case:

I have paddocks in the shape of a rectangle with the entrance gate and watering point at the same end; it is here that the cattle most often pass by or stand. Here, therefore, the excrement return is very much greater than at the other end of the field. While potash hardly seemed to act at all on the approaches to the gate and to the watering point, its action at the opposite end was extremely marked. I have therefore made a habit of increasing potash dressings on the half of my paddocks which receives less excrement return.

Use of nitrogen is of no interest except where grassland is rationally grazed

Numerous trials, even in recent times, ended by concluding that the use of nitrogen on pasture led only to very slight improvements in production and did not justify the application of nitrogenous fertilisers. These conclusions were correct under the conditions of the experiments, which were very far removed from those pertaining under a rational grazing system.

Let us quote, as an example, the experiment carried out by Woodman and Underwood (145), which produced, in 1932, the following conclusion:

"In the course of the two years of the experiment, the production of plots receiving sulphate of ammonia was 10% greater than that of plots not receiving it . . ."

Now, the conditions of the experiment were as follows. *The plots were not grazed*: they were cut with a mower at a height of ½–1 in. (*12·25* mm.) above the ground. The cuts were made always *after the same interval of time*, namely, every month, whatever the season. The quantities of sulphate of ammonia applied *were always the same*, 72 lb./acre [*80* kg./ha.] and did **not** vary: the dressings were applied once at the onset of growth, then subsequently after every cut except the last one.

On the whole, then, the large amount of 106 lb./acre [*118* kg./ha.] of nitrogen (corresponding to 500 lb./acre [*560* kg./ha.] sulphate of ammonia) was applied in order to obtain a 10% increase in yield.

We have just seen, however, that one can double the production of a rotational grazing by applying 108 lb./acre [*120* kg./ha.] (Table 10, p. 36).

It is evident that with the experimental methods of Woodman and Underwood (to whom we are grateful for other fine work elsewhere) one could only attain an increase in yield which was laughable. To tell the truth, the result might well have been a diminished yield.

In fact, Woodman and Underwood ascertained that nitrogenous fertiliser under the conditions described depressed white clover. It is evident that if one accelerates still further by dressings of nitrogen the growth of the grass in May–June, and if one cuts the grass at the end of 30 days (instead of 18, the normal rest period at this time), *the grasses will smother the clover*. And white clover is a marvellous factory for the free production of nitrogen, as will be seen in Chapter 7 of this section.

Chapter 6

AN IMMENSE ARMY OF LILLIPUTIAN PLOUGHMEN BURY THE PHOSPHO-POTASSIC FERTILISERS APPLIED TO THE PASTURE

Fears about the penetration of the fertiliser into the pasture

To justify the ploughing in of worn-out pasture one often hears it said that this is the only way to provide the soil with the fertilising elements it lacks. In addition, phospho-potassic fertilisers, which quickly become insoluble, are often badly distributed in the vertical sense, for the deeper layers seem to be poorer than the upper ones. This fact has been, and still is, attributed to the dressings of fertiliser affecting only the surface of the pasture. The deduction which it has been thought necessary to make from this is that it is essential to enrich the deeper parts of the arable layer, i.e., those between 4 and 8 in. [*10* and *20* cm.], with assimilable fertiliser elements by dint of mixing the fertilisers with the soil. This can be easily done by ploughing in the herbage. Sometimes, it was even thought that this was a means of supplying a reserve of fertiliser sufficient to satisfy the needs of the plants comprising the new herbage for several years to come.

Schulze's experiments at Rengen

When we study the influence of rational grazing on the improvement of the flora, trials carried out under the direction of Professor Klapp with the aim of improving common grazings on the Rengen estate (Eifel, Germany) will be discussed (pp. 277–83). The pastures of Rengen were in a state of complete dilapidation (they had probably never been marled and had never received mineral fertilisers). The results of different methods of fertiliser application were therefore very interesting. This is what Schulze did (91) under Professor Klapp's direction.

We know that the turf layer of an old pasture carried a felt about 4 in. [*10* cm.] thick that can be rolled out like a carpet. Schulze makes use of this fact in employing five methods of fertiliser application:

    1. The fertiliser is applied to the surface of the herbage as it stands.

    2. The turf layer is lifted and the fertiliser spread on the soil *underneath* it.

3. The turf layer is lifted, replaced and then fertiliser is scattered over its surface. It is the same method of fertiliser application as (1), while the turf is disturbed in the same manner as (2).

4. The herbage is ploughed under, the fertiliser is perfectly mixed with the soil in the course of the cultivations and a new sowing of grass is made.

5. The herbage is turned under, fertiliser scattered on the surface and then a new pasture is sown. This is the same method as (4) without the fertiliser being closely mixed with the arable layer 8 in. [*20* cm.] down. In 1941, before the experiments were started, the soil analysis of the grazings was as follows:

Depth of soil	pH	K_2O, milligrams	P_2O_5, milligrams
0–4 in. [*0–10* cm.] . . .	4·3	20·4	2·2
4–8 in. [*10–20* cm.] . . .	4·3	11·3	0·9

N.B. K_2O and P_2O_5 content measured in Neubauer figures (for 100 gm. of dry soil).

After each of the five treatments described above the following quantities of reserve fertiliser elements were added:

P_2O_5 216 lb./acre [*240* kg./ha.]
K_2O 432 lb./acre [*480* kg./ha.]
CaO   4320 lb./acre [*4800* kg./ha.]

The average results of analyses carried out in 1942 after these fertilising elements had been applied to the soil are contained in Table 12.

TABLE 12

Fertiliser application trials: on the surface and in the soil, with and without ploughing up (1942)

Method of fertiliser application	Depth of soil		pH	K_2O, milli-grams	P_2O_5, milli-grams
	in.	[cm.]			
1. *Fertiliser applied on surface* (without lifting turf)	0–4	[*0–10*]	5·6	33·4	6·6
	4–8	[*10–20*]	4·9	14·7	0·7
2. *Fertiliser applied in soil* (turf lifted)	0–4	[*0–10*]	5·4	23·1	4·7
	4–8	[*10–20*]	5·2	12·6	3·3
3. *Fertiliser applied on surface* (turf lifted)	0–4	[*0–10*]	5·4	30·2	5·9
	4–8	[*10–20*]	4·9	12·6	1·4
4. *Fertiliser mixed in by working of the soil* (with ploughing up and re-seeding)	0–4	[*0–10*]	6·1	24·3	2·0
	4–8	[*10–20*]	5·0	16·5	3·1
5. *Fertiliser applied on surface* (after ploughing up, then re-seeded)	0–4	[*0–10*]	6·7	29·2	3·3
	4–8	[*10–20*]	4·9	26·9	3·0

From Schulze (91).

Schulze considers that these figures prove that deep dressings of fertilisers, by lifting the turf layer (Method 2) or by mixing, through ploughing under (Method 4), have no definite advantage as far as good distribution of fertilising substances at depth is concerned nor as regards de-acidification of the soil. Only in the case of phosphoric acid did deep mixing of the fertiliser (Methods 2, 4 and 5) enrich the deeper layer. On the other hand, the application of fertiliser to the surface without touching the herbage (Method 1) led to a marked enrichment, absolute and relative, of the upper layer in phosphoric acid.

Yields produced by phospho-potassic fertilisers either applied to the surface or buried

The interest of these figures is purely indicative and completely relative. What is important are the harvests obtained, and Table 13 shows the relative average yields of forage in the experimental years (1941, 1942, 1943), then about ten years later, 1950 and 1951. It should be noted, moreover, that in the course of these ten years, all the areas, whatever their initial treatment, received a surface dressing of complete fertiliser every year (except for two years after the end of the War).

We see that during the *first three* years of the experiments the forage production where the herbage was turned under (Method 4) was very slightly superior to where the dressing was applied to the surface without touching the herbage (Method 1).

TABLE 13

Comparison of the yields of grass in the fertiliser application trials
(1941–51)

Method of fertiliser application	Relative yields of grass	
	Years of trials (1942, 1943, 1944)	Succeeding years (1950–51)
1. *Fertiliser applied on surface* (without lifting turf)	100·0	100·0
2. *Fertiliser applied in soil* (turf lifted) .	77·0	85·2
3. *Fertiliser applied on surface* (turf lifted)	88·5	89·5
4. *Fertiliser mixed in by working of the soil* (with ploughing up and reseeding)	108·9	93·3
5. *Fertiliser applied on surface* (after ploughing up, then re-seeded) .	100·0	93·3

From Schulze (91).

In the two cases where the turf layer was lifted (Methods 2 and 3) the production of forage was clearly diminished both in experimental and post-experimental periods; this is the consequence of damage to the superficial layer of the turf which makes itself felt for a very long time afterwards.

Schulze emphasises, and with good reason, that ten years later it is Method 1 (surface application of fertiliser *without touching* the soil or turf layer) which gives the best results. He considers that this is due particularly to the fact that the majority of the roots of the turf of old pastures are found in the upper soil layer (*vide* Table 14); it is to that part therefore that the reserve of fertiliser substances must be supplied, and not to the lower layer, where it will be much less well utilised.

TABLE 14

Distribution of the root mass in an old permanent pasture

Depth of soil		Percentage of total root mass
in.	[mm.]	
0–2·0	[0–50]	87·5
2·0–3·9	[50–100]	5·7
3·9–5·9	[100–150]	2·4
5·9–7·9	[150–200]	1·2
7·9–20	[200–500]	3·2

From Klapp (69).

The enormous expense of ploughing in and re-seeding has been in this case, a dead loss.

Schulze concludes:

"There are no grounds for deep applications of fertiliser to improve the yield of pastures."

It is quite clear that fertiliser applied to the surface is rapidly transported downwards into the soil. The workers responsible for this transportation are the animals of the micro-fauna of the pasture. We are going to show their great importance in this chapter; but first of all let us examine the distribution in depth of the roots, of the assimilable mineral elements and of these Lilliputian ploughmen.

Concentration of grass roots in the surface

The work of the Lilliputian ploughmen need not proceed very far downwards.

I will not be able, within this work, to deal with the subterranean part of pastures (soil structure, root development, the prodigious life of the sub-soil); I will deal with this question in greater detail in a work in preparation on the "Dynamic Ecology of Pastures". The subject of this is extremely important. A grass, in effect, is created by its roots. Darwin said that if a tree had a brain it would have to be sought in its roots. We will say that the brain of a grass must also be found in its roots.

I will content myself with indicating briefly the distribution of the root mass under an old permanent pasture.

Table 14 shows that 87·5% of the total root mass is found in the upper layer between 0–2·0 in. [0–50 cm.]. Contrary to widely held opinion, this concentration of roots in the surface seems to be one of the advantages of old, permanent pastures.

Let us note in passing that analyses of pasture soils should be particularly concerned with this superficial layer, where almost 90% of the life of the roots is concentrated.

TABLE 15

Assimilable mineral elements at different depths of an old pasture

Depth of soil		Relative content of:	
in.	[mm.]	P_2O_5 (phosphoric acid)	K_2O (potash)
0–2·0	[0–50]	100	100
2·0–3·9	[50–100]	54	78
4·0–5·9	[100–150]	36	70
6·0–8·0	[150–200]	21	61

From Klapp (69).

Surface concentration of mineral elements assimilable by the herbage

It is noteworthy that the assimilable mineral elements are concentrated especially in this upper layer down to 2 in. [50 mm.] in depth, as is shown in Table 15 (based on Neubauer analyses of the soil).

Surface concentration of the micro-fauna of pastures

In addition, the intense sub-soil life of the pasture (*life which creates its wealth*) is also to be found concentrated in this surface layer, in particular as is shown in Table 16, p. 45, which deals with the two great species of Oligochæta, the Lumbricids (earthworms) and the Enchytræids.

Enchytræids are very tiny white worms weighing on the average 300–400 times less than earthworms. They have been studied in detail at the Research Institute at Brunswick-Völkenrode (Germany), and I was truly impressed by the results shown me by Fräulein Trappmann (109) in her laboratory during my visit to Völkenrode.

Remarkable harmony prevails between the distribution of the roots, the assimilable elements and the micro-fauna. This helps us to understand why fertiliser laid on the surface of our pastures is efficacious.

TABLE 16

Decrease in the number of Oligochæta with depth of soil in permanent pastures

		Lumbricids (earthworms)					Enchytræids (small white worms)				
Soil Number		1		2		3		1		2		3
Depth of soil		Population per										
in.	[mm.]	sq. yd.	[m².]	sq. yd.	[m².]	sq. yd.	[m².]	sq. yd.	[m².]	sq. yd.	[m².]	sq. yd.	[m².]
0–1·6	[0–40]	268	[320]	308	[368]	298	[356]	3,394	[4,060]	25,147	[30,080]	8,400	[10,048
2·0–3·5	[50–90]	94	[112]	54	[64]	107	[128]	1,150	[1,376]	3,759	[4,496]	2,113	[2,528]

N.B. The soil was alluvial. 1 is grey soil. 2 and 3 are brown soils.

From Franz (24, p. 295).

We will understand it even better when we have some idea of the communal labour provided by that immense army of active Lilliputian ploughmen.

The livestock under the pasture is twice as heavy as that above

Table 17 shows the number and average weight of earthworms found in the course of numerous surveys made by Finck (21) on *rich* soils. One can suppose that these pastures, managed in the normal manner, were carrying $\frac{3}{4}$ of a beast to the acre (two beasts to the hectare) let us say 900 lb./acre [*1000* kg./ha.]. We see therefore that the "*subterranean stocking*" of earthworms is TWICE *as heavy as the cattle stocking.* We can therefore say that we have at our disposal, in the soils of our pastures, an army of Lilliputian ploughmen **whose total weight is** *twice* **that of the livestock the pasture is feeding.**

TABLE 17

Number and average weight of earthworms in pastures and arable land

	Number		Weight	
	Per acre	Per hectare	lb./acre	[kg./ha.]
Pasture . . .	1,200,000	3,000,000	1800	[2000]
Arable . . .	400,000	1,000,000	450	[500]

From Finck (21).

The prodigious tillage work carried out by the micro-fauna of permanent pastures

Table 18, based on soils quite mediocre in quality, has the merit of illustrating the very much more important rôle even of the Enchytræids in pastures.

TABLE 18

Annual production of wormcasts by earthworms and Enchytræids on arable land and on permanent pastures

Micro-fauna	Place	Individual weight in mgs.	Number of worms per		Annual production of wormcasts	
			sq. yd.	[m².]	tons/acre	[metric tons/ha.]
Earthworms	Arable	700	34	[41]	9	[22]
	Grassland	600	81	[97]	21	[52]
Enchytræids	Arable	2	1700	[2,000]	1	[3]
	Grassland	1	8800	[10,500]	5	[13]

From Graff (32).

We see that, even with this not very high number of Oligochæta the total excrement of earthworms and Enchytræids represents *per annum* 21 + 5 = 26 tons/acre [52 + 13 = 65 metric tons/ha.]. This is approximately the amount of farmyard manure which we apply *every six years* to our ploughed land.

Development of the micro-fauna in relation to the age of the pasture

It is interesting to compare these German figures with those of the English worker, Evans (Table 19).

TABLE 19

Weight of wormcasts as a function of the age of the pasture

Age of pasture in years	Weight of wormcasts	
	Tons/acre	[kg/ha.]
1	1½	[3,800]
7	2½	[6,300]
70	11	[28,000]
300	25	[63,000]

From Evans (19).

The latter tells us:

"Permanent pasture carries approximately 800 lb. to 1000 lb. of earthworms per acre representing 600,000 to 750,000 earthworms. During the first year after ploughing there is but little change, the ploughed-in grass supplying adequate food. After the first year, however, there is a rapid decline and by the fifth year only about 50 lb. to 100 lb. per acre—representing about 100,000 small earthworms—are present as a result of the great reduction in food supply.

Leys were found to carry 200 lb. to 700 lb. per acre according to age, and the proportions of the various species of earthworms present *differed* from those of permanent pastures and arable fields. Permanent pastures carry a high proportion of large species; arable fields a high proportion of small species; and leys are intermediate."

And Evans concludes:

"There is still a great deal to learn about the ways in which earthworms affect the soil by which we live. It would, however, appear that their value is much greater under conditions where man does not disturb the soil than where he does so."

Chapter 7

WHITE CLOVER, A FACTORY FOR FREE NITROGEN PRODUCTION

General considerations

THE nodule bacteria of legumes (and in particular, of white clover in pastures) fix the nitrogen in the soil.

Our concern is to know to what extent this nitrogen of white clover influences the associated grasses. Nitrogen applied in this way to grasses can originate either from the excretion of the intact nodules or from the decomposition of roots and nodules. In every case the nitrogen is organic, and it is interesting to ask what value in terms of an artificial mineral fertiliser has this organic fertilizer of white clover.

Unfortunately opinions on this point differ greatly, as is understandable in the light of all the reasons we have outlined above which produce a tremendous variation in the effectiveness of nitrogenous fertiliser on herbage according to the methods of management and of experimentation.

An experiment by Johnstone-Wallace

Professor Johnstone-Wallace (49 and 50) grew smooth-stalked meadow grass and white clover at Cornell, both separately and in association.

Smooth-stalked meadow grass grown in a pure stand produced an annual yield of 880 lb./acre [978 kg./ha.] dry matter.

White clover in pure stand yielded 3060 lb./acre [3400 kg./ha.] dry matter.

The total yield of these two plants when grown separately was thus 880 + 3060 = 3940 lb./acre [978 + 3400 = 4378 kg./ha.], but when grown in association they yielded a total of 4986 lb./acre [5540 kg./ha.], which is an increase in dry-matter yield of 1046 lb./acre [1162 kg./ha.] or 26·5%. (Note that the increase in the yield of crude protein was even more marked.)

My calculations indicate that the white clover provided the grass with 76 lb./acre [85 kg./ha.] of nitrogen, which corresponds to 364 lb./acre [405 kg./ha.] sulphate of ammonia and 477 lb./acre [530 kg./ha.] nitrate of lime.

Johnstone-Wallace, who did his calculations differently from mine, even considers that the white clover supplied the grass with the equivalent of 100 lb./acre [1225 kg./ha.] sulphate of ammonia (approximately 225 lb./acre

[*250* kg./ha.] nitrogen). Personally I think this figure is too high. However, the two experiments reported below appear to confirm both sets of figures.

Two experiments in the U.S.A.

Wagner (135) at Beltsville (U.S.A.) grew white clover, cocksfoot (orchard grass) and fescue separately; then he grew white clover in association with cocksfoot and again in association with fescue.

He compared the increase in yield of cocksfoot (or fescue) brought about by association with white clover with that due to the application of varying quantities of nitrogen.

It was estimated in this way that white clover supplied a net quantity of nitrogen equal to 63 lb./acre [*70* kg./ha.] a figure very close to my deduction from Johnstone-Wallace's work.

On the other hand, in North Carolina (Voisin, 117, Vol. II, p. 380) it is estimated that Ladino white clover furnishes the soil with 198 lb./acre [*220* kg./ha.] each year; this corresponds approximately to the figure calculated by Johnstone-Wallace.

White clover supplies the same amount of nitrogen as 450 lb./acre [*500* kg./ha.] nitrate of lime

While being very cautious we can, however, say that, under the conditions of rational grazing, **white clover supplies the grasses in the pasture with nitrogen equivalent to approximately 450 lb./acre [*500* kg./ha.] nitrate of lime.** It must be emphasised that this figure is variable according to many circumstances; but it represents, nevertheless, a minimum value under normal average conditions.

Clover provides an organic nitrogen, very probably possessing many different qualities not found in mineral nitrogen.

It has recently been discovered that the bacteria (Rhizobium) of the nodules of legumes produce Vitamin B_{12}, the anti-anæmia vitamin.

For reasons we are just beginning to grasp, legumes are great improvers of the soil.

The favourable influence exercised on the performance of the animal by the association of clover and grasses

The association of different plants, including even those wrongly described as weeds, is an essential element in the nutritive and health-giving value of the grass of our pastures.

Not only does the association of white clover with grass increase the yield of the latter, as we have just seen, but it produces a sward which allows the animal either to yield much more milk or to gain more live weight. We will refer only to the results obtained at the Welsh Plant Breeding Station (142).

Tethered sheep were grazed on swards consisting of grass only (pure sowing or a mixture) or of grass in association with white clover. The live-weight gains of the sheep over one year are contained in Table 20.

British workers, commenting on this table, write:

> "More than this, the average condition of the sheep in respect of fatness was decidedly better on the with-white clover than on the without-white clover plots.
>
> "This evidence as a whole, so favourable as it is to both wild white clover and to perennial rye-grass, affords an interesting commentary on the known performance of old permanent pastures of the Leicestershire type to which these two species so absolutely contribute . . .
>
> "The same observation is being made at present on all the swards where management is developing and maintaining the association of rye-grass and white clover. This illustrates the advisability, from the point of view of animal performance, of utilising species which complement and support each other."

TABLE 20

Influence of the association of wild white clover with grass species on the feeding value of grassland

	Live-weight increase of sheep			
	The grass species *without* wild white clover		The grass species *with* wild white clover	
	lb./acre	[kg./ha.]	lb./acre	[kg./ha.]
Perennial rye-grass . .	263	[292]	313	[347]
Cocksfoot (orchard grass) .	240	[267]	279	[309]
Perennial rye-grass and cocks-foot (orchard grass) . .	260	[288]	283	[315]

From the Welsh Plant Breeding Station (142).

Phospho-potassic fertilisers are the principal nitrogenous fertilisers for pastures

Let us recall that phospho-potassic fertiliser considerably aids the development of white clover in pastures. All farmers know (and scientists have confirmed) that these fertilisers result in a very evident modification to the flora of a pasture, namely, the development of white clover. This development is sometimes spectacular: after phospho-potassic fertilisers have been applied over two or three years, white clover is seen to appear and flourish in pastures where not a single clover plant had been before. I will refer only to the figures (Table 21) quoted in that excellent brochure on the *Manuring of Grassland* published by the Potash Society of Alsace (84).

TABLE 21

Influence of phospho-potassic fertiliser on the composition of flora

	Without fertiliser	With phospho-potassic fertiliser
Grasses.	36·6%	44·0%
Clovers.	7·1%	28·0%
Other plants	56·3%	28·0%

From *Potasses d'Alsace* (84, p. 9).

Apply nitrogen to a pasture without injuring the clover

To apply nitrogen to a pasture in such an unskilled manner as to favour the grasses and smother the white clover would be an uneconomic undertaking. We would be spending money on buying nitrogen to destroy a free source of this fertilising agent. What was said by a foreign agriculturalist at the Sixth International Grassland Congress in 1952 needs therefore to be treated with reservations:

> "Legumes certainly supply the soil with nitrogen: but it must be remembered that the rôle of the legume today is much less essential in this connection than it was at a time when no commercial nitrogenous fertilizers were available."

It is certainly not economic to neglect the maintenance of white clover, a free source of nitrogen (not to mention its other qualities), in a pasture just because we have an artificial and expensive source of this fertiliser at our disposal.

Contrary to an opinion which is too widely held, the use of nitrogenous fertilisers on grassland does not necessarily lead to the withdrawal of white clover. Everything depends on the management of the pasture. It is impossible to separate the use of nitrogenous fertilisers on grass from the question of the way in which the grass is managed.

To illustrate that the use of mineral nitrogenous fertilisers does not suppress the free nitrogen factory represented by white clover, I will cite the example of my own pastures.

White clover in the Voisin grazings

Table 22, p. 52, shows the percentages of white clover, grasses, weeds, etc., in my pastures.

One part of the paddocks analysed comprised old and very worn permanent pasture, the other part a pasture seeded in 1947 (the seeding was perfectly successful). These two sections in each plot received the same fertiliser dressings and were always identically managed. The amounts of nitrogen applied annually in the form of nitrate of lime never exceeded 58 lb./acre

[*65* kg./ha.] until 1954 (the year of this survey): these dressings were spread over the whole season. The pastures were always rationally managed, and phospho-potassic fertilisers were regularly applied.

It is interesting to note in passing the superior quality of the flora of the old permanent pasture as compared with the seeded sward. For example, the percentage of mosses, daisies and bare patches is 26·1% in the sown pasture against 6·6% in the old permanent grass.

But what is of interest here is the fact that the percentages of white clover were 26·7% in the old pastures and 17·1% in the pastures seeded seven years previously. In both cases, then, the percentage of white clover is high despite relatively large dressings of nitrogen.

It can be concluded that, *if grazing is rationally managed,* if the applications of mineral nitrogen are well spread out and if the requisite dressings of phospho-potassic fertiliser have been applied, there is no danger of a regression of white clover, that free producer of nitrogen.

TABLE 22

Flora in three paddocks of the Voisin pastures according to whether seeded or old permanent grass (in 1954)

Paddock		Percentage of:				
No.	Type	White clover	Rye grass cocksfoot, timothy	Agrostis, red fescue, Yorkshire Fog, smooth-stalked meadow grass	Meadow and creeping buttercup, dandelion, celandine, sorrels, nettles	Moss, daisies, bare patches
A. Average of four readings for each of three paddocks						
11 {	Permanent	28·8	55·4	8·5	2·2	5·1
	Seeded	17·9	43·1	11·9	1·4	25·7
12 {	Permanent	26·8	57·5	8·3	0·7	6·7
	Seeded	17·7	46·4	11·6	1·1	23·2
17 {	Permanent	24·6	58·6	8·0	0·8	8·0
	Seeded	15·6	41·9	10·1	2·9	29·5
B. Average of four readings for the above three paddocks together						
{	Permanent	26·7	57·2	8·3	1·2	6·6
	Seeded	17·1	43·8	11·2	1·8	26·1

From readings made by Monsieur Hédin (of the Grassland Research Station of Rouen, Normandy).

Chapter 8

AT WHAT HEIGHT SHOULD GRASS BE GRAZED?

Optimum height of grass for grazing

THE optimum height of grass for grazing is that which allows for maximum grass yield (as we have seen above) while at the same time making it possible for the animal to harvest the greatest quantities of herbage with optimum nutritive value.

It has been shown that the maximum yield of fresh grass is obtained by observing the rest periods which give greatest productivity per day. In the case of permanent pastures this result is attained with a sward 6 in. [15 cm.] high. It will be seen later that it is at this same height that the cow harvests the greatest quantities of grass. It therefore seems important to examine the relationship between the height of the grass and the quantity present. But, as will be seen, the factors entering into play are so numerous that each case presents its own particular problem which cannot be solved by any rule or mathematical formula.

What is the height of the grass?

The height must be determined from the soil level. The soil, however, is not as easy to define as one might think. It is, in fact, covered over with a carpet of vegetation consisting of a tangled mixture of roots and stems, with the result that it is not easy to see where exactly the zero level lies from which the height of the grass should be measured. When one places a measure vertically on the soil, one often wonders what pressure to apply so that the measure is resting on the soil itself and not on the surface layer of turf. It is impossible to give any exact rules, but what must be stressed is that errors, or, to be more exact, differences in estimation of up to $\frac{2}{5}$–$\frac{4}{5}$ in. [1–2 cm.] can easily arise with the same, or between different, measurers according to the way in which the zero level was chosen.

Let us look now at the other extremity, the upper part of the turf. No two grasses are the same height, even if the sward comprises one species only. The only definition that can be given therefore is of an average height: to say that a grass is 6 in. [15 cm.] high means, mathematically, that the total of the heights of all the grasses, divided by the number of grasses, is 6 in. It is

53

evident that this is an intellectual pastime and that in practice, what one is making is a visual estimate.

Density of grass

For the same height of grass, assuming that this is perfectly defined, the quantity of fresh grass present varies according to the density of grass, that is, according to the extent which the grass has thickened in.

Pastures recently re-sown are much more inclined to grow to height than old pastures, with the result that, at the same average height, the quantity of grass present on a *new* pasture is much less than on an *old*. Waite (136) has shown that on very rich pastures where the average height of the grass was 6 in. [*15* cm.] 10,489 lb./acre [*11,655* kg./ha.] fresh grass were present on an old pasture against 5076 lb./acre [*5640* kg./ha.] in a sward recently re-seeded with a mixture containing cocksfoot.

It can be stated therefore that in order to get the same quantity of grass present in both cases a *new* sward must be grazed at a greater average height than an *old* permanent pasture. It is probable, however, that the rest periods in both cases will be the same.

The quantity of grass present and the quantity of grass harvestable

Our interest is not in the quantity of grass actually *present* but in the quantity *harvestable*, that is in the amount which will be removed by the stock.

What is the relationship between the total amount of grass actually present and the amount of grass harvestable? Hardly any precise figures are available on this point, and even if they were, the relationship would vary greatly according to the conditions prevailing and the management applied. If one "scrapes" the pasture to the uttermost limit, the proportion of grass harvested in relation to that present will be greater. On the other hand, this proportion will be reduced if the periods of occupation employed are too long, with the result that poaching and soiling with excreta renders a large part of the pasture "unharvestable".

Basing my statement on Waite's work (136), although it was carried out under very specific conditions, I should say that where the average height of the grass is 6 in. [*15* cm.] a permanent pasture of medium quality will contain a total amount of grass equal to 5130 lb./acre [*5700* kg./ha.], 84% of which is removable; this represents 4320 lb./acre [*4800* kg./ha.] harvestable grass where grazing is well conducted and the degree of defoliation suitably adjusted. In the case of a pasture recently re-seeded, 6 in. [*15* cm.] should probably be replaced by 9 in. [*22* cm.].

These are average figures and, unfortunately, very variable. We will see that in Normandy for rotations during the period of strong growth and with dense, rich grass, a figure of 9000 lb./acre [*10,000* kg./ha.] utilisable fresh grass can be attained where old permanent pastures are grazed at an average height of 6 in. [*15* cm.] (*vide* Table 69, p. 313).

It is evident, on the other hand, that these same pastures, during a period of feeble growth (especially the last rotation of the year) yield barely 3600 lb./acre [*4000* kg./ha.] fresh grass. It is true that it is difficult at this time of the season to get a pasture averaging 6 in. [*15* cm.] in height. But even in May a poor pasture where the average height of the grass is more or less 6 in. [*15* cm.] will have difficulty in providing 3600 lb./acre [*4000* kg./ha.] fresh grass.

We will see later (p. 76) that Johnstone-Wallace found that with a sward 4 in. [*10* cm.] in height the green matter present totalled 4500 lb./acre [*5000* kg./ha.], whereas with the grass 10 in. [*25* cm.] in height the green matter totalled 4950 lb./acre [*5500* kg./ha.] (no precise details are given as to the definition of the height or the nature of the herbage).

It must be stressed that all references to the quantity of fresh grass present in an acre (hectare) concern the amount of grass harvestable with medium grazing intensity. The intentional looseness of these indications will be noted, for the figures vary from place to place and season to season.

It would be interesting if the Research Institutes would study this question more closely, although any such work would provide only a partial aid from the practical point of view. In the long run it is *the eye of the grazier*, supported by his experience, which is the judge.

Figures and the visual estimate

No amount of description or columns of figures will ever explain to a farmer when his grass or clover are ready for *cutting*. Generations of experience have shown him that his forage is fit for cutting when it has reached a certain stage of development and has assumed a certain appearance. He has been told time and time again that he cuts his lucerne when $x\%$ of the stalks are in bud, $y\%$ in flower, etc. No farmer in the world has yet spent his time working out such a calculation before cutting; his eye tells him to cut at the optimum time, which, in general, corresponds with the $x\%$ in bud and $y\%$ in flower advocated in lectures and text-books.

The clover has still made headway

The other day I asked my foreman:

"Do you think this red clover is ready for cutting?"

"No," he said, "I looked at it four days ago, and it has made a lot of headway since then. . . . And it has not hardened to any extent."

Obviously an agronomist would have replied:

"The clover is still providing a lot of daily growth and has not yet reached the second level part of the growth sigmoid (Fig. 2). The protein content is only very slightly diminished. . . ."

My foreman knows nothing of sigmoid curves and protein contents, but he knows nevertheless when it is most expedient to cut the clover; that is, he knows when the plant is no longer "making much headway" and is beginning to "harden too much".

The decision must be Man's and not the animal's

At the moment, with *continuous* grazing, the time for defoliating a pasture is determined by chance, dictated by the climatic conditions of re-growth and the goodwill of the animal.

Man must decide when the pasture is to be sheared by the animal; that is, he must determine the time which corresponds to maximum plant and animal performance. The grazier who practises rational grazing will in time acquire the same experience as for cutting his grass or clover: in other words, he will come to know the point, in each rotation, when grazing should commence.

My head cowman, an Italian, has been with me for seven years, and he now knows, as well as I do myself, when a paddock in my rotation should be grazed in order to obtain maximum yield under the prevailing conditions.

To what height should the grass be grazed?

In Fig. 2 we saw the curve of grass re-growth; this is extremely slow at the beginning, for the grass is solely dependent on the reserves in its roots. Then little by little, thanks to these reserves, it builds "chlorophyll workshops" and at this moment the re-growth is accelerated.

The logical idea is therefore not to graze the sward too closely so that the plant will be left with sufficient green surface, the chlorophyll of which will be able, right from the start, to carry out its work of synthesis and immediately aid re-growth. In this way the duration of the initial period of slow re-growth is reduced. From the plant physiology point of view one might say that the low, level part of the S curve of re-growth is reduced.

Unfortunately we see here again perfectly sound, theoretical and scientific considerations running foul of practical obstacles which could not be foreseen *a priori*.

Cows (or animals in general) have the habit of first grazing down the parts they prefer before going on to the herbage they like less. It would therefore be very difficult, or rather impossible, to get a pasture which has not been over-grazed without it containing a large area of herbage completely untouched by the stock. In consequence, these refused areas, if not mown, grow and mature, giving rise to deterioration of the flora and loss in yield. But it must not be forgotten that a sward which is insufficiently grazed down can equally be retarded in its re-growth due to insufficient stimulation of the crowns.

Over-grazing must be avoided, but so equally must under-grazing. There is an optimum degree of utilisation which allows optimum re-growth of the grass under the practical conditions realisable in the conduct of the grazing.

Chapter 9

THE COMPOSITION OF GRASS

Brief particulars

I SHALL restrict myself here to those details of the general composition of grass required for the purposes of the present volume.[1] The only question I will expand upon is that of the proteins (or more exactly, *false* proteins) of grass. This will enable the reader to better understand the accidents which can be caused by ill-conceived systems of intensive grassland management.

Limits of chemical analyses

The farmer has a profound respect for the analysis of the feedingstuffs he uses and imagines that the chemist can see, absolutely and perfectly, what these contain. For this reason these analyses have been misused to get oneself out of an awkward situation. I have often wondered even if analysis does not play in our modern agricultural science (like all sciences of life, so little advanced as yet) the same rôle as Latin did for seventeenth-century doctors. Both analysis and Latin are endowed with an air of mystery which always inspires great respect. Let us be under no illusion as to what analysis of our grass (or foodstuffs) can provide: vague indications which are an aid to the research worker or farmer but which cannot, in any circumstances, replace the experimentation of the scientist or the spirit of observation in the farmer.

It must be remembered that, until recent times, chemical analyses of herbage were always concerned with analytical groups and not with the chemical bodies determined. It is only in the course of the past few years that, thanks to electrophoresis, chromatography, etc., we have begun to understand a little better certain of the individual elements which constitute grass. But in spite of this recent progress, our analytical methods allow us to understand only some aspects of the composition of simple, ordinary feeding-stuffs and not very many aspects of the composition of grass.

[1] The author has work in progress, the results of which he will publish in due course, on "The Composition of Grass".

Analyses of grass must take the type of management into account

It has already been said with regard to the measurement of yield and growth as well as for determining the value of a new variety of herbage plant, that it is the management above all that must be taken into account. The most urgent problem ought to have been (and still is) the determination of optimum methods of grassland management.

It was stressed that, to obtain maximum yields of grass, the rest periods must be varied according to the season and that all the experiments up till now had been carried out with equal rest periods whatever the time of year (the experiments always being discontinued in winter). The same has been true of the analysis of grass: it is the average composition of herbage cut every week, every two weeks, etc., that has been determined.

We cannot do better than cite, in this connection, the work carried out by Woodman at Cambridge before the last war. In the course of this extensive research which lasted several years, Woodman (145 and 146) examined the composition of grass defoliated by means of a cutting instrument every 1, 2, 3 and 4 weeks respectively, that is to say, always after the same interval of time.

When Geith (27), in Germany, wanted to determine the composition of grass, he grazed it with cows. This represented an enormous step forward by comparison with the laboratory method, which to-day is still that normally used and consists of cutting the grass with a scythe or shears instead of employing the grazing animal. Unfortunately Geith always had his animals return to the same plot after an identical interval of about 16 days (maximum variation of 14–20 days). These analyses, although an improvement, still do not correspond to the conditions which must prevail in a well-conducted rational grazing.

It is understood, therefore, that every indication of the composition of herbage must be accompanied by a definition of the conditions in which the grass was managed. The scope of the present work cannot include these details, and it is therefore preferable not to give figures which in themselves are of little significance and run the risk of misleading the reader.

Composition of grass cut at different, fixed intervals

Studies of this nature, although carried out under conditions far removed from continuous grazing and still further removed from those of rational grazing, are nevertheless instructive to some extent.

Table 23, p. 59, contains the results obtained by the Swiss Geering (26) cutting the herbage every 1, 2 and 8 weeks in the course of the season. In Table 24, p. 60, I have collected some of the results extracted from Woodman's monumental work. These two tables, 23 and 24, illustrate clearly some of the points which I will have occasion to refer to again in examining methods of grassland management themselves:

Young grass, cut every week, is:

1. Very rich in crude protein (that is to say, nitrogenous compounds, protein and other substances).

2. Contains little crude fibre and ballast (non-digestible, organic portion).

3. Is relatively rich in potassium and phosphorus and relatively poor in calcium.

4. The nutritive ratio (ratio of digestible crude protein to starch equivalent) is very narrow; that is to say, the proportion of protein (or so-called protein) is much too high in relation to the nutritive units.

TABLE 23

Composition of dry matter cut at different fixed intervals

	Cut every				
	Week	2 weeks	4 weeks	6 weeks	8 weeks
Composition of dry matter of the *grass:*					
% digestible crude protein	22·8	18·0	14·0	10·9	10·0
% starch equivalent	63·6	61·4	58·7	58·1	53·5
% ballast	16·0	17·1	18·8	20·7	22·3
% potassium (K)	2·91	2·74	2·41	2·24	2·08
% calcium (Ca)	0·71	0·93	1·14	1·14	1·29
% phosphorus (P)	0·52	0·48	0·44	0·39	0·35
Annual yield:					
Dry matter— lb./acre	875	4610	6350	7420	7790
[kg./ha.]	[980]	[5170]	[7110]	[8310]	[8730]
Digestible crude protein— lb./acre	199	831	888	809	779
[kg./ha.]	[223]	[931]	[995]	[906]	[873]
Starch equivalent— lb./acre	550	2830	3720	4240	4170
[kg./ha.]	[620]	[3170]	[4170]	[4750]	[4670]
Nutritive ratio	1 : 2·8	1 : 3·4	1 : 4·2	1 : 5·3	1 : 5·4

N.B. 1. The nutritive ratio is the ratio of digestible crude protein to starch equivalent.
2. Ballast = Non-digestible organic matter.

From Geering (26).

It is thought that these indications are sufficient for the purpose of the present work, but it is considered necessary to add a few words on the protein or, to be more accurate, the nitrogenous compounds of grass.

TABLE 24

The feeding value of the dry matter of pasture grass cut at different (fixed)
intervals

	Weekly	Fortnightly	Three-weekly	Monthly
Crude protein . .	24·74	23·48	12·14	20·23
Ether extract . . .	5·29	6·53	6·04	6·51
Crude fibre . . .	15·39	15·94	17·16	16·92
Nitrogen-free extractives .	44·79	44·53	46·68	47·58
Ash	9·79	9·52	8·98	8·76
Starch equivalent . .	67·74	69·87	69·39	67·04
Digestible crude protein .	19·97	18·75	16·66	14·76
Nutritive ratio . .	1 : 3·39	1 : 3·72	1 : 4·17	1 : 4·54

N.B. 1. Figures are percentages in dry matter.
2. The nutritive ratio is the ratio of digestible crude protein to starch equivalent. This ratio was calculated by the author and was added to the table.

From Watson (139), p. 19, following Woodman's original work.

The so-called proteins of grass

In the hundred years that have passed since the beginnings of scientific investigation into the feeding of livestock, the nitrogen in the grass (or feeding-stuff) has been analysed and multiplied by a factor close to six; the result of this multiplication has been considered to correspond to the crude protein. This confusion between nitrogen and protein is one of the most serious errors in our feed tables, as well as in the analyses carried out by the different laboratories. We will see in this work to what catastrophes such confusion can lead. For the moment I will cite only the opinions of three scientists in three very different countries (U.S.A., Britain and New Zealand) on the so-called proteins of grass.

A "crude" which merits its name

Albrecht is a scientist from the University of Missouri (U.S.A.) to whom we are indebted for perhaps the most original studies on the relationship between the nature of the soil and the composition of living matter. He writes (1):

"It is very unfortunate for us that what we call protein is in fact merely the ensemble of diverse organic combinations of nitrogen. . . . Up to the present we have made no clear distinction on the quality of the nitrogen in our foodstuffs.

"We must no longer be satisfied to burn the foodstuff in sulphuric acid, collect the nitrogen thus obtained (Kjeldahl method), multiply this nitrogen by a factor between 5·75 and 6·28 and consider the result as representing the protein. . . . Such a protein is called crude protein and certainly merits its name."

Albrecht continues:

> *"Such protein is only 'crude' protein and certainly one all too crude to be taken as a basis for complete nutrition."*

The opinion of a Nobel prize-winner on the practical value of feedingstuffs analyses

Synge, the inventor of paper chromatography, received the Nobel Prize for Chemistry in 1952. Fortunately for us, he is a member of the staff of the Rowett Research Institute in Scotland, which studies almost exclusively the scientific aspects of digestion in animals (*vide* Part Two, Chapter 5). More fortunate still from our point of view is the fact that Synge very modestly considers that the science of feeding livestock has advanced so little as yet that one must be more than careful in handing out the smallest piece of advice based on analyses to a farmer.

He writes (Synge, 104):

> "It is sheer presumption to recommend changes in farm practice on the basis of a few Kjeldahl figures."

It is a comfort to see such a great scientist so wary with regard to feedingstuffs analyses and to read his conclusion that the empirical methods employed by the farmer in feeding his livestock must be looked upon with respect.

We must learn more about the protein fraction of the herbage

Melville, the Director of the Grassland Research Station, Palmerston North, New Zealand, writes (75):

> "When we talk so glibly of 'carbohydrates' and 'protein' what do we really mean? To take the latter first, the reported protein content of the feed, in virtually all animal nutrition studies, is nothing more than its nitrogen content multiplied by some factor between 6·0 and 6·25. In the first place some 5% to 25% of the total nitrogen of the leaf occurs not in protein but in low molecular weight, water-soluble compounds. It must not be forgotten that this non-protein nitrogen fraction also contains other nitrogenous compounds which can exert a considerable influence on the health of the animal. . . .
>
> "But leaving out of account the non-protein nitrogen fraction, the lumping together, under the generic term 'protein', of the whole complex array of individual proteins which exist in the leaf cell, has always been a source of grief to one who was once a protein chemist.
>
> "If we are in earnest about our search for pasture quality, we *must* learn more about the protein fraction of the herbage.
>
> "And to show that nitrogen metabolism studies should not be the monopoly of the plant biochemist, I should like to ask the animal physiologists a simple question. Why does the herbivore, unlike any of the omnivores thus far studied, excrete up to 15% of its urinary nitrogen as amino-nitrogen (both free

and combined) and of this amino-nitrogen why is it practically all in the form of simple amino acid glycine?"

Such is the opinion of this great research worker and grassland specialist. We repeat with him:

"If we are in earnest about our search for pasture quality we *must* learn more about the protein fraction of the herbage."

Melville, moreover, emphasises the point concerning the constitution of the nitrogenous substance in the grass which must have been the cause of many blighted hopes and also, it must be said, of many catastrophes in the intensive management of pastures. The New Zealand scientist reminds us in effect that: "5–25% of the total nitrogen of the leaf occurs not in protein but in low molecular weight, water-soluble compounds".

Too many of the problems of animal nutrition have been obscured by a mathematical formula

As in many instances in animal nutrition, the hope was to camouflage the problem by using a mathematical formula, and the protein equivalent has been defined as equal to:

$$\frac{Digestible\ nitrogenous\ substances\ +\ Digestible\ protein\ substances}{2}$$

which goes back to the supposition that, in every case, non-protein substances are utilised in half.

This fiction could hardly be allowed to continue, especially in the case of grass, where the non-protein nitrogenous substances often attain such importance, and where there is a marked lack of balance between the nitrogen and the carbohydrates.

A high percentage of non-protein nitrogen in the grass may endanger the health of the animal

In fact, as Sullivan (103) p. 11, of the Pasture Research Laboratory, Pennsylvania, has stressed, the percentage of non-protein nitrogen in young grass can sometimes represent half the total nitrogen. He considers that if the percentage exceeds 20% and if the animal is fed exclusively on this grass, serious digestive disturbances will result. This is what we shall have occasion to see in studying the digestion of protein in the ruminant (pp. 117–122).

Approximate composition of grass as grazing advances

Having made all these reservations to our knowledge of the quality of grass, it is sufficient finally to point out the essential elements contained in the grass as grazing advances (Table 25—based on Geith's work). We will see later

(Table 31, p. 82) what this division of the periods of occupation actually means.

TABLE 25

Approximate composition of grass of an average height of 6 in. [15 cm.] at different stages of grazing

	1 lb. of fresh grass contains			1 kg. of fresh grass contains			1 lb. of dry grass contains		1 kg. of dry grass contains	
	Dry matter, gm.	Digestible crude protein, gm.	Starch equivalent, lb.	Dry matter, gm.	Digestible crude protein, gm.	Starch equivalent, kg.	Digestible crude protein, gm.	Starch equivalent, lb.	Digestible crude protein, gm.	Starch equivalent, kg.
1. Composition of whole eatable part . . .	9·5	12·2	0·132	21	27	0·132	59	0·63	130	0·63
2. *Period of occupation divided into two parts:* A. Portion grazed during first part.	9·1	12·7	0·132	20	28	0·132	64	0·62	140	0·62
B. Portional grazed during last part .	10·0	11·8	0·138	22	26	0·138	54	0·63	120	0·63
3. *Period of occupation divided into three parts:* A. Portion grazed during first part.	9·1	13·6	0·132	20	30	0·132	68	0·61	150	0·61
B. Portion grazed during second part . .	9·5	12·2	0·132	21	27	0·132	59	0·63	130	0·63
C. Portion grazed during last part .	10·4	11·3	0·142	23	25	0·142	50	0·62	110	0·62

Paragraph 1 of Table 25 corresponds to all the grass present on a turf 6 in. [15 cm.] high on the average. This is the total composition, but, as we shall see in studying the tastes of the cow, the animal put out to graze a sward of this nature tends to select the most tender grass. The result is that at the beginning of grazing a paddock with grass of the composition described in paragraph 1 the cow is going to select a grass more tender and more rich in proteins (2A and 3A for example).

The action of this selection, and consequently the difference in the quality of grass selected is the more marked when one divides the total time of grazing a paddock into a larger number of fractions or, what amounts to the same thing, if one divides the herd into more groups.

It is emphasised that these are approximate figures only, their aim being to serve as guides in certain considerations later.

PART TWO
THE COW

Chapter 1

HOW THE COW HARVESTS THE GRASS

The "harvesting" of the grass

FIRST and foremost I should like to insist on the word *"harvesting"* of the grass by the cow. Indeed, I believe it indispensable to distinguish clearly between the following two actions:

> (1) feeding in the stall with grass *previously cut* (or any other food that one carries to the animal), and
> (2) feeding on the pasture with the grass *underfoot*.

In the first case I say, by way of definition, that the cow **eats** the grass; in the second, that the cow **harvests** the grass. In effect, these are two very different actions, and I think it preferable to differentiate between them by not using the same verb in both cases.

All the works on grass and forage crops devote numerous chapters to methods of harvesting crops for drying or the making of silage. But, if there is much preoccupation with *man's* methods of harvesting the animal's food for four months in the year, not a single word is to be found regarding the *animal's* method of harvesting its own food during eight months of the year. Take as an example an excellent American work of 720 large-sized pages on green forage crops. One hundred pages are devoted to methods of harvesting and conserving forage crops either by drying or ensilage, but not a single line describes the cow's own harvesting methods.

Let us try, therefore, to see how the cow proceeds to harvest her food. When we know what methods she employs we will be able to help her in her work, that is to say to get a better yield from our pastures.

The cow's programme of work

At the University of Cornell (U.S.A.) between 1940 and 1943, Professor Johnstone-Wallace carried out some remarkable work on the dividing up of the different operations performed by the cow each day in order to harvest her food. Cows and suckling calves were chosen for the investigation. An observer was installed on a raised platform and for 24 hours of the day the behaviour of the cow was observed.

If one takes as "grazing" the whole of the operation consisting of moving about in search of food and then shearing the grass (that is, browsing) once it has been found, **the time spent in grazing in the course of the 24 hours of a day was a little less than 8 hours. This time was never exceeded.**

As for the browsing operation itself, it lasts a little less than 5 hours per day.

With the average temperature prevailing at Cornell, 60% of the grazing took place by day and 40% by night. As the temperature rises, the proportion of the harvesting work done by night increases.

The distance travelled in the course of 24 hours was about 2½ miles [4 km.]. But the strange thing was that 80% of the travelling takes place during the day, although the grazing time by day represents only 60% of the total grazing time. Professor Johnstone-Wallace considered that this increased efficiency of night grazing was due to the fact that the cows, at night, were not being disturbed by all the things that take place round about them during the day: these, by arousing their curiosity, incite them to move. Personally, I have wondered if flies and insects might not equally be the explanation.

The time spent in ruminating is approximately 7 hours. These investigations showed that this time was greatly influenced by the crude-fibre content of the grass: but unfortunately, the question was not pursued further.

Part of the rumination takes place while the cow is lying down and part while she is standing up. A cow lies down for a period slightly less than 12 hours. This total period is normally divided into nine rest periods of unequal length, the minimum being 1 hour and the maximum 6 hours.

The grass was very tender and had a high water content, and although the cows had free access to water they took only one drink per day.

On the average the cows defæcated twelve times and urinated nine times in the course of the day.

Cows are union members

It was stated above that the grazing time, that is the harvesting time, was the total of the times spent moving about and browsing, and that this amounted to approximately 8 hours a day. It is important to stress a few points in this connection:

This time remains remarkably constant, and as Professor Johnstone-Wallace humorously observes, it appears that the Union of Cows has imposed upon its members very strict rules which none would dare to infringe. The time remains the same whether the cow is grazing a tender, lush meadow or a bare and poor pasture. In the latter instance, however, as we shall see, the cow will gather barely 44 lb. [20 kg.] of poor-quality grass, which is insufficient to meet her maintenance requirements. But all Professor Johnstone-Wallace's experiments indicate that the **cows refuse to work any hours of overtime.**

It must be admitted that by the end of 8 hours the animal, in travelling and shearing, has utilised an amount of energy representing the maximum effort of which she is capable. To understand how enormous this effort is, let us examine more closely the mechanics of the cow's grazing.

The mechanics of grazing

Browsing consists in the animal severing the leaves and stems with the aid of its two jaws the width of which, in a fully grown beast, varies between 2·4–2·6 in. [*6–6·5* cm.]. The row of teeth in the lower jaw comes into contact with the muscular pad of the upper jaw, thus producing the action of shearing. The relative position of the teeth and this pad makes it impossible for the cow to browse closer than $\frac{1}{2}$ in. [*12* mm.] from the soil.

During browsing the tongue is continually in action: it emerges from the mouth and moves from one side to the other. It is used to gather a certain amount of grass and get it into the mouth.

In the course of grazing the cow travels forward, moving her head and neck regularly from one side to the other within an arc of 60–90° and taking between 30 and 90 bites per minute.

It is immediately evident that the rhythm per minute can vary from simple to triple time. Professor Johnstone-Wallace observed that a cow which is very hungry and is grazing a prime-quality sward makes 90 jaw movements per minute, although in other cases, which unfortunately are not detailed, this rhythm can fall to 30 beats per minute.

The time during which the cow bites uninterruptedly varies greatly. The maximum time observed has been 30 minutes and it is interesting to note that this was on a sward 4–4$\frac{1}{2}$ in. [*10–12* cm.] in height of almost perfect quality.

We have spoken of the uninterrupted rhythm of shearing. But a factor intervenes to stop the shearing without always stopping the meal—the length of the grass. If a cow is grazing very long grass, let us say 10–14 in. [*25–35* cm.] high, she can sever the upper layer to about 2$\frac{1}{2}$–3 in. ("the cream"), but she can also plunge her muzzle into the mass and tear off a mouthful about 12 in. [*30* cm.] long. In this case, the cow will not be able to absorb such a large and long mouthful without submitting it to a great deal of manipulation. She will then raise her head and it will take her about 30 seconds to masticate and swallow the mouthful. Now, during these 30 seconds a cow grazing a sward 4 in. [*10* cm.] high will have swallowed 30 mouthfuls containing infinitely more grass and of a much greater value than the single mouthful.

In conclusion, it should be noted that Professor Johnstone-Wallace perfected a device for recording the movements of the cow's jaws. Unfortunately, it has hardly ever been used except with cows eating *cut* grass in their stalls.

To-day, one puts a harness on the back of the grazing cow which, having attained greater perfection, permits many and diverse recordings.

Division of the work of harvesting the grass in the course of a day

Grazing and ruminating alternate in periods during the course of the day. As the diagrams published have become more and more profuse, we will be content with reproducing only the one contained in Fig. 7, below. These observations were made in England in May 1949 at the Grassland Research Station, Stratford-on-Avon.

The total time devoted to harvesting the grass varied between 6 and 8 hours, and 93% of this harvest was gathered by day. Rumination, a large part of which took place by night, lasted from $5\frac{1}{2}$ to $9\frac{1}{2}$ hours according to the animals.

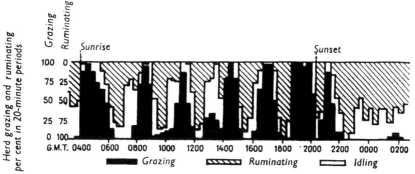

FIG. 7. Periodicity of grazing and cudding in a herd of eight bullocks, Stratford-on-Avon, May 1949.

From Tayler (105a).

Fig. 7 shows the percentage of animals in the herd which for every period of 20 minutes is, in the course of grazing, ruminating or idling. Grazing commences with the rising of the sun and alternates with periods of ruminating and idling. It is the first period of grazing which is the longest; this is the big morning meal. The next big meal will be eaten before sunset. These meals are regular in all temperate climates, but the other periods of grazing during the day can vary greatly according to external conditions.

Various studies on the time the cow spends in grazing

Studies on grazing time, or, to be more exact, harvesting time, necessarily multiplied following on the work of Johnstone-Wallace and Kennedy. Numerous factors were seen to be influencing the harvesting time: temperature, density and fibrousness of the grass, etc., etc.

In general, the harvesting times varied in the region of the 8 hours or 480 minutes noted by Johnstone-Wallace and Kennedy (51). This is well illustrated by Table 26, taken from Tribe (110). But there was one factor which renders these results inconclusive and often difficult to explain—the fact that time spent in harvesting is a hereditary characteristic in the cow.

TABLE 26

Number of minutes per day spent grazing by cows

Worker	Grazing time, minutes
Hodgson (40)	411–439
Hancock (35)	410
Castle, Foot and Hailey (13)	390
Waite, MacDonald and Holmes (137) . .	552
Cory	461
Shepperd	495
Johnstone-Wallace and Kennedy (51) . .	452

From Tribe (110).

Grazing time and hereditary character

A study of grazing times was made at the Ruakura Animal Research Station in New Zealand with six pairs of monozygotic twins, that is both springing from the same egg, and therefore having exactly the same inheritance. This gave rise to the surprising discovery that the time which an animal devotes to grazing is a distinctly *hereditary* characteristic, as is shown by Table 27:

TABLE 27

Grazing time for monozygotic twins

Date	Time grazing, minutes											
	T 1	T 2	T 11	T 12	T 13	T 14	T 17	T 18	T 23	T 24	T 31	T 32
Nov. 8, 1946	448	370	472	456	376	427	465	494	284	325	399	430
Dec. 12, 1946	427	413	520	531	410	448	488	489	421	442	443	455
Jan. 11, 1947	403	424	468	440	410	405	502	461	387	397	358	422
Feb. 5, 1947	386	414	412	396	362	353	447	437	296	281	355	337
Feb. 25, 1947	366	384	408	448	363	337	428	456	312	314	356	318
Apr. 10, 1947	357	349	444	411	372	377	464	434	318	291	365	391
May 8, 1947	433	433	496	475	416	452	576	539	363	346	419	410
Average per heifer	403	398	460	451	387	400	482	473	337	342	400	395
Average per monozygotic pair	400		455		393		477		339		397	

N.B. 1. Brackets indicate a pair of monozygotic twins.
2. It should be noted that these observations were made in the Antipodes where summer corresponds to winter here.

From the Ruakura Animal Research Station, New Zealand (36).

In fact, for one pair of monozygotic twins at a certain date (that is, under certain climatic conditions), the grazing time is identical to within $\pm 2\%$, which is insignificant. By contrast, if we consider the averages of the grazing times of each pair of twins, we find that between the lowest and the highest figures there is a difference of 40%, which is truly enormous for beasts which have been grazed strictly under the same conditions.

It is obvious, therefore, that *time spent in grazing is an essentially individual characteristic in the cow*, a function of her hereditary complex. We will meet this same hereditary influence later in the quantity of grass harvested by the cow.

The average rhythm of the jaw movements appears to be constant

The studies made at Ruakura Animal Research Station must also provide us with valuable information on the number of jaw movements made by the cow each day in harvesting her food. The results obtained from two pairs of monozygous twins are contained in Table 28. From these, surprisingly, it is obvious that whatever the hereditary character, the average number of jaw movements made per minute remains more or less the same: that is, very near to 50, although, as we saw from Johnstone-Wallace's work, the rhythm can vary from 30 to 90 beats per minute.

TABLE 28

Daily number of bites for monozygotic heifer twins

Number of the heifer	First pair			Second pair			Average for the 4 heifers
	T 1	T 2	Average for pair	T 1	T 2	Average for pair	
Daily grazing time, minutes	354	404	379	576	534	555	467
Total number of bites:							
Grazing	17,025	20,538	18,781	30,955	27,346	29,150	23,966
Ruminating	15,525	18,657	17,091	18,686	15,600	17,143	17,117
Grazing and ruminating	32,550	39,195	35,872	49,641	42,946	46,293	41,083
Average rate of bites per minute	48·1	50·8	49·4	53·7	51·2	52·4	51·3

From Hancock (36).

The total number of jaw movements per day is a hereditary character

As stated above, daily grazing time is governed by the inherited aptitude of the animal. As a result, the *always identical* average number of jaw movements per minute multiplied by an *increased* number of minutes spent in harvesting, in the case of the cows most suited (by nature) for harvesting grass, gives a higher total number of jaw movements per day.

A heifer of the pair with the greatest aptitude was capable of 29,150 jaw movements per day for browsing, while a heifer of the least-suited pair made only 18,781. The former was capable of devoting 555 minutes (9 hours 15 minutes) to harvesting, and the latter only 379 minutes (6 hours 19 minutes). In other words, *an "average" heifer from the most able pair of twins made 55%* *more jaw movements and gathered grass for a period 46% longer than the heifer* *from the pair with the less favourable hereditary characters.* Of course, as we have just seen, this rhythm per minute varies in the course of the same day or in the course of a "browsing passage".

There are no "quick" grazers, but there are "long" grazers

When a cow, having rested and ruminated, starts to browse again, she shows a tendency to browse more quickly to begin with, but the rhythm slows down gradually until finally she stops altogether. For a long time it was thought that there must be "quick grazers" and "slow grazers", the character being hereditary. According to the New Zealand work, it would appear that, on the contrary, all grazing cows have the same average rhythm but that *heredity produces "grazers with* LONG *harvesting times" and "grazers with* SHORT *harvesting times".*

Life in the herd and individual behaviour

Cows normally live in a herd, and so it is important to know to what extent their individual behaviour is influenced by this communal life.

We have all noted that our cows have a tendency to gather together in groups before lying down. If one beast is not in the group we can be certain that she has not been in the herd for long and has not yet acquired the "freedom of the city". Apart from these exceptional cases, cows graze together in groups, and their browsing is almost rhythmical. There is no doubt that the cows in a herd tend to adopt certain identical attitudes and that fixed "laws of the herd" govern the behaviour of each animal.

John Hancock, of the Ruakura Research Station in New Zealand, has carried out some very interesting studies on "herd instinct" (35). On the basis of his work and my own observations, this is how that instinct works.

Cows have in general a tendency to graze, ruminate or rest simultaneously and together. When a group of cows is in the act of grazing, if a minority begins to ruminate and, in spite of this appeal, the rest continue grazing, the minority starts grazing again with the majority. On the other hand, if the signal is taken up by a majority, then the whole herd will finally start ruminating.

Are there cows which play the part of leader and whose will is imposed on the mass? This is difficult to affirm, but it does seem probable, as one can easily see in the case of mountain herds.

John Hancock carried out a very unusual experiment on this subject also.

He took a herd of *twin* cows which he divided into two groups, the pairs of twins being separated into each of the groups. The two groups were placed in two adjacent pastures, which, in fact, were only one single pasture divided into two, with the result that the two could be considered identical and there was no particular soil or herbage influence on the attitude of either group.

After milking, the two groups were put out simultaneously on the pasture, but it was repeatedly observed that there was nothing simultaneous about the behaviour of the two groups: one, for example, being in the act of grazing while the other was in the act of ruminating. One might think, therefore, that there must be, within a herd, one or more cows that impose their will on the group as a whole.

Cows tend to graze together or to travel in accordance with certain rules: on a long, narrow grazing the group moves from one end to the other, generally following the same path. On a square pasture they tend to travel in circles.

The external temperature has a very strong influence on the tendency of the cows to form open or close groups. When it is very warm, and especially if it is very heavy, the cows gather very closely together, just as the group is relatively more scattered at cooler temperatures. We have all seen cows in stormy weather lying so closely together that they are almost touching.

After great excitement the cow needs a grass cordial

There is an observation which anyone can make: take a dog into or round the outside of a pasture. All the cows will rush towards this unwanted animal, abandoning whatever they are doing or the position they happen to be in at that moment. When the object of their anxiety has disappeared they will return to the pasture, and whether they were ruminating or grazing when they rushed towards the dog, *they will always start* GRAZING *when they go back.* You will never see them lying down or starting to ruminate, as if they are too much on edge to give their mind to these occupations requiring calm. One gets the impression that they want to recover from their great excitement fortifying themselves with a grass cordial.

Herd instinct and division into groups

It would obviously be interesting to know more about the causes and effects of the herd instinct in cows. But from what has just been said, a probable conclusion can already be drawn: namely, that it is desirable to form homogeneous groups of cows so that they can draw up as easily as possible their common programme of occupations (*vide* p. 155). One might fear that an exceptional cow, capable of a longer grazing time, will be hindered because the other cows will start to ruminate. In fact, it is probable that due to its herd instinct this exceptional cow will not complete the effort she was in the course of making to harvest her grass, but will adopt the attitude of the herd as a whole.

PHOTO 1

*Farm Worker of the Elorn
Valley (Finistère)
mowing a plot which produces
53 tons per acre
(120 tons per ha)
fresh grass per annum*

Photo Voisin

PHOTO 2

*One of Professor Ivins'
isolating cages at Nottingham
University (England)*

Photo Voisin

Photo 3

Dr. Schutzhold beside an
isolating cage.
Grassland Research Station
(Cleves, Germany)

Photo Voisin

Photo 4

Rengen:
Professor Klapp beside an
isolating cage

Photo Voisin

A TABLE, shewing what Plants are eat or neglected by the five most common Domestic Animals, Oxen, Goats, Horses, Sheep, and Swine.

§ 1.

		O.	G.	Sh.	H.	Sw.
1	Veronica *officinalis.* Male speedwell P. V.	1	1	1	1	0
2	——— *scutellata.* Narrow leaved water speedwell, P VI.	1	1	1	1	n
3	——— *agrestis.* Germander Speedwell, or chickweed, A V.	1	1	1	1	n
4	——— *hederifolia.* Ivy leaved Speedwell, or small hen-bit, A IV. V.	1	1	1	1	n
5	——— *triphyllos.* Trifid Speedwell, A V. VI.	1	1	1	1	n
6	Anthoxanthum *odoratum.* Vernal or spring grass, P VIII. Tab. 9.	1	1	1	1	n
7	Milium *effusum.* Millet-grass, A VI. VII.	1	1	1	1	n
8	Aira *caespitosa.* Turfy hair grass, P VII. VIII.	1	1	1	10	1
9	—— *flexuosa.* Small hair grass, P VII. VIII.	1	n	1	1	n
10	—— *montana.* Mountain hair grass, P VII. VIII.	n	n	1	n	n
11	—— *caerulea.* Purple hair grass, P VIII.	n	1	1	1	n
12	—— *canescens.* Grey hair grass, P VII.	1	1	n	n	n
13	Poa *aquatica.* Reed meadow grass, P VII.	10	1	1	10	n
14	—— *compressa.* Creeping meadow grass, A VI. Tab. 8.	1	1	1	1	n
15	—— *annua.* Annual meadow grass, or Suffolk grass, A VI.—IX.	1	1	1	1	1
16	—— *pratensis.* Great meadow grass, A VI. Tab. 7.	1	1	10	1	1
17	—— *angustifolia.* Narrow leaved meadow grass, A VII.	1	1	1	1	1
18	—— *trivialis.* Common meadow grass, P VI. VII.	1	1	1	1	1

APPENDIX.

365

* Sheep I have found delight much to pasture on fields that abound with the purple fescue grass, and eat them quite bare.

† This grass is not much liked either by oxen or horses.

PHOTO 7

Palatability of 583 plants for 5 animal species, determined by Swedish botanists in 1740

From James Anderson's work, 1777 edition
Photocopy made by National Institute of Agricultural Botany, Cambridge

Chapter 2

THE QUANTITIES OF GRASS HARVESTED BY THE COW

Methods of measuring these quantities

THE methods of measuring quantities of grass harvested by the cows represent a technical and very special subject which I will leave aside here. However, since the method of the isolating cage was used by Johnstone-Wallace, whose work throws light on our own ideas to no little extent, I will state briefly of what this method consists.

Before the stock are put on to a grazing, different sections are isolated by means of isolating cages, all of the same surface area. The grass under one or more cages is cut, and from that yield the quantity of grass per acre at the disposal of the livestock is assessed. At the end of the grazing period an unprotected, grazed section equal in area to that under the cage is cut. The difference between the quantity of grass on the grazed area and that on the isolated area is indicative of the quantity eaten by the stock. Analysis of the herbage on the grazed and on the isolated areas will likewise provide information on the quantity of nutrients absorbed by the animal (numerous formulæ have been suggested to improve these calculations; *vide* Brown (10)).

In Photo 2 (facing p. 74) I have reproduced a photograph (which I took at Nottingham University) of the isolating cages used by Professor Ivins, who has made such an important contribution to the progress of our methods of grassland utilisation (*vide* also Photos 3 and 4).

The amount of dry matter "eaten" by the cow

The majority of, although not all, feed tables for cattle, state that a beef or dairy cow weighing approx. 10 cwt. [*500* kg.] consumes daily about 29 lb. [*13* kg.] of dry matter which is sufficient to satisfy it (*vide* p. 97). It must be stated in this connection, however, that the question of "satisfying" ruminants has been very little studied and that books on nutrition do not tell us on what experiments (numerous and repeated) they base this figure of 29 lb. [*13* kg.] of dry matter as satisfying a 10-cwt. [*500*-kg.] beast.

75

It should be added that the quantity of dry matter required to satisfy varies from author to author. For example, we find:

Dry matter absorbable by cows weighing 10 cwt. [500 kg.]

Wolff-Lehmann .   .   . 27–33 lb. [12–15 kg.]
Kellner   .   .   . 24–38 lb. [11–17 kg.]
Morrison .   .   .   29 lb.   [13 kg.]
Armsby   .   .   . 22–33 lb. [10–15 kg.]

Even in the stall, however, the quantities of dry matter absorbed by cows of 10 cwt. [500 kg.] are very different from the supposed average of 29 lb. [13 kg.]. This was shown as much as fifteen years ago by Fissmer's study, my French translation of which was published in the "Annales de l'Institut de la Recherche Agronomique" (22): his work seems to have attracted very little attention.

In studying the work done by Johnstone-Wallace as mentioned above, I was forced to come to certain fundamental conclusions as regards the quantities of grass harvested by the cow; conclusions which were going to allow me to shake the idol of 29 lb. [13 kg.] of dry matter satisfying a cow of 10 cwt. [500 kg.] which, up till that time, was dominating the text-books on nutrition and feed tables (Voisin, 118).

Let us look first of all at the figures obtained at Cornell (U.S.A.) in 1942 by Johnstone-Wallace and Kennedy (51, 53 and 54).

Professor Johnstone-Wallace's results

Professor Johnstone-Wallace, in collaboration with Kennedy, measured by means of isolating cages the quantities of grass harvested by cattle. The cows were put on to pasture where the average height of the grass was 4 in. [10 cm.]; this was considered as optimum. During the first 3 days the cows gathered an average per day of 150 lb. [68 kg.] of fresh grass containing about 32 lb. [14·5 kg.] of dry matter.

In the course of the next 3 days the cows harvested only 90 lb. [41 kg.] of fresh grass containing about 20 lb. [9 kg.] of dry matter from this pasture, which had already been stripped. Finally, during the last 3 days of grazing this same sward, which was now almost bare, the cows harvested only 44 lb. [20 kg.] of fresh grass per day containing about 10 lb. [4·5 kg.] of dry matter (far removed from the so-called indispensable 29 lb. [13 kg.] of dry matter if the cow is to be satisfied).

These cows were also grazed on a pasture where the grass had been allowed to grow to a height of 10 in. [25 cm.]. Here they harvested, per day, 68 lb. of fresh grass containing approximately 16·5 lb. [7·6 kg.] of dry matter.

The quantities of fresh grass and dry matter per acre available to the stock were also measured. With regard to Table 29 (p. 77), the following remarks are of interest:

When the herbage reaches a height of 10 in. [*25* cm.] the quantity of fresh grass present is 5000 lb./acre [*5550* kg./ha.] as opposed to 4500 lb./acre [*5000* kg./ha.] for a height of 4 in. [*10* cm.]. In other words, when the grass is two and a half times as high (that is, 10 ins. [*25* cm.] as compared with 4 ins. [*10* cm.]) the quantity of fresh grass increases by 11% 5000 lb./acre [*5550* kg./ha.] against 4500 lb./acre (*5000* kg./ha.) and the quantity of dry matter by 13% (1200 lb./acre [*1320* kg./ha.] against 1000 lb./acre [*1165* kg./ha.]).

TABLE 29

Results obtained at Cornell (U.S.A.) by Professor Johnstone-Wallace

	Quantity available to cows of				Percentage of dry matter in herbage available to cows	Amount grazed per cow per day of			
	Herbage		Dry matter			Herbage		Dry matter	
	lb./ acre	[kg./ ha.]	lb./ acre	[kg./ ha.]		lb.	[kg.]	lb.	[kg.]
Grass 10 in. high	5000	[5550]	1200	[1320]	23·8	70	[32]	20	[7·5]
Grass 4–5 in. high at start of grazing:									
First 3 days of grazing .	4500	[5000]	1000	[1165]	21·3	150	[68]	32	[14·5]
Second 3 days .	2200	[2440]	500	[535]	21·9	90	[41]	20	[9·0]
Third 3 days .	1100	[1220]	250	[275]	22·5	45	[20]	10	[4·5]
					Average	95	[43]	20·7	[9·3]

From (51).

It has already been noted earlier that a pasture of 6 ins. [*15* cm.] average height (in the case of permanent pasture) represents 3569 to 8922 lb./acre [*4000–10,000* kg./ha.] consumable grass according to the circumstances. Johnstone-Wallace's figures, therefore, lie within these limits, although the quantity of green matter present on a pasture 10 ins. [*25* cm.] high seems little enough, unless it is a case of a young pasture recently re-sown.

It is evident once more, therefore, that quantities of harvestable grass for the same, or approximately the same, height can vary to a great extent.

Be that as it may, with 11% more grass present per acre, the cow, per day, has only harvested 70 lb. [*32* kg.] of the 10 in. [*25* cm.] grass compared with 150 lb. [*68* kg.] of the 4 in. [*10* cm.] grass. We have already seen how the mechanics of grazing explain this difference. In other words **it is not on very long grass that the cow harvests the maximum quantity, but on the pasture of medium height which allows the animal to carry out its harvesting operation with maximum efficiency.**

Cows do not work overtime

Throughout these trials the cows were observed in the manner previously described. It was confirmed that, in the four cases, they continued to devote only eight hours to the operation of "grazing", that is to say, to travelling and browsing (it was mentioned above that they appeared to have formed unions). Even during the last 3 of the 9 grazing days (on a pasture with an initial height of 4 in. [*10* cm.]) the cows put in not one single hour of overtime. Although the 44 lb. [*20* kg.] of grass harvested hardly satisfied their maintenance requirements (more will be said on this point later), the cows were nevertheless incapable of any additional effort to get themselves more food.

One of the most remarkable facts brought to light in these experiments was the following. **No increase in the grazing area led to an increase in the amount of grass harvested by the cow,** even if it was a pasture of poor quality with very small quantities of dry matter harvested (9–15 lb. [*4–7* kg.]). (We suppose, of course, that the area of pasture added was of the same quality as the original grazing.)

Grass harvesting is an enormous task for the cow

Professor Johnstone-Wallace illustrates this fact by the following picture:

"The cow's jaw is about 2½ in. [*65* mm.] wide. Let us suppose that we have to cut grass with a mower 2½ in. wide: we can guess the work entailed in cutting some tens of pounds of grass during an eight hour day with such an implement. It is understandable that, if we can mow 154 lb. [*70* kg.] of grass 6 in. [*15* cm.] high, this figure will be much more difficult to attain, for the same effort, if the grass is only 2 in. [*5* cm.] high."

And the Professor concludes:

"The pasture could be 100 acres instead of one, but this would not help us to mow, within 8 hours, 154 lb. [*70* kg.] of grass, 2 in. [*5* cm.] high, with a mower 2¼ in. [*65* mm.] wide."

Using the same image, I will add that, if the grass is very long, our 2½-in.-wide mower will continually be becoming choked, in the same way as the cow gets mixed up in the long grass. As a result, we will constantly be having to stop working, and at the end of our 8-hour day we will be far from having cut 154 lb. [*70* kg.] of grass. We see, then, that the Union of Cows has acted wisely in forbidding its members to continue the arduous work of gathering grass for more than 8 hours per day.

When Professor Johnstone-Wallace carried out his first investigations in 1942–43 there were not many other sides to the question, but observations made abroad on the quantities of grass harvested by cattle have increased somewhat, and I should like to compare some of these later results with those obtained in the course of Johnstone-Wallace's pioneer work. Brief mention will be made of two only.

A Scottish and a German observation on the quantities of grass harvested by the cow

Waite (136), in Scotland, found the following average results: On a pasture with 50% grass less than 6 in. [15 cm.] high, the cattle harvested an average of 24 lb. [11 kg.] dry matter: when the height of the grass increased, the quantity fell to 21 lb. [9·5 kg.]. Here we encounter once more one of Professor Johnstone-Wallace's results: it is not where the grass is longest that the cow gathers the maximum quantity. There is an optimum height (around 6 in. [15 cm.] average height) which seems to allow the cow to work most efficiently at harvesting her grass (naturally account must also be taken of the *density* of the grass).

Schmidt (90), at the Hohenheim Institute in Germany, measured, over five years, the quantities of grass harvested by three different breeds of cows. These figures are not the measurements taken in one or several experiments, but rather the statistical results over several years.

TABLE 30

Average quantities of fresh grass harvested by animals of three different breeds

Breed	Average live weight of animals		Quantity of fresh grass harvested per animal per day	
	cwt.	[kg.]	lb.	[kg.]
Fleckvieh . . .	12·4	[630]	137	[62·2]
Braunvieh . . .	11·4	[580]	125	[56·9]
Hinterwälder . .	8·5	[430]	96	[43·5]

N.B. The trials were carried out during the five years 1946–51.
From Schmidt (90).

The results are contained in Table 30. The heaviest breed (12·4 cwt. [630 kg.] live weight) harvested 137 lb. [62·2 kg.] fresh grass daily and the lightest breed (8·5 cwt. [430 kg.] live weight) 96 lb. [43·5 kg.].

In a very recent study at the Hohenheim Institute, Mehner and Grabisch (74) found that cows weighing approximately 8 cwt. [400 kg.] gathered about 88 lb. [40 kg.] fresh grass.

These results are therefore in the region of Johnstone-Wallace's figures and it can be said that, on the average, **when an animal of 10 cwt. [500 kg.] live weight grazes a pasture of 6 in. [15 cm.] average height (which it must scrape bare) it harvests quantities of fresh grass in the region of 99–110 lb. [45–50 kg.], corresponding to about 22 lb. [10 kg.] dry matter.** But this is a point of such importance that we will proceed to examine it in more detail.

Diminution of the amounts of grass harvested in the course of progressive scraping of the pasture

Let us look back at Table 29 (p. 77). When stock are put out to graze a sward with an average height of 4 in. [*10* cm.] they harvest:

	Fresh grass		Dry matter	
	lb.	[kg.]	lb.	[kg.]
1st third of period of occupation .	150	[*68*]	32	[*14·5*]
2nd third of period of occupation .	90	[*41*]	20	[*9·0*]
3rd third of period of occupation .	44	[*20*]	10	[*4·5*]
Average . . .	95	[*43*]	20·7	[*9·3*]

This illustrates the enormous fall in efficiency which takes place when cows are forced to gather grass on a sward which has been more or less scraped bare. Unfortunately, we cannot be content with taking "the cream" off the grass, for a grass sward which is only half grazed would produce a miserable yield on re-growth.

We will see later (as is evident, moreover, without making calculations) that, *given equal milking abilities*, the cow that gathers 141 lb. [*64* kg.] of grass produces more milk than the cow that gathers 88 lb. [*40* kg.]. In the same way, the bullock that harvests 141 lb. [*64* kg.] of grass will gain more weight than one that harvests smaller quantities on a pasture already grazed bare or which has made but little re-growth.

We have just seen that, according to the observations and statistics available, Professor Johnstone-Wallace's figures should be raised slightly, the more so as the grass used by him at the outset seems not to have been very *dense*, has a somewhat low average height and seems to have been "scraped" too far. We will assume therefore that a cow of 10 cwt. [*500* kg.] is put on to a grazing where the average height of the grass is 6 in. [*15* cm.] and that she is forced to graze to the ground, harvests, on the average, 100 lb.[1] [*48* kg.] of fresh grass per day. (6 in. [*15* cm.] applies to permanent pasture; for a re-seeded sward 9 in. [*22* cm.] should be substituted.)

What an eighteenth-century French pioneer had to say about the quantity of grass the cow harvests

In 1786 the illustrious ancestor of the *Academie d'Agriculture*, the *Société Royale d'Agriculture*, made the following the subject of a competition: 'What temporary pasture species can be grown with most advantage in the region

---

[1] The actual equivalent of 48 kg. is 106 lb. British system, but as explained in the preliminary note, in order to achieve clarity, we have decided to convert into round numbers. We realise that there is a difference of 6%, but this is well within the margin of error allowed nowadays in discussing a question of this kind.

of Paris and what is the best cultivation?" The prize was awarded to Gilbert, professor at the *Ecole Royale Vétérinaire* now the *Ecole Nationale Vétérinaire d'Alfort*, a school which to-day remains one of the glories of French science.

In a communication to the *Académie d'Agriculture* (131) I pointed out the great interest of this remarkable memoir and how it helps to make clear how the tithe postponed the replacement of the fallow with temporary pastures composed of legumes.

Among the valuable data contained in this memoir we find the following statement with regard to the harvesting of grass by cattle: "A bullock of average size eats 53–106 lb. [*24–48* kg.] of grass per day."

It is quite remarkable that the highest figure corresponds to the maximum amount of grass a cow can harvest when she is forced in the end to scrape the pasture bare.

We can only affirm, as we have often had occasion to say elsewhere, that the agronomists of the Age of the Enlightenment always kept *in touch with reality*. It is striking that Gilbert carefully notes that the quantity of grass harvested by a bullock is *variable*, although the feed tables were later based on the arbitrary affirmation that, whatever happens, a beast of average size eats 29 lb. [*13* kg.] dry matter, which means about 132 lb. [*60* kg.] of grass.

The quantities of grass harvested in the course of the different fractions in a period of occupation

Table 31 (p. 82) shows the quantities of grass harvested by cows of 10 and 12 cwt. [*500* and *600* kg.] live weight during the complete occupation period on one plot or when this period of occupation is divided up either into two or three parts. It should be noted that, in order to arrive at the quantities harvested by a cow of 12 cwt. [*600* kg.], we multiplied the figures obtained for a cow of 10 cwt. [*500* kg.] by the factor of 6/5.

In other words, we assumed that the quantities harvested are proportional to the weight of the animal. This is obviously an arbitrary hypothesis, which corresponds, moreover, to a rule applied in our feed tables. It must be looked upon as merely indicative in the hope that future study of the question will clarify this point.

As will be seen later (Table 37, p. 107), it was ascertained in the course of palatability trials with single species that bullocks weighing 6 cwt. [*300* kg.] could harvest 31 lb. [*14·3* kg.] of dry matter per day on plots of pure brome grass. One might obviously wonder if a bullock of 6 cwt. [*300* kg.] live weight is not capable of harvesting just as much grass as one weighing 12 cwt. [*600* kg.]. One can equally wonder whether or not these were bullocks with exceptional hereditary grazing qualities; but, as we shall see, no consideration was given in these experiments to the various elements which might influence the quantities of dry matter harvested. It should also be noted that, in the course of the same series of trials, bullocks that had harvested 31 lb. [*14·30* kg.] of dry matter on brome grass only gathered 11·5 lb. [*5·20* kg.] of dry matter on fescue.

TABLE 31

Quantities of grass harvested during progressive grazing

	10 cwt. [*500* kg.] live weight				12 cwt. [*600* kg.] live weight			
	Fresh grass		Dry matter		Fresh grass		Dry matter	
	Average quantity harvested per day							
	lb.	[kg.]	lb.	[kg.]	lb.	[kg.]	lb.	[kg.]
1. *Single period of occupation* (or single group) .	100	[*48*]	21·0	[*10·1*]	120	[*58*]	25·2	[*12·2*]
2. *Double period of occupation* (or two groups): A. 1st half of period (or 1st group) .	116	[*56*]	23·2	[*11·2*]	140	[*67*]	28·0	[*13·4*]
B. 2nd half of period (or 2nd group).	84	[*40*]	18·5	[*8·8*]	100	[*48*]	22·0	[*10·6*]
3. *Treble period of occupation* (or three groups): A. 1st third of period (or 1st group) .	133	[*64*]	26·6	[*12·8*]	161	[*77*]	32·2	[*15·4*]
B. 2nd third of period (or 2nd group) . .	92	[*44*]	19·2	[*9·2*]	110	[*53*]	23·1	[*11·1*]
C. 3rd third of period (or 3rd group) . .	75	[*36*]	17·3	[*8·3*]	89	[*43*]	20·5	[*9·9*]

As mentioned in the preliminary note and in the footnote to p. 80, it was decided to use the round figure of 100 lb. as the equivalent of 48 kg. All the other British figures of this table are thus slightly modified, having been converted on the same basis.

Dividing the herd into groups and quantities harvested

We will see later (pp. 152 and 157) in studying the practical application of rational grazing that the herd can be divided into groups. It is easily understood, then, that the first case above corresponds to the harvest realised by one single group, obliged, in itself, to graze the pasture to the bottom (average of 100 lb. [*48* kg.]) of grass per day for a cow of 10 cwt. [*500* kg.] live weight.

The second case corresponds to two groups: the first being in a position to harvest greater quantities (116 lb. [*56* kg.] for 2A) than the second group, whose job it is to scrape the pasture clean and which gathers amounts less than the average (84 lb. [*40* kg.] for 2B).

The third case is that of three groups. The first group is able to produce record harvests (133 lb. [*64* kg.] in 3A), but the third finds itself confronted with a pasture already very much grazed down and is therefore able to gather only the small harvest of 75 lb. [*36* kg.] (case 3C); this figure can fall even further if scraping is prolonged.

Principles governing the harvesting of grass by the cow

From the experiments of Johnstone-Wallace and Kennedy the following rules can be deduced:

1. The cow devotes only a strictly limited time to harvesting grass. The result is that she harvests food only during a certain period, which is in the neighbourhood of 8 hours.

2. It appears that the cow is incapable of exceeding this time limit, even if she has harvested only quite a small quantity of grass, barely sufficient for her needs.

3. The fundamental factor determining the quantity of grass harvested by the cow is the height of the herbage (under conditions of equal density). In view of the structure of the cow's jaw, the average height of grass allowing the maximum harvest is about 6 in. [*15* cm.].

4. If the grass is longer or shorter than this optimum, the quantity of grass harvested is diminished.

5. With grass of optimum height and density, the quantity harvested by a cow of 10 cwt. [*500* kg.] live weight reaches its maximum at 100 lb. [*48* kg.] fresh grass and 21 lb. [*10·1* kg.] dry matter, **if the cow is forced to scrape bare the plot on which she is grazing.**

6. At the end of the grazing of a plot (or of the first group, where the herd is divided into groups) a cow of 10 cwt. [*500* kg.] live weight that is taking the full mouthfuls at each bite, can harvest 133 lb. [*64* kg.] fresh grass or 26·6 lb. [*12·8* kg.] dry matter approximately each day.[1]

7. *If the dry matter absorbed is a very doubtful criterion of satisfaction where the cow in the stall is concerned, it leaves even more to be desired as a criterion for the grazing animal.*

8. Increasing the grazing area (assuming, of course, that the pasture added offers the same degree of grazing or re-growth) does not lead the cow to make an added effort to harvest a greater amount of grass, even if the quantity gathered is hardly satisfying her maintenance requirements.

9. Milk yield, stage of lactation or gestation do not affect the appetite of the cow to any appreciable extent or, to be more correct, do not bring her to make a supplementary effort to harvest a greater quantity of grass.

This last conclusion is so surprising at the first glance that it merits closer examination.

[1] See footnote to Table 31.

Does milk yield influence the amount of grass harvested?

Johnstone-Wallace and Kennedy, as has just been stated, came to the conclusion that milk yield, or the advance of lactation, did not appear to exert any influence on the quantity of grass the cow harvested. In Scotland, Waite and his co-workers (136) reached the same conclusion. Other research workers, however, considered that a cow that produces more milk harvests increased quantities of grass. Personally, I think that both parties were right. In effect, it was a vicious circle. In my opinion one must not say "The cow that produces more milk harvests more grass" but "The cow that harvests more grass produces more milk". It is even clearer to say: "*The cow that is* CAPABLE *of harvesting more grass produces more milk.*"

It had in fact to be recognised that the quantity of grass a cow is capable of harvesting is, as we are about to see, a *hereditary* capacity.

The hereditary character of the cow determines the amount of grass she harvests

In the preceding chapter we saw that the experiments at Ruakura Animal Research Station with monozygotic twins showed that *the hereditary character of a cow determines the length of time she is capable of devoting to gathering grass and the total number of bites which she takes each day to harvest that grass* (Table 28, p. 72). It was later observed that the quantities of grass harvested were equally a function of hereditary character.

In fact, the following were the observations made by the New Zealander, Wallace (Hancock, 37), in the course of further experiments:

1. Between two beasts of the same pair of twins, the quantity of dry matter harvested daily differed by not more than 1·20 lb. [540 gm.].

2. *The average quantity of grass dry matter harvested daily by a twin of the pair gathering the largest quantity was 34 lb. [15·5 kg.] against 20·5 lb. [9·3 kg.] for the pair harvesting the smallest quantity (about 163 lb. [74 kg.] fresh grass against 97 lb. [44 kg.]).*

This shows therefore that:

(a) The most accentuated difference in the dry matter harvested was *between* two pairs at 13·2 lb. [6·2 kg.] that is to say 12 times greater than the highest difference (1·2 lb. [0·540 kg.]) existing *within* a pair.

(b) *The heifers in a pair, whose inheritance had endowed them with a greater capacity for harvesting, gathered 163 lb. [74 kg.] fresh grass per animal against 97 lb. [44 kg.] for heifers less favoured from the point of view of inheritance.*

In other words, *a cow, having favourable hereditary characteristics, harvests 63% more grass than one whose inheritance in this charasteristic is less good.*

Fundamental consequences for stock-rearing of the New Zealand observations

It is, of course, not certain that the cow *capable* of harvesting the greatest quantity of grass is the one which possesses *the best milking qualities*. But it is a well-known fact that the high yielders are the big eaters. There is an old peasant dictum which says: "Milk is made through the mouth."

Moreover, it must not be forgotten that in the most important stock-rearing areas, the animals eat grass for eight months in the year. At the moment, champion cows, the mothers of our own champion bulls, produce record yields, thanks mainly to a high consumption of concentrated feeding-stuffs: these are *cows particularly suited to the utilisation of oil-cake and grain*. It is not at all certain that the cow which is particularly able to *"eat"* cake and grain efficiently is also the cow most capable of *"harvesting"* the greatest amount of grass and, in consequence, producing the highest yield of milk from grazing.

The correlation, or antagonism, between the *efficiency* of the animal adopting these two methods of nutrition has not been studied as yet so far as I am aware.

Good grazers must be selected

It therefore appears essential to draw the attention of the animal experts, whether research workers or breeders, to this extremely important point. If further studies confirm the results of the New Zealand workers, we must select bulls which procreate cows with the highest *grazing* qualities: that is, capable of harvesting very large quantities of grass while, of course, possessing the best possible milking qualities.

It might even be asked whether it is prudent to wait and whether it would not be wise, from now on, to make provision for regulations (governing showing and breeding) to the effect that cows, at least during the grazing period, should be fed on grass which they have harvested themselves, either exclusively or with the addition of strictly limited supplements. These animals would be the mothers of bulls destined for areas where grazing is the chief source of nourishment.

It seems, too, that our ancestors selected cows with a "wide and long mouth" to allow for record harvesting. This was the case, for example, with our old Cauchois [1] breed. I remember having such a cow, Caillotte by name, which produced sensational milk yields, especially when out on grass, where for several months she gave $5\frac{1}{2}$ gal. [25 litres] of milk without any supplement whatever. But she had a proper "goat's head" with an extra long neck (Photo 5, between pages 74 and 75).

[1] Cauchois = from *Pays de Caux*, a part of the Norman *département* of Seine-Maritime.

This question of good grazers is the more important now that artificial insemination is multiplying the defects or qualities transmitted by a bull.

It therefore seems essential to select, above all, cows capable of utilising efficiently the grass on which they graze. Grass was always the cheapest foodstuff, and to-day is even a bigger bargain in view of the progress which must be made, thanks to rational management of grazings, which allows us to multiply the yields of our swards.

Chapter 3

QUANTITIES OF MILK PRODUCED BY THE COW WHEN SHE HARVESTS HER OWN GRASS

The cow's expenditure of energy in harvesting grass

IT has already been noted that a cow devotes 8 hours per day to the harvesting of her food: then, as if exhausted by this effort, she refuses to work one more hour, even if the quantity of grass she has gathered is barely sufficient for her maintenance requirements. It seems therefore that the cow is capable of supplying only a certain amount of energy for the harvesting of her food and is incapable of any supplementary contribution towards this end.

It would be of particular interest to know what this expenditure of energy is. It may perhaps be a personal factor in the cow, contributing to no mean extent to her good or bad milking qualities.

Moreover, all the measurements of the cow's requirements of energy or nutritive substances to maintain her production have been assessed in the cow-shed. Before it is possible to deduce the requirements of the animal at grass it is essential to know how much supplementary energy is necessary to cover the expenditure incurred by the animal in the cause of harvesting her food. Unfortunately, as in so many other aspects of the feeding of livestock in general and the cow at grass in particular, our knowledge of this expenditure of energy by the cow in the course of her harvesting grass is as good as non-existent.

The figures put forward are the result of guesswork with not the slightest experimental evidence. I have examined this question in a study to which I refer the reader (116). In order not to overload the present work, I will state simply that, on the basis of various investigations, I believe I have been able to deduce, by means of indirect calculations, that the *expenditure of energy incurred by a cow of 10 cwt. [500 kg.] live weight to harvest a grass sward must be in the region of 3000 calories of net energy*, representing 2·8 lb. [*1·3* kg.] of starch equivalent. To be cautious, I take a round figure, 2·20 lb. [*1·0* kg.]. Moreover, it seems probable to me that a cow does not expend any more energy on a tight than on a lush sward, but for an equal expenditure of energy she will harvest much less grass on the second than on the first.

What must be emphasised is that the figure quoted is purely hypothetical.

87

It is both desirable and an urgent necessity that measurements and investigations be undertaken to determine the expenditure of energy by the cow in the course of her harvesting operations. To quote the words of Allan Fraser (25):

> "We have no exact measurement of the energy expended by the animal in grazing, although it is fundamental. . . ."

And he goes on, not without humour:

> "Our feed tables would certainly be more easy to use if our cows would be kind enough not to graze. **But our cows do graze, indeed, it is their main activity. . . . And so, if we want our tables to have a practical value and continue to be used, we must make them applicable to animals that graze.**"

For lack of anything better while awaiting more precise data we will use the figures quoted above, but without any illusions as to their accuracy.

Requirements of the cow at pasture in order to achieve different milk yields

Tables 32 and 33 show the amounts of digestible crude protein and starch equivalent required by cows of 10 cwt. [*500* kg.] and 12 cwt. [*600* kg.] live weight in order to supply varying yields while grazing. In both cases an expenditure of 2·20 lb. [*1·00* kg.] of starch equivalent has been allowed for the gathering of the grass; this has arbitrarily been assumed to be the same for both live weights.

TABLE 32

Requirements of a 10-cwt. (500 kg.) cow when grazing for different levels of milk production

Milk production 3·4% butter-fat		Digestible crude protein		Starch equivalent					
				In the manger		To harvest the grass		Grazing	
lb.	[kg.]	oz.	[gm.]	lb.	[kg.]	lb.	[kg.]	lb.	[kg.]
0	[0]	8·8	[250]	5·51	[2·50]	2·20	[1·00]	7·72	[3·50]
6·6	[3]	15·7	[445]	7·17	[3·25]	2·20	[1·00]	9·37	[4·25]
13·2	[6]	22·6	[640]	8·82	[4·00]	2·20	[1·00]	11·02	[5·00]
19·8	[9]	29·5	[835]	10·47	[4·75]	2·20	[1·00]	12·68	[5·75]
26·5	[12]	37·0	[1050]	12·13	[5·50]	2·20	[1·00]	14·33	[6·50]
33·1	[15]	43·2	[1225]	13·80	[6·25]	2·20	[1·00]	15·98	[7·25]
39·7	[18]	50·1	[1420]	15·43	[7·00]	2·20	[1·00]	17·64	[8·00]
46·3	[21]	57·0	[1615]	17·09	[7·75]	2·20	[1·00]	19·29	[8·75]
52·9	[24]	63·8	[1810]	18·74	[8·50]	2·20	[1·00]	20·94	[9·50]
59·5	[27]	71·1	[2015]	20·39	[9·25]	2·20	[1·00]	22·60	[10·25]
66·1	[30]	77·6	[2200]	22·05	[10·00]	2·20	[1·00]	24·25	[11·00]
72·8	[33]	84·5	[2395]	23·70	[10·75]	2·20	[1·00]	25·90	[11·75]

N.B. Calculations based on 0·250 lb. starch equivalent and 1·0 oz. crude digestible per lb. of milk at 3·4% butter-fat (0·250 kg. starch equivalent and 65 gm. crude digestible protein per kg. of milk at 3·4% butter-fat).

TABLE 33

Requirements of a 12-cwt. (600 kg.) cow when grazing for different levels
of milk production

Milk production 3·4% butter-fat		Digestible crude protein		Starch equivalent					
				In the manger		To harvest the grass		Grazing	
lb.	[kg.]	oz.	[gm.]	lb.	[kg.]	lb.	[kg.]	lb.	[kg.]
0	[0]	10·6	[300]	6·61	[3·00]	2·20	[1·00]	8·82	[4·00]
6·6	[3]	17·5	[495]	8·27	[3·75]	2·20	[1·00]	10·47	[4·75]
13·2	[6]	24·3	[690]	9·92	[4·50]	2·20	[1·00]	12·13	[5·50]
19·8	[9]	31·2	[885]	11·57	[5·25]	2·20	[1·00]	13·78	[6·25]
26·5	[12]	38·1	[1080]	13·23	[6·00]	2·20	[1·00]	15·43	[7·00]
33·1	[15]	45·0	[1275]	14·88	[6·75]	2·20	[1·00]	17·09	[7·75]
39·7	[18]	51·9	[1470]	16·53	[7·50]	2·20	[1·00]	18·74	[8·50]
46·3	[21]	58·7	[1665]	18·19	[8·25]	2·20	[1·00]	20·39	[9·25]
52·9	[24]	65·6	[1860]	19·84	[9·00]	2·20	[1·00]	22·05	[10·00]
59·5	[27]	72·5	[2055]	21·50	[9·75]	2·20	[1·00]	23·70	[10·75]
66·1	[30]	79·4	[2250]	23·15	[10·50]	2·20	[1·00]	25·35	[11·50]
72·8	[33]	86·2	[2445]	24·80	[11·25]	2·20	[1·00]	27·01	[12·25]

N.B. Calculations based on 0·250 starch equivalent and 1·0 oz. crude digestible protein per lb. of milk at 3·4 butter-fat (0·250 kg. starch equivalent and 65 gm. crude digestible protein per kg. of milk at 3·4% butter-fat.

Quantities of nutritive elements harvested and possible yields of milk

Taking into account the quantities of grass harvested as shown in Table 31 (p. 82) and the composition of the herbage at different stages of growth (Table 25, p. 63), we have calculated (Tables 34 and 35, pp. 90 and 91) the quantities of digestible crude protein and starch equivalent that a cow of 10 cwt. [500 kg.] or 12 cwt. [600 kg.] harvests per day during the different stages of occupation of a paddock or, what comes to the same thing, as a function of the number of groups into which the herd has been divided. Thereafter, the yields of milk possible for these two types of cow were calculated.

It should be noted in passing that, relative to ordinary feed tables, the tables here (34 and 35) have increased the requirement of starch equivalent, that is of energy equivalent, due to the fact that they are allowing for the expenditure of energy by the animal in harvesting her grass feed. As a result *it is the starch equivalent harvested which is always the limiting factor.* In other words, there is always too much digestible crude protein relative to starch equivalent in the grass gathered. It would obviously be interesting if experiments could show more exactly whether this situation indeed corresponds to reality.

Table 36 (p. 91) reviews the quantities of grass harvested and the milk yields possible therefrom.

Maximum milk yields of a cow at grass

On examination of Table 36 it becomes evident that, even in a well-conducted rational grazing and with the herd divided into three groups (which has certain disadvantages as will be seen later), it is difficult for a cow harvesting her own grass and not having to scrape the pasture to exceed the following milk yields:

1. Cow of 10 cwt. [*500* kg.] live weight: 4 gal. [*18* litres].
2. Cow of 12 cwt. [*600* kg.] live weight: 5¼ gal. [*24* litres].

If the cow is forced to "scrape" the sward (Case 1) it is difficult for her to exceed:

1. Cow of 10 cwt. [*500* kg.] live weight: 2½ gal. [*11* litres].
2. Cow of 12 cwt. [*600* kg.] live weight: 3¼ gal. [*15* litres].

Any producer wanting to increase his milk production knows very well that he must feed the cow a supplement of concentrates if she is to provide higher yields of milk.

TABLE 34

Possible theoretical production of milk from 10-cwt. [500 kg.] cows when grazing

| | Quantity of fresh grass harvested | | Quantity harvested of | | | | Possible production of milk according to the intake of | | | |
| | | | Crude digestible protein | | Starch equivalent | | Crude digestible protein | | Starch equivalent | |
	lb.	[kg.]	lb.	[gm.]	lb.	[kg.]	gal.	[litres]	gal.	[litres]
1. *Single period of occupation* .	100	[*48*]	2·57	[*1296*]	13·2	[*6·33*]	3·5	[*16*]	2·4	[*11*]
2. *Double period of occupation:*										
A. 1st half of period .	116	[*56*]	3·08	[*1568*]	15·3	[*7·39*]	4·4	[*20*]	3·3	[*15*]
B. 2nd half of period .	84	[*40*]	2·08	[*1040*]	11·6	[*5·52*]	2·6	[*12*]	1·5	[*7*]
3. *Treble period of occupation:*										
A. 1st third of period	133	[*64*]	3·80	[*1920*]	17·6	[*8·44*]	5·7	[*26*]	4·0	[*18*]
B. 2nd third of period	92	[*44*]	2·37	[*1188*]	12·2	[*5·80*]	3·1	[*14*]	2·0	[*9*]
C. 3rd third of period	75	[*36*]	1·79	[*900*]	10·6	[*5·11*]	2·2	[*10*]	1·3	[*6*]

N.B. 1. For the quantities of fresh grass harvested, see Table 31, p. 82.
   2. For the composition of the grass, see Table 25, p. 63.
   3. For the requirements of the cows, see Table 32, p. 88.
   4. The possible production of milk is, in every system, always calculated according to the figures of intake of the metric system.

Of course, a cow can produce 6½ gal. [*30* litres] of milk from grass alone; but to do so she is taking it out of herself, with the result that production will fall all of a sudden, and the health of the cow may even be endangered.

TABLE 35

Possible theoretical production of milk from 12-cwt. [600 kg.] cows when grazing

	Quantity of fresh grass harvested		Quantity harvested of				Possible production of milk according to the intake of			
			Crude digestible protein		Starch equivalent		Crude digestible protein		Starch equivalent	
	lb.	[kg.]	lb.	[gm.]	lb.	[kg.]	gal.	[litres]	gal.	[litres]
1. *Single period of occupation* .	120	[58]	3·22	[1566]	15·85	[7·75]	4·2	[19]	3·3	[15]
2. *Double period of occupation:*										
A. 1st half of period .	140	[67]	3·81	[1876]	18·50	[8·84]	5·3	[24]	4·2	[19]
B. 2nd half of period	100	[48]	2·60	[1248]	13·80	[6·62]	3·3	[15]	2·2	[10]
3. *Third period of occupation:*										
A. 1st third of period	161	[77]	4·82	[2310]	21·20	[10·16]	6·8	[31]	5·3	[24]
B. 2nd third of period	110	[53]	2·95	[1431]	14·52	[6·99]	3·7	[17]	2·6	[12]
C. 3rd third of period	89	[43]	2·21	[1075]	12·62	[6·10]	2·6	[12]	1·8	[8]

N.B. 1. For the quantities of fresh grass harvested, see Table 31, p. 82.
2. For the composition of the grass, see Table 25, p. 63.
3. For the requirements of the cows, see Table 33, p. 89.
4. The possible production of milk is, in every system, always calculated according to the figures of intake of the metric system.

The personal character of the cow upsets all our figures

It is considered necessary to give some figures as guides for the farmer or research worker, but this is done with some hesitation, for, in many cases,

TABLE 36

Possible theoretical production of milk from 10- and 12-cwt. cows when grazing

	Quantity of fresh grass harvested by a cow weighing				Possible production of milk from grass harvested by a cow weighing			
	10 cwt.	[500 kg.]	12 cwt.	[600 kg.]	10 cwt.	[500 kg.]	12 cwt.	[600 kg.]
	lb.	[kg.]	lb.	[kg.]	cal.	[litres]	cal.	[litres]
1. *Single period of occupation* .	100	[48]	120	[58]	2·4	[11]	3·3	[15]
2. *Double period of occupation:*								
A. 1st half of period	116	[56]	140	[67]	3·3	[15]	4·2	[19]
B. 2nd half of period	84	[40]	100	[48]	1·5	[7]	2·2	[10]
3. *Treble period of occupation:*								
A. 1st third of period . .	133	[64]	161	[77]	4·0	[18]	5·3	[24]
B. 2nd third of period . .	92	[44]	110	[53]	2·0	[9]	2·6	[12]
C. 3rd third of period . .	75	[36]	89	[43]	1·3	[6]	1·8	[8]

N.B. Figures have been calculated from Tables 34 and 35.

these figures may mislead the reader. It must be remembered that a multi-plicity of elements can modify all of these tables. Mention has already been made of the reservations which must be applied to the so-called proteins of grass. In addition, it was noted that no attention is being paid at all as yet to the energy expended in the course of harvesting grass, etc., etc.

But there is one point to bear in mind. We have seen that a cow with the inherited capacity of being a *good grazer* can harvest 63% more grass than a cow with unfavourable hereditary qualities. She will produce more milk, just as a bullock with the hereditary capacity of harvesting more grass will show greater increases in weight. It is believed that the tables drawn up here correspond to average, ordinary cows and that, naturally, the figures should be altered one way or the other according to whether the cow possesses the aptitudes of a good or bad grazer.

Conduct of the grazing and climatic conditions enter into play

It must likewise be stated that these figures correspond to average con-ditions of management. I can only cite the example of my last group.

We will see later that in 1954, the year that I was working with three groups, my third and last group made an average daily gain throughout the season of 1·43 lb. [*650* gm.] (Table 64, p. 308). In 1956, when I was working with only two groups, my second last group was put out under conditions far more favourable from the point of view of grass harvesting and, in con-sequence, should have shown greater weight gains. In this instance, the average daily gain was only 0·7 lb. [*320* gm.]. This was due to the fact that throughout that year conditions on the average were very unfavourable for grass re-growth.

In order not to make the grazing rotations too rapid (what I will later call "untoward acceleration") I was obliged to "over-scrape" my paddocks, with the result that the last group necessarily found themselves in circumstances similar to those under which the experimental cows of Professor Johnstone-Wallace were working (Table 29, p. 77) at the end of their scraping opera-tion: that is to say, in a position to harvest little more than 44 lb. [*20* kg.] of grass.

We can only repeat therefore: these figures are *indicative*.

Simultaneous variation in the quality of the grass and the amount harvested

If I decided to give these few figures, it was not so much for the figures themselves as to show how, in the future, when we have more precise informa-tion, we will have to proceed in trying to forecast the possible performance of the animal harvesting its own grass and, in this way, assist the practical conduct of grazing.

Up till now the starting-point has been the erroneous idea that, in every case

*and in all circumstances, the cow of 10 cwt. [500 kg.] harvests 29 lb. [13 kg.]
dry matter (corresponding more or less to 132–154 lb. [60–70 kg.] fresh grass).*
The grass has been analysed, and by multiplying the constituents found in
1 lb. or 1 kg. of grass by the 132–154 lb. [60–70 kg.] harvested, the quantities
of nutritive elements absorbed by the cow have been calculated. Numerous
recent studies could be quoted where this procedure has been adopted. But
there are two factors which vary with the nature of the pasture:

1. The composition of the grass.
2. The quantity of grass that the cow (given equal grazing capacities)
is capable of harvesting.

*Up till now only the first factor has been taken into consideration, the second
being left as a constant.* If we continue on this assumption we cannot possibly
progress in our knowledge and understanding of the phenomena observed
among animals at grass.

Efficiency of the cow in rational grazing

Examining Table 36 (p. 91), we understand even better how, by applying
rest periods of sufficient length, which increase the productivity of the grass
to a remarkable degree (Figs. 3 and 4), we are contributing to the efficiency
of the cow in the case of rational grazing as compared with continuous grazing.

In rational grazing, with a sward 6 in. [15 cm.] in height at the disposal
of the cow throughout the season, we are allowing her a large harvest (100 lb.
[48 kg.]) even if she is forced to scrape the pasture to the limit. In addition,
we are better able to distribute the "rations" according to the production and
therefore to the feeding requirements of the stock. Just as we give a bigger
ration to a cow producing a lot of milk, so also, thanks to the division into
groups, we afford the most productive cows the *possibility* of harvesting more
grass if they are capable of it. This aim is achieved by putting the highest
yielders into the leading group, where they will harvest a greater amount of
richer herbage.

We see, then, that a cow of 10 cwt. [500 kg.] put on to a sward 6 in. [15 cm.]
high *which she herself is to graze bare*, can produce only 2½ gal. [11 litres],
whereas in the leader group she can produce 3¼ gal. [15 litres] when there are
two groups and 4 gal. [18 litres] when there are three groups: this of course,
on condition that her milking and grazing capabilities (as well as her stage of
lactation) allow her to do so.

Continuous grazing and rational grazing

With continuous grazing there is a short period in the month May–June
during which the cow may probably find herself in conditions akin to those
of a first group. But it must be stressed that this is only for a short period
and at the price of the wastage of grass which is not sufficiently grazed down

(in fact, at this time the cow is "creaming" the grass). During this short space of time a cow of 10 cwt. [*500* kg.] on continuous grazing produces almost 4 gal. [*18* litres] of milk without taking anything out of herself.

But for the rest of the season the cow will find only a thin sward of meagre height corresponding, according to circumstances, to the more or less bared sward of a rational grazing. Here she will only be able to harvest restricted quantities of grass allowing her to produce only the very much lower yields corresponding to cases 2B, 3B and 3C in Table 36.

Supplementation at grass can only be determined empirically

As has just been stated, the 132–154 lb. [*60–70* kg.] of grass (supposedly harvested in all circumstances) was multiplied by the quality of the grass analysed. Then using the feed tables, the quantities of nutritive elements thus harvested were compared with the requirements of the cow relative to her milk yield. From this the supplements of food were deduced which would have to be fed in order to obtain such and such a yield.

If the calculation of rations by means of feed tables is very risky even with stall feeding, it becomes doubtful in the extreme when supplements for cows at grass are being estimated. I believe that for a long time (if not always) supplementary feeding of grazing stock will have to be regulated empirically, the most that feed tables and calculations can do being to provide certain general principles. The milk yield and particularly any drops in production of the cow at grass will have to be watched. This will show to what extent supplementation is affecting milk production and, more particularly, checking any drops in yield.

In this present work the long calculations and the theories which have developed regarding supplementary feeding of cows at pasture will be omitted. I believe that this is a question which can only be answered on the spot, by the skill of the farmer himself.

Figures do not govern the cow's world

It must be borne in mind that the figures quoted are merely guide figures and must on no account be taken as absolute values, for they are a function of numerous variable factors. As Goethe said: *"Figures do not govern the world; but they help us to understand how the world is governed."* No more do figures govern the world of the cow, but they help us to understand it very much better.

Now that we have come to understand that it was not only the composition of the grass that made the quantities of nutritive elements absorbed by the cow vary, but that the variation in the quantities of grass harvested according to the nature of the sward must be taken into consideration, we are in a position to explain certain phenomena which seemed so strange to us in the past.

The phenomenon of the peasant woman's cow is by no means a tale of sorcery but purely and simply a technical problem.

The knitting peasant and her cow

Even to-day in parts of Europe one can still see a good old peasant woman sitting knitting while keeping her eye on her cow grazing by the roadside. The abundance of milk produced by such a cow is always a source of surprise: the milk pail seems ready to overflow, and this not only in May but throughout the season. The conclusion reached is that this is a cow with extraordinary milking qualities, and we make a point of remarking to the happy owner: "Have your cow served by a good bull and if she produces a heifer, we will give you a good price for it." But, even if the bull is good, the heifer, when it becomes a cow, will under normal conditions of continuous grazing, produce milk yields which are quite commonplace.

In actual fact, the cow that grazes along the roadside is always a cow in the first group of a rational grazing system. She only creams the grass, for nobody is the least concerned whether this insufficient grazing down will retard re-growth and diminish the grass yield. Basing our estimate on the figures in Table 36 (p. 91), we will say that the knitting peasant woman's cow is in a position to harvest regularly and continuously 161 lb. [77 kg.] of grass per day, which obviously allows her to give excellent performance and, in particular, to produce (Table 35, p. 91) at the beginning of lactation, daily milk yields of 5¼ gal. [24 litres] if her milking qualities are the least bit suitable at all.

Fluctuations in milk production when the cows stay too long on one paddock

The simultaneous variation in the quality of the grass and the quantity harvested enables us to understand the progressive decline in milk production which takes place when cows on rational grazing stay too long on the same paddock.

The Centre National d'Experimentations Agricoles at Courcelles-Chaussy (Moselle) has made the following observations (63), p. 36.

"With a period of stay of 2–7 days, the percentage milk production for one day in relation to that of the preceding day is as follows:

Production of 2nd day in relation to 1st day 101·7%
,, 3rd ,, ,, 2nd ,, 96·1%
,, 4th ,, ,, 3rd ,, 97·2%
,, 5th ,, ,, 4th ,, 95·2%
,, 6th ,, ,, 5th ,, 98·6%
,, 7th ,, ,, 6th ,, 96·3%

"If one takes the milk production on the first day of grazing as 100, the relative productions on subsequent days are as follows:

Production of 1st day	100
2nd	101·7
3rd	97·7
4th	95
5th	90·4
6th	89·1
7th	85·8

"Production falls rapidly, therefore, from the third day onwards; and has already diminished by almost 10% from the fifth day and by 15% from the seventh day.

"It should be noted that these results were obtained in the months of May and June, when grass growth is very vigorous. The reduction makes itself felt moreover in proportion as the season advances. . . .

"It is evident that when the period of stay is prolonged, cows harvest *less and less grass of poorer and poorer quality*: this leads to the fall in production observed at Courcelles-Chaussy. In the case of continuous grazing the cow remains permanently on the same pasture throughout the grazing season. Certainly the nature of the grass she eats is not the same from 15th April until 15th November, but the evolution of the quality of the grass and of the quantity harvested *takes place slowly* and so, almost always, if one considers the herd as a whole, does the evolution of total milk production, the average production falling when the grass becomes more scarce."

Rhythmic production of milk when the period of stay is too long

In the case of rational grazing, where the period of stay of each of the groups is too long, we see emerging, in place of this continuous evolution, an alternative, rhythmic evolution. In the leader group comprising milch cows *a periodic fluctuation in the quantity of milk produced takes place: this finally finds expression in a much more rapid drop of the lactation curve.*

On an Isère (French Alps) farm I had the opportunity of seeing a very large herd of cows gathered together in one single group, grazing each paddock for *seven* days. Total herd milk production was recorded for each day, and it could be seen quite plainly that the production curve was represented by a sinusoid with a periodicity of seven days: that is to say, a curve with its alternating maxima and minima separated by an interval of 7 days corresponding to the period of stay. Such fluctuations are obviously very detrimental to the total production of a cow during one lactation (for the year).

The cow's mechanism for halting fluctuations in milk production

Rhythmic fluctuations in milk production would be even more accentuated if the cow did not *take it out of herself*, that is to say, if she did not utilise the reserves in her own body to maintain a high milk yield; this represents the requirements that the quantities of grass (which the nature of the sward allows the cow to harvest) are insufficient to cover. This is her primary mechanism for wiping out fluctuations in milk production.

But there is an even more marked regulatory factor constituted by the mass of nutritive substances in the alimentary canal of the animal, and principally in its rumen.

When, in the stall, we alter the feeding of our cows (for example, by the addition or suppression of concentrates in the ration) this modification of the ration begins to influence production after 36 hours, and is exercising its full effect from after about 4 days. In consequence, if the animal remains for 24 hours on the same paddock, the change in quality and quantity of food, the effects of which begin to make themselves felt after 36 hours, will have no influence on performance. If, on the other hand, the animal remains on the same paddock more than 3 days we are reaching the limit of the delay of 4 days, after which a change in food exercises its full effects. Finally, if the animal remains 1–3 days on the same plot, the influence of a change in food will be very incomplete indeed. These considerations will lead us later to establish the second Universal Law of Rational Grazing.

Let us consider another point where the figures (uncertain as they are) help us to understand a technique of our ancestors, still very much in use to-day, sometimes in new and ultra-mechanised form.

Must the cow be spared the work of harvesting?

If fairly tender, cut grass is carried to the animal, either in its stall or in some exercise park or other, a cow of 10 cwt. [*500* kg.] live weight will consume about 143 lb. [*65* kg.], that is to say, 29 lb. [*13* kg.] of dry matter. This observation has no doubt led to the error, with such serious consequences, that has already been stressed: namely, the theory that, in all circumstances, the cow eats and is satisfied by 29 lb. [*13* kg.] of dry matter (*vide* p. 75).

It can be said, therefore, that if one cuts, in advance, a tender forage crop, be it grass or red clover, and carries it to a 10-cwt. [*500*-kg.] cow, she will *eat*, every day, 143 lb. [*65* kg.] and sometimes even more, although on the average if this cow has to forage for herself, she will *harvest* a maximum of 100 lb. [*48* kg.] of grass on rational grazing, and probably, with difficulty, 88 lb. [*40* kg.] on continuous grazing. It is quite evident that, all other things being equal, the cow that *eats* 143 lb. [*65* kg.] grass will supply a much greater yield of milk than the cow that "harvests" 88 lb. [*40* kg.] of grass. In the same way, a bullock *eating* this grass will fatten more quickly than if he is *harvesting* it.

Allowance must also be made for the amount of energy expended by the cow in harvesting her food (*vide* pp. 78 and 87).

Other factors entering into the picture are: the health of an animal living an enclosed existence, cost of man-power, qualities (hormones, œstrogens, antibiotics, etc., lost almost immediately by mown grass), non-return of excrement to the pasture, etc.

In the chapter on grazing "rationed in time" we will see how, particularly in countries where the days are very hot, they try to combine grazing and soiling (in the stall or in a sheltered exercise park). In this present connection I will cite only one French example.

Soiling in Finistère

Mention was made above of the peasants in the Elorn valley (Finistère) (Part One—Chapter 3)(Photo 1). In this valley the grassland, which is irrigated with water laden with organic substances, is capable (in the exceptionally favourable climate pertaining here) of providing eight to nine cuts of grass. To this grass the farmer never fails to give, for each "rotation", the appropriate period of rest, with the result that its productivity is high. It was said that this method of management might be called a "mown rotation" (*vide* p. 24–25).

This method of previous cutting and feeding green is very much in use in Finistère, be it grass, red clover, etc., that is involved. All, or almost all, the operations are manual: little mechanisation is employed in view of the plantiful supply of man-power in this region. The Agricultural Advisors have tried, not without reason, to turn the Breton farmers against this method of feeding green and to persuade them to graze much more than they are doing. In the course of a few days spent touring this area before giving a paper at a Grassland Conference I had long talks with these farmers. All were agreed in giving the same answer to the Agricultural Advisors who were accompanying me: "Our bullocks fatten quicker and our cows give more milk when we feed them cut, green grass than when we allow them to graze. . . ."

It is evident, as was said above, that this answer is very probably correct, for in this way animals *eat* almost 154 lb. [*70* kg.] of grass instead of the 88 lb. [*40* kg.] they *harvest*, the latter amount being often perhaps even less with ordinary continuous grazing.

It should be noted, moreover, that the Breton farmers are careful to let their cattle *graze for a few hours each day* on pastures situated close to the farm. This is a prudent measure (and no doubt indispensable) to maintain the good health of the stock (*vide* p. 245).

This example is a good illustration of the fact that traditional farming methods are never without their reasons. Before criticising and modifying the method, one must first try to understand the reasons for it.

In examining the use of soiling as a means of compensating for seasonal fluctuations in grass production, mention will be made of the ultra-mechanised methods of the Americans, a sharp contrast with the manual, but highly refined, methods of the Breton farmers, who, unfortunately for themselves, have not the benefit of cheap petrol (*vide* p. 173).

Chapter 4

THE COW IS A GOURMET

Palatability is the link between the grass and the grazing animal

MANY definitions can be given of palatability. The following passage taken from Ivins provides a definition that is both precise and original (47):

> "Palatability is the sum of the factors which operate to determine whether, and to what degree, the food is attractive to the animal; it can thus be held to constitute the *connecting link* between grass and the grazing animal and is regarded by various authors to be of greater importance than nutritive value. Necessarily relative, it is influenced by such variables as the animal itself, stage of growth and development of the herbage, alternative foods and the management and manuring of the herbage."

This definition of palatability and the considerations which accompany it provide a good introduction to our subject of study, namely, the cow considered as a gourmet.

The cow has very decided tastes

The cow, like ourselves, has certain preferential tastes which it would be of great value to us to know. In fact, if a cow in the course of her travels on a paddock does not come upon food to her taste she will snuffle and sniff and spend her time and energy uselessly in looking for food she likes and which suits her. In the end, for the limited time and effort devoted to harvesting her food, she will harvest a greatly reduced quantity of grass, which will lower her production.

Unfortunately, up to the present, pastures have been studied particularly from the point of view of the plant and very little from the point of view of the animal. Without doubt, analyses have been made of the herbage: protein and fibre contents have been determined, but no one has thought to find out which grass the animal prefers, that is, which it considers to be the most palatable. For instance, chemical analysis has shown that the nettle has a high nutritive value, but no cow on the hoof ever touches a *green* nettle.

Tastes and physiological requirements

Before examining the tastes of the cow I should like to pose the following, preliminary question: Does the cow choose certain foods because they are more or less suited to her physical, chemical or physiological equilibrium? Sir R. G. Stapledon replies explicitly (101):

> "The cow has an appetite instinct which allows her to select the foods which best satisfy her physiological requirements."

J. W. Gregor (3), on the other hand, imposes these reservations:

> "Certainly it is convenient to imagine that the grazing habits of livestock on a mixed vegetation are in some way related to their dietary needs. Yet it would be quite wrong to think that our domestic animals are endowed with an instinctive appreciation of what is best for them i.e. of dietetics. No doubt it sometimes happens that when animals are given a choice of foods their feeding habits conform to the rules of 'good husbandry', though it does not in the least follow that their likes and dislikes are in any way directly related to their needs."

I believe that there is reason to adopt a more cautious attitude to the instinct of the cow. *A priori* Stapledon is very probably right, but with one fundamental reservation: the cow has an excellent "appetite instinct", but *on condition that she finds herself in circumstances corresponding with those to which this "appetite instinct" has become adapted over thousands of years.*

Traditional habituation

To explain: in days of old the cow adapted herself to the conditions of her natural environment. If these conditions are changed, the animal will be completely lost and is liable to make a number of blunders. Here are two striking examples:

If the instinct of the cow did not warn her against poisonous plants, all the cows would have been dead long ago. For example, the cow never touches poison-sumac (poison-ivy), that dangerous plant so common in certain regions of the U.S.A. (Voisin, 117). Now man has been clever enough to discover hormone herbicides, and when sumac is treated with a hormone its appearance is so changed that the cow no longer recognises her mortal enemy, eats it without a qualm and dies.

The instinct of the cow is adapted to the green plant; but have 4000 years of feeding with dried crops allowed her to adapt her instinct to dried food?— I doubt it. In fact, certain poisonous plants which are carefully left aside in the green state are unfortunately eaten in dried forage. This is the case, for example, with colchicum, which is encountered on some new alluvial deposits of the Lower Seine and which causes very serious accidents in dried forage crops.

There is one way to baffle the cow's instinct: starve her. Obviously if a cow finds herself on a sward grazed to the ground the unfortunate beast will eat everything she can lay her teeth upon: that is to say, a multitude of weeds, if not poisonous plants, which she would not look at in normal circumstances.

But apart from these extremes, how many intermediate circumstances are there in which Man's intervention confuses the instinct and tastes of the cow? To what extent does fertiliser application alter the taste of plants and upset the animal's instinct? Are re-seeded pastures not very different from the permanent pastures to which cows were accustomed? Do the strains or species selected by Man correspond in taste or smell to those to which the cow was accustomed in the days of our forefathers?

These are important questions to which it is often difficult to find an answer in the present state of our knowledge. Sometimes, however, the cow provides a very definite clue, as in the following experiment at Rengen.

Cows prefer indigenous to selected grasses

A pasture at the Rengen Grassland Research Station (Germany) was sown down to grass five years ago in the following manner:

(*a*) one half was re-seeded with selected strains of grass;
(*b*) one half consists of grass which has re-grown *spontaneously* on a ploughed field, and consequently comprises ordinary species indigenous to that place (*bodenständige Pflanzen*).

The cows on this pasture can choose freely between the two kinds of herbage and show a distinct preference for the part comprising the indigenous species and strains which have grown naturally. The difference in grazing is very marked: the part with the local varieties is already grazed bare when the other part, with selected species, is still hardly touched by the animal's teeth.

For reasons unknown, the cow prefers species and varieties that grow normally and naturally under local conditions. Reference will be made later to Ivins' experiments on the subject (Table 38, p. 108). For the present let us look more closely at the reasons for the cow's choice.

What sensory activities guide the cow in her choice?

As a rule four sensory activities can guide the cow in her selection of grass: taste, touch, sight and smell.

As far as taste and touch are concerned, I will say that I have never seen a cow spit out food once she has got it into her throat. I therefore assume that other sensory activities guide her in her choice of plants which would have a bad taste for her palate or irritate her throat. Sight may certainly help her to avoid eating such prickly plants as thistles, but it is very probable that the principal and basic sensory experience guiding the cow in her choice of food is smell. It is enough to watch a cow at the feed-trough, how she sniffs with disgust at a mouldy forage or a rotten silage, to understand how she is guided above all by her sense of smell. But it is even more evident if one watches a cow grazing and sees her walking about smelling at the grass before selecting her food. Moreover, so far as I know, the cow is not nyctalopic, which means that she does not see at night. But by night as well as by day, she avoids poisonous plants and the thick grass of the dung pats.

The palatability of foodstuffs must be correlated, above all, with their smell. How does this smell tie up with the chemical composition of the plant? How, to take the question even farther, does the smell tie up with the "appetite instinct" described by Stapledon which allows the animal to harvest the ration best suited to her physiological equilibrium? These are certainly questions which it would be interesting to have clarified.

But to these four sensory activities, one of which, smell, appears to play the fundamental rôle, I should like to add another: pursuit of the *pleasure of rumination*. The ruminating animal undoubtedly has a truly satisfied air; and so it does not seem to me too presumptuous to suppose that the animal will tend to create for herself a ration which allows her to find the maximum pleasure in ruminating.

Quest of the pleasure of rumination

On a very young pasture the cow tends to seek out the mature grass. On a mature sward she harvests, for preference, the youngest grass. *What is the factor which guides her in this systematic choice?* Nutrition experts generally reply that she acts in this way so as to harvest the best balanced food from all points of view. This may be true, but how is her aim achieved?

The cow has not taken a course in animal nutrition, and since she cannot make use of a slide-rule, we must assume that she has certain criteria more simple and more within her reach for determining the balanced ration most suited to her needs. Unfortunately, we are reduced to hypotheses on this point. Here are some of these regarding the fibre content.

It is possible that the cow, like ourselves, observes the law of minimum effort and, having ascertained that a mature plant demands great effort in shearing and mastication, prefers to gather a less tough plant, just as we prefer a tender to a tough vegetable. This seems quite natural. But it is much more difficult, on the other hand, to explain why the cow's instinct compels her to seek out a grass rich in fibre when she is surrounded with grass of very low fibre content. I believe that this fact can be explained by the following hypothesis: *the animal wants to experience the maximum possible pleasure in digesting her food and particularly in ruminating.*

If, after harvesting a plant that is too young and poor in fibre, the cow does not have the pleasure of a sufficiently prolonged rumination, and if, moreover, this young plant, as a result of scouring, has caused her abdominal pains, one can imagine that the animal will hesitate before gathering this plant again.

I persist in believing that the rôle of rumination is fundamental. It is certain that rumination is a great source of pleasure to the cow; a popular expression describes a person as "having the satisfied air of a cow chewing the cud".

Now the time spent in ruminating diminishes with the fibre content of the ration, and may even become more or less non-existent with foods of low fibre content, such as cattle cakes or an *exclusive* diet of very young grass. In studying the problems of applying rational grazing we shall see, for example, that under certain systems called "rotational", cows which had only grass 2–3 in. [5–8 cm.] high at their disposal were incapable of ruminating unless fed a supplement rich in fibre, such as straw. It is understandable that these cows tended to browse along bankings and in the hedgerows: they were compiling for themselves a ration with a sufficient fibre content to allow them at least a little rumination, the latter being not only a pleasure to them but a vital necessity. With cows, as with men, not all pleasures are unwholesome.

My point can be summarised as follows:

The cow does not like its grass to be too mature, for it then requires too big an effort to shear due to its high fibre content. By contrast, when grazing a very tender sward, she seeks out grass which is richer in fibre so as to increase her pleasure in rumination. But if these hypotheses are correct, it still remains to be understood how the cow estimates the fibre content of a plant. Perhaps there is some relationship for her between the *smell* and the fibre content. It can also be supposed that the animal looks for an average ration with not too high a water content, and thus provides herself with a diet neither too high nor too low in dry matter.

"Eatable water"

W. Ellison (17) reports the following experience:

Part of a hill pasture was tilled and sown with a mixture of different ryegrasses. At the beginning of the autumn the cows were allowed to graze this new sward simultaneously with an old pasture. It was quickly noted that the cows spent much more time on the old pasture, although this did not seem very palatable.

Samples of grass from the two pastures were analysed, and it was found that the rye-grass mixture, if relatively rich in protein in relation to dry matter, had a low percentage dry matter, namely 15%. The old pasture, on the other hand, although of a low nutritive value, had a very high dry matter content, about 30%.

The cows were therefore endeavouring to obtain a balanced ration from the point of view of the protein as well as from that of "*eatable* water".

It is important to dwell for an instant on this expression "eatable water". This is the water contained in the food itself, as distinct from the water contained in the ration, which is the sum of the "eatable water" contained in the food and the water *drunk*, be this water in a trough, or dew, or rain lying on the plants.

This, Stapledon (101) explains in these words:

"Animals, if given the chance, graze with discretion. They move from one type of vegetation to another not only 'in search' of particular minerals shall we say, but because *they sense the need of a particular balance between 'eatable' water (wet matter) and dry matter.*"

Here we certainly have a very interesting and important hypothesis, but research still has not provided us with data on the percentage of "eatable water" most appreciated by the cow and the factors which can influence that percentage, such as quantity of water *drunk*, fibre content, protein content, etc.

The instinct of the cow can never be set down as an equation

It would obviously be interesting to have knowledge of all these points, and it is certain that future research will provide us with valuable data. But let us have no illusions: the instinct of the cow will never be set down in equations. As a philosopher has said, "The animal which acts by instinct seems unaware of the end, with a view to which it does what it does."

After these general considerations, let us look at some measures of palatability.

Swedish precursors of the Encyclopædist era.

At the time when, in the eighteenth century, all human knowledge was being gathered together into vast encyclopædias at the instance of the French philosophers, a group of Swedish botanists undertook, in 1747, the study of the tastes of domestic animals for different grasses. 583 species were studied in this connection.

Sheep ate 422 of these 583 species. Cattle ate only 328 of the 515 species offered to them. Of all the domestic animals, it was the goat that had the least exacting tastes, eating 470 of the 545 species offered to it. But the investigators were not content with seeing whether an animal ate a plant or refused it. Each plant was given an index of palatability.

James Anderson (with whom I shall deal at length in Chapter 1 of Part Six) has reported these Swedish experiments for us. Photo 7 (facing p. 75) shows one of the pages of the 1777 edition of this Scottish author containing the indices of palatability of some plants for the five animal species. Anderson's work contains the complete index list for the 583 species tested.

I have not been able to find any searching study since Anderson's time of the index of palatability of our pasture plants. It fell to Professor Ivins, that friendly and active research worker of Nottingham University (England), to carry out some remarkable work on this palatability index (*vide* Table 38, p. 108), but it is only in very recent times that Barbara Mott, a co-worker of Dr. Klapp's (77), has provided an index of palatability as complete as that drawn up by the Swedish workers of the Encyclopædist era.

Relationship between palatability and quantity of grass harvested

It has often been considered that the quantity of grass (or, more generally, of forage plants) harvested by the animal could be looked upon as a measure of the palatability of that grass. It should be noted, moreover, that the same thought has been expressed in stall-feeding trials. When a 10-cwt. [*500*-kg.] cow ate more or less than the 29 lb. [*13·5* kg.] of dry matter (which it was considered must satisfy her) this was attributed to the fact that the ration fed was more or less palatable.

In my opinion, this is only correct, in the case of stall feeding, if one includes in the term "palatability" all the many characteristics of the food: taste, extent of chopping, degree of toughness, fibre content, etc. I believe, indeed, that the cow is capable of allotting only a definite and limited amount of energy to mastication (cf. Voisin, 118). Even if she likes the taste of the food, she will absorb only it in an amount which, relatively, is not very great if she has to devote much effort by way of mastication in order to eat this delicious food. Irrational as Man may be, I doubt whether a person who loves nougat eats it in large quantities: his jaws would quickly tire.

In the case of the grazing cow there is another factor which makes the estimation of the quantity of grass harvested by the cow even more tricky. As we have seen, height and density of the herbage have an enormous effect on the amount harvested. If we suppose that two grass swards have the same chemical composition and the same taste, but one is 6 in. [*15* cm.] high and the other 2 in. [*5* cm.] we have seen that, as a result of her jaw mechanism, the cow will harvest 150 lb. [*68* kg.] approximately in the first instance and 45 lb. [*20* kg.] approximately in the second.

In all the experiments concerning palatability of pasture plants with which I am acquainted, the investigator has been content to measure the quantities of these plants harvested by the cow: I have never found any reference whatever to the height and density of the plants at the time when they were being harvested by the animal.

Palatability trials are further complicated by differences in earliness and different rates of "re-growth" after cutting. Rarely have I found any data on these factors in accounts of palatability experiments.

Such trials, therefore—and they are often contradictory into the bargain— must be treated with the greatest reserve, but since they throw light on some interesting aspects of the question, we will examine a few in more detail.

The Middleburg experiments (West Virginia, U.S.A.)

As stated at the beginning of this work, Research Institutes and Stations have studied pasture plants on small plots in particular. In 1949 a station was set up at Middleburg in West Virginia with the express aim of studying pastures and the plants they comprise *under actual conditions, that is to say, grazed by stock.* I had the opportunity to visit this station in 1951 (Voisin, 117, Vol. I, pp. 43–45).

One of the studies in progress is concerned with palatability of plants. This is measured in two ways:

1. The quantity eaten by the animal.
2. Direct observation of the animal.

So that the cow's preferences can be better observed directly, plots have been sown within the pastures comprising either one species only or a simple mixture of species. These have been named "cow cafeterias"; one might say that they are self-service restaurants for cows.

In the case of pure stands, the two methods of measurement classified the grasses in the following order:

1. Cocksfoot (orchard grass).
2. Fescue.
3. Smooth-stalked meadow grass.
4. Timothy.
5. Brome grass.

Where the menu consisted either of grasses alone or of grasses and white clover in mixture with grass, the cows indicated their preference as follows:

1. White clover and cocksfoot.
2. Cocksfoot alone.
3. White clover and fescue.
4. White clover and smooth-stalked meadow grass.
5. Fescue alone.

The cow prefers a varied diet

This classification, however, is valid only when cows are continually on grazings offering them five "dishes" to choose from. Indeed, the sensational observation was made that the *preference exhibited by the cow depends on what she was eating previously.* The cows were allowed to graze for several weeks on pastures comprising only white clover and cocksfoot; then they were put on to a cafeteria grazing, where the five different diets mentioned above were at their disposal. The beasts neglected the mixture (white clover and cocksfoot) completely, and this was placed last. It is evident, therefore, that cows like a change. A man can be a great lover of fillet steak, preferring it to sausage; but if he is made to eat fillet steak at each meal every day for several weeks and then is offered a choice between fillet steak and sausage, there is no doubt that he will choose the sausage, only too glad of a change of food. The cow, too, is a gourmet and likes a little variation.

Stimulating the cow's appetite

Everyone knows that the addition of salt stimulates the cow's appetite for a food that she has previously been slow to eat. Cattle-cakes eaten with distaste are gobbled up when mixed with pulp. A truly curious observation on the behaviour of a cow at grass has been made in California:

Cows were grazing a miserable pasture, very poor in protein. They were given a complementary feed of 16–32 oz. [*500–1000* gm.] of cotton cake. The result was that the cows, as if their appetites had been stimulated by this high-protein food, *consumed far more of the grass from this wretched pasture.*

J. W. Gregor (33), reporting this instance, concludes that if cows were made to graze a small area of herbage rich in protein, for example, young white clover, for a certain period every day, they would eat far more grass from the mediocre pasture comprising their normal grazing. This would obviously be of particular interest with poor grazings: it would be sufficient to create a few small areas of very rich pasture to help the animals to eat a greater amount of the herbage from the poorer-quality hill grazings. Gregor carried out numerous trials with cattle and sheep, but could not reach any definite conclusions on this very important point.

Obviously there are grounds for wondering if it is the feeding of a certain amount of protein (in the cake or young grass) that has stimulated the cow's appetite; or if, as has just been seen, the cow that likes variation in her diet appreciates the old grass (sausage) more after having eaten young grass (fillet steak). It is also possible that this ration of protein allows the development in the rumen of a micro-flora more abundant and of better quality which is capable of digesting foodstuffs rich in fibre, that is to say, grass of low quality.

Palatability trials at the University of Kentucky

At the University of Kentucky the quantities of dry matter contained in different plants harvested by bullocks weighing 6 cwt. approximately [*300* kg.] have been measured (Voisin, 117, Vol. I, pp. 235–240). Some of the results obtained are contained in Table 37. Unfortunately, the average height of the plants has not been noted, and so the results obtained are somewhat hazardous.

TABLE 37

Quantity of dry matter harvested daily by bullocks during the grazing season in Kentucky

	Quantity of dry matter harvested			
	1st series of trials (1949)		2nd series of trials (1950)	
	lb.	[kg.]	lb.	[kg.]
Lincoln brome grass	25·09	[*11·38*]	31·53	[*14·30*]
Ladino white clover	21·12	[*9·58*]	26·01	[*11·80*]
Lucerne (alfalfa)	16·67	[*7·56*]	23·96	[*10·87*]
Cocksfoot (orchard grass)	14·70	[*6·67*]	21·98	[*9·97*]
Smooth-stalked meadow grass	13·40	[*6·08*]	19·22	[*8·72*]
Fescue (Kentucky 31)	11·51	[*5·22*]	11·46	[*5·20*]

N.B. The bullocks used weighed between 5·4 and 6·2 cwt. [*272* and *317* kg.]

This is sufficient to explain certain anomalies: as, for example, the considerable variations, in the course of the two series of trials, in the quantities of dry matter harvested by a bullock from certain plants (difference of 50%

for lucerne), whereas there was no variation to speak of in other plants: fescue, for example.

If we compare the Kentucky with the Middleburg palatability trials we find totally different classifications. For example, brome grass, which at Middleburg was classed as the least palatable diet, heads the list in the Kentucky trials. The two brome grasses were perhaps very different, but my belief is that the amount of grass harvested is only a measure of its palatability if all the factors which can influence the animal in its harvesting are taken into consideration, namely height and density of the grass, hereditary grazing capacities of the cows used, etc.

Professor Ivins' index of palatability

Professor Ivins of Nottingham (England) sowed an area of approximately 4 acres [*1·60* ha.] in plots of simple mixtures of a grass and white clover. The plots were sited at random and replicated six times. The cows were under continuous observation while grazing the experimental area, and counts were made at frequent intervals of the number of animals grazing a particular species. In this way an *index of palatability* was calculated, using a special method which need not be described here: the results are contained in Table 38.

The cow must be asked for her opinion

A very strange fact is that the most palatable plant is a weed, or rather so-called weed, very common on pastures, namely plantain. What is more worrying, however, is that the least palatable grass is a strain of cocksfoot, S.143. Ivins writes:

> "Taking into account differences in seasonal growth, S.143 cocksfoot was never outstandingly palatable. This fact is mentioned by Thomas (108) who refers to the criticisms by (British) farmers of the poor palatability of S.143."

TABLE 38

Relative palatability of different herbage plants

Strain	Average percentage of cows grazing
1. Plantain (ribgrass)	17·6
2. S.125 meadow fescue	11·1
3. S.48 timothy	9·8
4. S.100 white clover	8·6
5. S.23 perennial rye-grass	7·9
6. American timothy	7·4
7. Montgomery L.F. red clover	7·4
8. Danish cocksfoot	6·7
9. Irish perennial rye-grass	6·1
10. Cockle Park Mixture	5·8
11. S.143 cocksfoot	5·5

Compare Table 39, p. 111.
From Ivins (47).

This proves the danger of launching on to the market strains which have been obtained by cultivation on small plots in experimental gardens and cut with a cutting instrument. The errors which can arise from this practice from the point of view of the plant have already been alluded to: here we see the mistake from the animal's standpoint. **It is essential to seek the cow's opinion: that of the scientist is not enough.**

In large vineyards tasters of wine are employed: Plant Breeding Stations must likewise employ tasters of grass.

Stapledon (100), as Ivins points out, recognises that bred strains of grass are generally less palatable than ordinary, natural grasses (cf. the Rengen experiments, p. 102 above, in this connection).

Soil and palatability

Barbara Mott (77) recounts some of the observations she has made concerning the influence of the soil on the palatability of grass. In general, cows prefer to graze plant communities in dry, as opposed to wet, places. This is perhaps the reason why, when stocking a field where there are dells, cows prefer to graze on the highest parts. B. Mott concludes by saying: "Cows have a discriminating tongue for choosing the herbage that has grown in the drier places."

Perhaps it is also because of the dampness that cows on a pasture graze less readily in the shade than on sunny spots.

Be that as it may, differences in humidity do not explain all the animal's tastes. B. Mott observed that in the same paddock, cows preferred to graze where the gley [1] character of the soil was less pronounced. Soil of this nature can make itself felt, indirectly, by modifying the flora, for, on the parts with less gley, the percentage of white clover was higher and that of Yorkshire Fog (*Holcus lanatus*) lower.

Altitude may also have some effect. B. Mott (77) reminds us that wavy hair grass (*Aira cæspitosa*) is eaten in quantity by cows in mountain regions, but is not touched at all in the plains, due to the fact that, at these low altitudes, it is harder and much tougher.

Influence of phospho-potassic fertilisers on palatability

If mineral fertilisers give the grass a bad taste they could not be used even if they increase the yield tenfold. Fortunately not only is this not the case but, on the contrary, at the rates at which these fertilisers are normally applied they give herbage plants a taste much appreciated by the cow.

Twenty-five years ago Tacke (105) had already noted that on pastures in marshy places cows always grazed those parts of the grazing that had received potash. The same observation was made at Rengen in 1953 by B. Mott.

---

[1] Gley soil = soil with high ground water and iron-oxide accumulation in the region of the water table.

Klapp (70) found that of a pasture on poor land, the half that had received a complete fertiliser (phosphoric acid, potash and lime) was grazed in preference to the half that had received *no* fertiliser.

As yet we have no precise data in our possession allowing us to state whether this improved palatability of the herbage as a result of phospho-potassic fertiliser is due to modification of the flora or to a particular taste afforded by the fertiliser to the plant. The same uncertainty prevails with regard to the effect of nitrogenous fertilisers.

Influence of nitrogenous fertiliser on the palatability of grass

In the course of his investigations determining the indices of palatability mentioned above, Ivins applied a complete fertiliser (2 cwt./acre [*254* kg./ha.] of 12–12–15) over the whole experimental area. A quarter of each replicate, however, received the equivalent of 2 cwt./acre [*254* kg./ha.] nitrate of lime, distributed by hand (47). Whether it was June or October, the cows spent 80% of their time on the part that had received the nitrogen supplement, while only 20% of the cows grazed the untreated area. Ivins writes:

> "In fact, very little grazing took place on the unmanured areas until the manured areas had been grazed right down. Untreated cocksfoot was hardly touched whilst treated cocksfoot, of both strains (*vide* Table 38, p. 108) was eaten to a considerable extent."

Obviously one may wonder whether the nitrate of lime had not affected the palatability of the plants in modifying their composition. Ivins expresses this opinion:

> "It cannot be concluded that palatability is linked up with a high content of crude protein and a low percentage of crude fibre or dry matter for although this seemed to be the case when manured herbage was compared with unmanured, no such correlation exists when strains are compared one with the other—the animals did not necessarily choose the strains with highest protein content."

B. Mott (77) likewise observed at Rengen that where a grass had received a phospho-potassic fertiliser the parts that had received nitrogen at the high rate of 216 lb./acre [*240* kg./ha.] in several dressings were much more readily grazed than those parts that had received no nitrogenous fertiliser.

One might, however, wonder if there is not an optimum dressing of nitrogen to exceed which will diminish the palatability of the grass. This is what they tried to find out at the West Virginia Pasture Research Station.

Quantity of nitrogenous fertiliser and optimum palatability

At the Middleburg Research Station, West Virginia (Voisin, 117, Vol. I, pp. 45–46), varying dressings of nitrogenous fertiliser were applied to "cafeteria" grass paddocks: these comprised, in terms of pure nitrogen:

20 lb./acre [*22* kg./ha.]
60 lb./acre [*66* kg./ha.]
119 lb./acre [*132* kg./ha.]

In spring the 60 lb./acre [*66* kg./ha.] dressing apparently produces the greatest palatability while during the heat of the summer it is the highest rate (119 lb./acre [*132* kg./ha.]) that provides the grass most appreciated by the cow. It is probable that this variation, according to season, in the optimum nitrogen dressing providing the greatest palatability may be linked up with seasonal variation in the palatability of grasses which we will now briefly examine.

Seasonal variation in the palatability of grasses

Ivins traced seasonal variations in the index of palatability. The figures in Table 39 are selected from his numerous results.

TABLE 39

Seasonal variations in index of palatability

Grazing period	April 13–23	May 8–10	May 16	June 8–12	June 28– July 2	July 17–18	Aug. 16–21	Oct. 9–12	Average over whole period
Strain					Index of Palatability				
Ribgrass (plantain) .	13·9	9·4	6·5	21·9	14·0	13·8	23·1	9·7	17·6
Rye-grass Irish perennial	22·1	7·2	12·4	1·6	1·3	1·7	3·0	9·7	6·1
S.23 perennial	0·6	9·4	8·5	7·5	3·3	2·4	5·9	8·7	7·9
Cocksfoot Danish .	26·5	4·4	7·5	2·1	4·7	6·2	9·2	6·7	6·7
S.143 . .	0·3	5·3	3·3	2·1	11·6	11·7	6·2	6·4	5·5
Timothy American .	16·8	11·6	14·1	2·7	4·0	2·7	6·5	10·1	7·4
S.48 . .	4·1	12·5	8·8	7·2	15·0	13·8	7·7	10·4	9·8
White clover S.100 . .	3·8	3·7	4·2	15·2	12·7	11·7	—	—	8·6

N.B. 1. Figures are for the year 1950.
2. Compare Table 38, p. 108.

From Ivins (47).

It is particularly interesting to follow the figures for one type of grass. For example, Irish rye-grass is much more palatable in April than S.23, but the opposite is true in June. The differences in April are without doubt due to the variation in earliness of the plants, but this makes the variable differences, and not always variable in the same way, difficult to explain. As Ivins says, research in this sphere is only in its initial stages, and it is still difficult to pick out general rules.

From Table 39, p. 111, it is evident that a grass can be more palatable in one season than in another, and that these variations can differ greatly according to the strain of a species and, in particular, between ordinary strains and pedigree strains.

The cow and weeds

Basically, a weed is a plant not eaten by the cow, and may even be injurious to her health. It is more exact, however, to say that *a weed is a plant less readily eaten than a good grass*. Even this may not be going far enough: we should add "on the average and in most circumstances". It is not at all certain that a cow will not eat a young and *tender weed* more readily than a *tough* GOOD *grass*; and, lastly, there are a number of grasses so much half-way between the two categories, that one does not know how to classify them: we need only name Agrostis as an example.

Another complication which arises is the fact that although the cow refuses certain weeds in the green state, she will devour them with relish when they are withered: a typical example is the nettle. Be that as it may, there are so-called weeds readily eaten by cows and even for preference. It has already been noted (Tables 38 and 39) that Ivins found plantain (ribgrass) to be far superior, from the point of view of palatability, to all other grasses and white clover, Ivins writes (47):

> "The manner in which the dairy cows selected and preferred plantain (ribgrass) was most marked. . . . The taste of the cow for a plant such as this may be due to the fact that she finds it to contain certain indispensable nutritive elements. This fact should be stressed in view of the modern tendency towards simplification of seeds mixtures and further restriction in the range of types of herbage plants available to the grazing animal. On the basis of productivity it is probable that nothing will be lost by the inclusion of plantain (ribgrass) in seeds mixtures, for Milton found that in the early years after sowing, plantain (ribgrass) outyielded perennial rye-grass. Moreover, Stapledon (30) refers to the fact that under the poor conditions at Cahn Hill lambs were satisfactorily fattened on ribgrass swards."

Weeds in their capacity as opponents of deficiency diseases

It is quite probable, as Ivins suggests, that the cow's preference for so-called weeds, like plantain, stems from the fact that the animal finds in them certain essential elements which ordinary analysis perhaps does not reveal. Plantain has always been widely used in pharmacy; and recent research (Voisin, 125) has shown that it is one of the richest plants in natural antibiotics.

Many workers have claimed that weeds supply large amounts of mineral elements, especially trace elements which avert deficiency diseases among stock. American observers noted that on pastures very poor in mineral matter, cows showed a much greater tendency to eat weeds, especially those with deep roots, which seem to carry up the minerals from the lower strata of the soil. Perhaps the most interesting observations in this connection are those made by Von Grünigen in Switzerland during the last war (34). On the ploughing up of permanent pastures cattle derived their nourishment mainly (if not exclusively in some places) from temporary leys re-seeded with mixtures containing only a few plants. The combined effect of ploughing and lack of variation in the flora gave rise to deficiency diseases, which were easily cured when the animals were supplied with hay from *natural* and permanent meadows, normally containing a certain percentage of weeds (*vide* p. 125).

It appears therefore that the cow's tendency to prefer gathering certain weeds, such as plantain, has an important nutritional basis. It is probable that, as our methods of analysis are perfected, we will gradually achieve a better understanding of her reasons.

The cow prefers to harvest one part of a grass plant

Apart from her individual tastes for certain plants, the cow has a tendency to:

(a) Choose the most succulent part of the plant; in general, the leafiest parts when the plant is rather tough. This is known as progressive defoliation.

(b) Harvest for preference on certain parts of the pasture, known as *creaming* the herbage.

Professor L. Saltonstall of the University of Cornell (U.S.A.) wrote to me as follows in 1951:

"The herbage eaten by a bullock was collected in the rumen by means of a fistula, before it had undergone certain digestive processes. One of the experimental animals was grazing a very mature sward; the other was being fed in the stall with cut grass from the same sward. The nitrogen content of the grass collected in the rumen of the stall-fed bullock was always lower than in the case of the grazing animal. In other words, the bullock at grass was choosing a diet richer in protein.

"The difference in nitrogen content between the rumen contents of the two bullocks increased with advancing maturity, which indicates plainly that the selection of grass by the animal became more marked as the grass, in maturing, provided it with a diet more and more removed from the average and balanced quality it is inclined to look for."

It appears, therefore, that the animal wants to harvest a food with certain optimum characteristics from the point of view of water content, protein, fibre, etc.

How the cow chooses the part of the plant she prefers

When the cow chooses one plant and neglects another, the mechanism of her choice is clear and evident. But when the cow chooses one part of a plant, how does she go about it? How does she divide and dissect it to get at the part she likes best? Reference has just been made to *progressive defoliation* of plants by the animal, and it is probable that a slow-motion picture might help us to understand this mechanism better. Such films have been widely used in the study of the *growth* of plants, including grasses. Would it not also be possible to use them for studying the *decline* of the grass, that is to say, the way in which the cow reduces it, more or less progressively?

Improvement of our knowledge in this connection would certainly be of value; it would allow us to better compare cutting with scissors (or the blade of the mower) and the work of the animal's teeth. It might then be possible to improve the method of isolating cages and thus make the measurement of the quantities of grass eaten by the cow more exact. It is probable, however, that the way in which a plant is defoliated also plays an important part in the rate of its re-growth.

"Progressive defoliation" and "creaming" of the sward

It was explained above that, to enable grass to achieve its maximum productivity, it must be granted sufficient rest periods to allow it to realise its "blaze of growth" (*vide* p. 14, Fig. 2) and reconstitute the reserves in its roots. We know, too, that some parts of this grass, with an average height of 6 in. [*15* cm.] will be more mature than others in the spring and at the beginning of summer, in particular, certain grasses will have gone to height.

The result is that in favouring grass productivity we would run the risk of furnishing high-yielding cows with grass too rich in fibre and not rich enough in *true* proteins. In other words, in satisfying the requirements of the grass there would be the risk of failing to meet the needs of one section of the herd.

Now, with rational grazing, it is possible to divide the herd (p. 152) and, by means of "progressive defoliation" and "creaming" the sward, the leader group will be able to *graze selectively*, harvesting a ration with a higher protein and lower fibre content than the sward as a whole. To refer back to Table 25 (p. 63): 1 lb. of grass 6 in. [*15* cm.] high contains 12·2 gm. crude protein [1 kg. contains *27* gm.]. When the herd is divided into two groups, the first "creams" a grass sward with 12·7 gm. of crude protein per lb. [*28* gm./kg.] and leaves the second with a herbage containing 11·8 gm./lb. [*26* gm./kg.]. The differences are even more marked when the division is into three groups.

This selective harvesting is very obvious to the observer. In spring I have often had to put my first group into a paddock where the cocksfoot had already gone to height. When, after a stay of 2 days, the first group was removed to accommodate the second, the long stems of the cocksfoot had obviously hardly been touched at all, while the rye-grass and white clover had already suffered a considerable amount of grazing. This did not prevent the second group from grazing all the long cocksfoot stems to the ground. It is probable that the last group must have had their teeth sharpened to allow them to cope with the harder stems, while the micro-flora of the rumen has modified simultaneously to such an extent as to be able to "digest", with ease, a ration relatively richer in fibre.

Before leaving the subject of the cow's tastes, let us see what she thinks of the grass which has grown on her excrement.

The cow and her dung pats

It is common knowledge that a cow will not eat grass growing where she has defæcated; such places are absolutely refused. But there is a fact to note: the cow will readily eat grass that has grown where horse dung has fallen, and the horse, which refuses to eat from horse-dung pats, eats grass growing on cow-dung pats. This shows the advantages of associating young horses and cattle, where possible, in the last group of rotation (cf. p. 154).

Where sheep are concerned, opinions are contradictory, and it would be of interest to have the position clarified. Be that as it may, while it is not possible to make an absolute statement of fact, it appears that the cow accepts grass grown near sheep's droppings but refuses to eat grass at a place where a sheep has lain.

Another remarkable and less well-known fact is that if one cuts the grass from a dung pat and transfers it to another clean spot in the pasture it will be readily eaten by the cow. It can be concluded that it is not chiefly sight but smell which guides the cow in her choice of food. The latter, moreover, is essential to her if she is to be equally selective at night, and everyone knows that by night as much as by day a cow will not eat grass that has grown in a place where there is cow dung.

A second conclusion is that the smell of the dung drives the cow away, but the dung has not communicated any bad smell or taste to the grass growing upon it.

Finally, it would be interesting to know for how long the dung retains this smell which is so repulsive to the cow. In fact, good grassland management demands regular scattering of dung pats so as to prevent the formation of refused areas, the development of certain weeds, such as *Ranunculus* (buttercup), on the dung pats and the progressive disappearance of white clover in places where there is an accumulation of droppings.

The scattering of dung pats over the whole area of a pasture could make the cattle anything but willing to graze. Fortunately, it is certain that the disagreeable smell will have disappeared after a certain time, its disappearance being greatly accelerated by breaking up large heaps of dung into small pieces so as to aerate it and expose it to the sun. In my own experience, 12 days or so are long enough in this case for the smell to disappear. At all events, the time required is certainly less than the rest period of 18 days, the minimum given in spring to grass paddocks in rotation.

The cow and her urine

The cow avoids grazing the grass growing on a dung pat. What is her attitude to grass from a patch where she has urinated? Dr. Alfred G. Etter investigated this question at Malvern Clopton Farm, Clarksville (Missouri, U.S.A.), with the following obviously unexpected result: **the cow seeks out grass which has grown where she urinated.** The author writes (18):

> "In the course of my research work on meadow grass (blue grass) I discovered that cattle and horses much prefer grass that has grown on a soil impregnated with their urine. They seek out the dark green grass which has grown where their urine fell, thus saving and recovering the nitrogen, potassium and trace elements it contained. They profit from the stimulating effect exercised by the mineral elements, hormones and vitamins found in the urine. . . . The animals seek out the grass fertilized by their urine with just as much care as they avoid the grass which has grown on their dung pats. . . ."

If this observation is confirmed, it would be of interest also to examine more closely what effect fresh urine can have on the composition of grass. Unfortunately any study of this question is unknown to me.

This attraction for the cow of grass growing where her fresh urine has fallen is strange, moreover, in view of the fact that she shows a marked repugnance for herbage over which *fermented* urine, that is liquid manure, has been distributed; a repugnance which leads to liquid manure being restricted, in general, to grass for conservation.

Chapter 5

CONVERSION OF NITROGENOUS SUBSTANCES IN THE STOMACH OF THE COW

The ruminant's particular method of digestion

LET us repeat: when we speak about grass we must never forget the cow; when we speak of the cow we must always think of the grass. To get a good grasp of the relations between grass and cow we must understand how that grass is digested by the cow (or more generally, by the ruminant). But this would necessitate studying the mechanics of digestion in ruminants, a vast subject sufficient to fill a whole book in itself, and so we will here examine only one aspect of that digestion which will help us understand better the accidents which can take place with intensive grazing due to certain erroneous, but nevertheless current, methods.

Nourishing the microbes of the stomach

The digestive mechanism of the ruminant is very different from that of monogastric animals. The micro-organisms of the stomach play a fundamental rôle in the former's digestion. It has even been said that *it is not the ruminant itself that is being nourished, but in actual fact the microbes in the rumen, which, in turn, nourish the animal which is their host.* This, then, is a case of fundamental symbioses between the ruminant and the microbes of its rumen. But even if one assumes that the ruminant is nourished exclusively by microbial proteins (which is probably not correct) one is only shifting one's ground without simplifying the problem. It is not certain that the problem is not made even more complicated and tricky by the fact that one is going to have to nourish the micro-organisms of the rumen in particular without feeding the ruminant.

In fact, bacteria are particularly sensitive to the presence and absence of an amino acid. Indeed, they react so obviously to the presence of amino acids that they are used to-day to determine the quantity of the latter present (amino acids are constituents of proteins).

Moreover, administration of nutritive substances alters the composition of the rumen flora.

117

Synthesis of proteins by bacteria

Fundamental research showed that the bacteria of the rumen were capable of synthesising proteins from non-protein nitrogenous substances (erroneously called amides), such as urea, for example. The deduction from this was that there was no point in devoting too much attention to the nature and composition of the protein, and that the bacteria of the rumen would manage to produce from all these nitrogenous substances microbial proteins of value, which in the long run would be the proteins on which the ruminant would have to feed.

This question of the utilisation of non-protein nitrogen by ruminants has been the subject of a mass of investigations, for which the reader is referred to the excellent review compiled by Thérèse Terroine (106). Almost all the work has been devoted to this positive side of protein digestion, which was, and is, of so much interest from the practical point of view. It was hoped that it would lead to simple and cheap nitrogenous substances, such as urea, being used in the feeding of ruminants. Some of the results obtained were encouraging, but equally many were deceiving. I will not expand on this here but merely state that in reading a multitude of works on this subject I always had the impression of not quite understanding all the varying results: indeed, I felt a little at sea. But everything seemed to become quite clear once I made the acquaintance of the admirable work published by the Rowett Institute (Scotland).

Two opposing microbial actions in the stomach: synthesis and degradation of nitrogenous substances

One of the most eminent of the Rowett workers and a Nobel prize-winner (1952) is Synge, who with his customary clarity and precision writes (104):

> "Protein of the food is in part broken down by micro-organisms and in part passes on unchanged to the abomasum (or true stomach); part of what is broken down by the micro-organisms is assimilated by them and built into microbial protein which likewise passes on to the abomasum. *Another part is converted to ammonia and is absorbed by the animal directly from the rumen.* Urea and other nitrogenous compounds of low molecular weight in the food, in the saliva or entering the rumen from the animal's blood may undergo the same treatments by the micro-organisms. It is also very generally assumed that at least a fair proportion of the protein digested by the animal in the abomasum and small intestine is microbial protein, and it has been shown experimentally that the protein of mixed rumen micro-organisms is a high-grade, well-balanced protein for rats. The fact that ruminants show none of the signs of amino-acid imbalance on being fed exclusively on proteins that are ill-balanced for rats, chickens and men is therefore taken to be explained by the contribution of microbial protein, and the *requirements for absorbed amino-acids are assumed to be not dissimilar to those of the animals whose requirements have been studied in detail.*

"Beyond this point nearly everything is controversial. Because the mixture of amino-acids absorbed by the ruminant is a well-balanced one, it does not follow that the overall utilizition of protein is always *efficient.* Two opposing tendencies can be seen at work: (*a*) non-protein nitrogen, poor in 'essential' amino-acids, entering the rumen is upgraded to microbial protein; (*b*) food protein of good amino-acid composition *is attacked in the rumen and converted into ammonia which is to a great extent absorbed and excreted as urea.*

"The important thing is to ascertain *whether processes of type (a) or of type (b) predominate.* This has very rarely been done unequivocally and it is worth pointing out here that the conventional digestibility studies, traditional to agricultural chemistry, are ill-adapted for the purpose; to know the net absorption of nitrogen from the intestine provides no evidence whether it has entered the animal as ammonia or as essential amino-acids. At this stage of development of the subject, practical experiments on growth, milk yield and so forth or else thorough-going nitrogen-balance experiments, in which urinary nitrogen is measured, seem preferable to digestibility experiments...."

When a Nobel prize-winner overturns the idols of our feed tables

Synge has been quoted at some length, for it is essential that we should consider and understand his every phrase, so full of meaning and with important consequences. Moreover, I was gratified to know that I had the support of such an authority in drawing attention to the fact that our feed tables are constructed on doubtful foundations and still have but a very slight practical value for the farmer of to-day. Next to failure to investigate the criterion of the animal's satiation, the weakest point in our feed tables is the question of so-called proteins. It has already been stated, and is again repeated, that there has too often been a tendency *to confuse nitrogen and protein and to attribute the same biological (nutritive) value to all proteins.* This was a serious error in the case of ordinary foodstuffs: it had disastrous consequences where grazing is concerned.

But to return to the digestion of nitrogenous substances. One of the most fundamental points to be grasped from Synge's exposition is this:

Two microbial actions are taking place concurrently in the rumen—

—Synthesis of nitrogenous substances, supplying proteins (anabolism).
—Breakdown of the nitrogenous substances (including the proteins) into more simple nitrogenous substances and finally into ammonia (catabolism).

Rate of protein breakdown in the stomach

We will again call upon Synge. Reporting on his experiments with urea supplements (referred to above) he writes (104):

"It seems conclusively established that supplementation of certain diets with urea can lead to good use being made of this supplementary nitrogen. *This is known to depend very much on the type and amount of carbohydrate simultaneously fed.* The best results have been with beef cattle and fewer favourable experiences with dairy cattle (which require a higher protein concentration in the diet) have been recorded. It is perhaps significant that these favourable experiences have mainly been in the U.S.A. where maize plays a much larger part in cattle feeding than it does in northern Europe. McDonald has studied the rates at which different proteins were converted to ammonia in the rumen of sheep and found zein, the chief protein of maize, to be converted *very slowly* indeed. He was able, by making use of its unusual solubility in aqueous alcohol and of its low lysine content, to study quantitatively the disappearance of zein and the simultaneous formation of lysine-rich microbial protein in the rumen. It would help if similar specific tests were available for studying the fate of other proteins.''

Yet another consideration can be added to those put forward by Synge, namely, that very rapid decomposition of the proteins (or of simple substances such as urea) can, as the result of too great and too rapid a production of ammonia, give rise to pronounced alkalinisation of the rumen contents, with serious consequences for the animal (Scharrer, 87, p. 83).

But let us now turn to the most brilliant experiment carried out by the workers of the Rowett Institute (Scotland) about the digestion of nitrogenous substances by the ruminant.

The ruminant utilises casein better when it does not pass through the rumen

It has become more and more evident that the value of urea or other simple nitrogenous compounds as sources of microbial protein depends, in part, on the rate at which the urea is converted to ammonia. If this process is rapid, much of the nitrogen of the urea is transformed into ammonia and passes through the kidneys to be lost in the urine (6). Thus it appears that *the value of proteins to the ruminant must diminish in proportion as the protein is rapidly broken down and de-aminated by the micro-organisms of the rumen.* It is possible that the slow decomposition of zein, the protein of maize, helps to increase its biological value for the ruminant.

Chalmers and co-workers (14) have recently observed that the rapid conversion of the nitrogen of casein (milk protein) into ammonia in the rumen was the cause of its being lost in the urine. They ascertained, in fact, that the feeding of casein supplements to pregnant ewes on a low-plane diet (negative nitrogen balance) did not result in the animal retaining the supplementary nitrogen. They carried out the following experiment with ewes using rumen fistulæ: during a control period of 8 days, the animals were fed each day with a basal ration containing 6·06 gm. nitrogen. 6·82 gm. nitrogen were excreted in the urine and 2·78 gm. in the fæces (negative nitrogen balance). Then a supplementary 6·15 gm. of nitrogen were administered in the form of casein, which was put through the fistulæ of the *rumen*. The resulting daily nitrogen excretion was 11·42 gm. in the urine and 3·16 gm. in the fæces. This method of feeding was maintained for 7 days. It is evident that the amount of casein nitrogen retained was low when this protein passed through the rumen.

Subsequently, for a period of 12 days, 6·15 gm. of casein nitrogen were introduced by means of a *duodenum* fistula, with the result that *the casein did not pass into the rumen.* The quantity of nitrogen excreted in the urine now fell to 9·44 gm. (instead of 11·42 gm. as previously) *while that excreted in the fæces remained more or less identical. It is clear that the quantity of casein nitrogen retained by the animal was therefore increased when the casein was put directly into the intestine without passing through the rumen.*

It must be assumed that the micro-organisms of the rumen were responsible for this diminution of the biological value of the protein (casein), for the amino acids constituting that protein were *too quickly* de-aminated and converted into ammonia, which was eliminated by the kidneys, its nitrogen consequently being lost to the animal.

Heating casein reduces its rate of breakdown

In vitro experiments (that is using laboratory apparatus) showed, moreover, that the rumen liquor, in 5 hours, de-aminated 2·4 gm. of the 13 gm. of casein nitrogen present. This production of ammonia *in vitro* was considerably reduced by replacing ordinary commercial casein with hardened casein obtained by heating ordinary casein in alkaline solution and drying it at 220° F. [*105*° C.]. The resulting product is almost as hard as horn.

The laboratory experiment was confirmed by an *in vivo* experiment (that is, using the animal) as startling as it is informative.

The same treatment can make the biological value of a food for monogastric animals and ruminants vary in opposite ways

The hardened casein described above was subsequently used to supplement the rations of the ewes and to study their nitrogen balance. The amount of nitrogen retained by the animal was found to be very much higher than when commercial, untreated casein was used. It is very strange that the treatment which hardened the casein reduced its biological value as a food for monogastric animals such as rats and dogs.

This is a fundamental experiment, probably the first of its kind, which demonstrates that *a treatment which* ENHANCES *the biological value of a protein for ruminants* REDUCES *the biological value of the same protein for monogastric animals.* It is difficult to find better proof that the biological value of proteins in monogastric animals (rats, etc.) can provide but little indication of their value for ruminants. It may even happen, as has just been seen, that the same treatment will cause the biological value to vary in opposite ways for the two animal species.

It should be noted that, in the course of these same experiments the workers at the Rowett Institute found once again that *the presence of easily assimilable carbohydrates (cereal meal or starch) reduced the quantity of ammonia produced when a ration very rich in protein was introduced into the rumen.* This confirms the hypothesis that a substance rich in energy equivalent stimulates the multiplication of rumen micro-organisms, which are then able to assimilate rapidly the amino acids resulting from the breakdown of the various nitrogenous substances (protein and non-protein) in the ration.

Knowledge of the digestion of grass by ruminants must be improved

It was thought necessary to deal at length with this aspect, one among many, of protein digestion in ruminants. This has shown the extent to which knowledge of the subject is still in the embryo stage, and also how essential it is for scientists to pursue research on all these points so as to provide some guidance on the feeding of livestock, more particularly during the grazing period. Special stress has been laid on the details indicating that one is setting out along a dangerous path in assuming that all forms of nitrogen are suitable for the ruminant, since the bacteria of its stomach make it their business to produce high-grade protein from common ordinary nitrogen. Recent experiments, especially those conducted at the Rowett Institute, have revealed the problem in its complexity and have demonstrated that, in fact, two contradictory phenomena are taking place in the rumen at any one time:

1. Non-protein nitrogen is being converted to protein of superior quality.

2. The protein in the ration, being of good amino-acid composition, is being attacked in the rumen and converted to ammonia, a large part of which is excreted in the urine in the form of urea.

As the New Zealand worker A. T. Johns writes (48):

"It is not the crude protein determination, or even the amino acid analysis that determines the value of protein to the ruminant. The solubility of the protein and its susceptibility to microbial breakdown, the amount of non-protein nitrogen and the nature and amount of the carbohydrate being simultaneously fermented, are obviously important factors."

To perfect the feeding of cattle, more must be known about all these mechanisms of digestion. Improvement of our knowledge, which at the moment is unfortunately very meagre where such matters are concerned, would lead to better methods of managing pastures and, in particular, would prevent many disasters from taking place.

Chapter 6

GRASS TETANY

GRASS tetany is a paralysis affecting animals at grass in particular; it can also attack animals kept indoors, but this less frequently. Taken as a whole, the symptoms are similar to those of milk fever, but the primary causes and remedies are apparently different.

Causes of grass tetany

In spite of many uncertainties and unknown factors, it has almost generally been agreed that grass tetany is due to an ionic imbalance, that is, the ratio of certain mineral elements is unfavourable, in particular the ratio of four elements:

$$\frac{\text{potassium, sodium}}{\text{calcium, magnesium}}$$

This question will be studied in more detail in my work on the composition of grass. As far as the ionic imbalance is concerned, the reader is referred to the excellent study made of the problem by Verdeyen (112). In the present context only the question of magnesium will be briefly examined.

Hypomagnesemia

The blood of animals attacked by grass tetany has an abnormally low magnesium content; intravenous injections of magnesium salt, if given in time, effect as spectacular cures as calcium salts in the case of milk fever. Almost all the research workers (Sjollema and Seekless in Holland in particular) are more or less of one accord in considering that the disease is not caused directly by a lack of magnesium in the diet, for the grass of pastures which produce tetany has often as high a content of magnesium as that of other pastures where tetany does not occur. It appears also that there is no correlation between the nature of the soil and the appearance of tetany (Allcroft, 3), although cases have been reported where application of magnesium to the soil has reduced the frequency of tetanies; this, however, is a highly controversial point.

If grass tetany does indeed correspond to a hypomagnesemia, that is to say,

123

a lowering in the magnesium content of the blood, there is very little certainty as to the cause of the drop.

Development of grass tetany with ley farming

Grass tetany appears to have developed particularly rapidly in Great Britain following upon the policy of ploughing up old pastures recommended during the War. Ruth Allcroft (2), of the Weybridge Veterinary Research Laboratory (England), writes:

> "Increase in grass tetany is associated with adoption of an intensive system of ley farming. . . . According to the information received at this laboratory, tetany is increasing from year to year. . . ."

As far back as 1948 Muir (78) was writing in the *Journal of the British Grassland Society*:

> "The opinion is widely held, at least among veterinary surgeons and farmers, that while parasitism is usually greatly reduced on leys, physiological disorders, *such as bloat and grass tetany, are much more common on leys than on permanent pastures.*"

In 1954 one British agricultural journal, *The Farmers Weekly*, reports (7):

> "Enquiries made by our correspondent in Northumberland show that mortality due to grass tetany has been higher than average this spring. . . . The disorder is spread throughout the county but is concentrated in particular on the best farms and *especially on farms where stock have been put out on young leys. . . .*"

Grass tetany and temporary pastures

Several hypotheses have been put forward to explain how ley farming (considered as a plough-up policy for old pastures) could lead to the development of grass tetany. As mentioned above (p. 113), deficiency diseases were found to put in an appearance in various countries, Switzerland in particular, when old pasture was ploughed up during the War. It has been thought that these deficiencies could be indirect causes of tetany.

But I consider Verdeyen's explanation the most satisfactory. In his vast study (112) of the influence of mineral imbalances on grass tetany, the Belgian scientist writes:

> *"The exteriorization of mineral imbalance resulting from the sowing of a pasture is characteristic: and stock fed exclusively from leys normally suffer more easily from tetany than stock fed exclusively from old pastures."*

The good mineral balance of old pasture grass

Verdeyen also gives a series of curves of certain mineral balances (especially ratios of K_2O and CaO) in leys and old pastures. From these he concludes:

"Old pasture provides a grass which, for 100 parts (artificial unit) dry matter, contains 85–105 milli-equivalents of K_2O and 30–35 milli-equivalents of CaO while the grass of newly-sown neighbouring paddocks contains, for 100 dry matter, 100–125 milli-equivalents of K_2O but the content of CaO in this instance drops to 22 milli-equivalents. . . . *These figures provide quite a natural explanation of the development of tetany on leys. . . .*"

In other words, ley grass contains an excess of potash for too little lime, giving rise to imbalance among the four elements whose reciprocal relationship was pointed out above.

Let us now consider the question of management.

Erroneous forms of intensive grazing and grass tetany

In studying the various forms of rational grazing we shall see that the Hohenheim system advocates grazing very young grass. The same is true of various supposedly intensive systems of grazing. It will also be seen that the current error made in rational grazing, what I have named "untoward acceleration", leads to the stock being supplied with grass that is much too young (p. 226). The yields from such grass are small. It re-grows slowly because it has not been able to accumulate sufficient reserves, and consequently cannot produce its "blaze of growth", since the period of re-growth allotted to it is insufficient.

These erroneous conceptions of rational grazing, while not satisfying the requirements of the GRASS, *fall so short of meeting the demands of the* COW *that they gravely endanger the latter's health.*

Dangers of very young grass, an unbalanced food

Very young grass is a highly unbalanced food. In continuous grazing the animal (except for exceptional cases) is confronted with clumps of tough grass, while in erroneous systems of rational grazing it feeds exclusively on this very young grass. Promoters of the Hohenheim and analogous systems, however, very soon saw that cows put on to paddocks of very young grass saturated with nitrogen showed a marked tendency to look for the old grass growing under hedges or on bankings.

Moreover, in studying the composition of grass it has been seen that this very young grass which the pioneers of rotation considered to be very rich in protein is in fact only very rich in nitrogen.

This, then, is a very unbalanced food because:

—The ratios of the various mineral elements are unfavourable.

—Nitrogen is excessive in relation to the energy equivalent of the ration (i.e., in relation to the carbohydrates).

—The small proportion of fibre does not permit good rumination.

—In the case of various abnormal external conditions, which are not well-known, the percentage of non-protein nitrogen can amount to as much as 50% of the total nitrogen.

All these circumstances militate against good utilisation of the grass by the ruminant, and may even seriously impair its health.

M. Latteur, Veterinary Inspector of the Belgian Ministry of Agriculture, has been able, on the above basis, to explain the cause of grass tetany. His is probably the most satisfactory explanation available at present.

Influence of excess ammonia on the state of the rumen

Latteur (71) has ascertained that in the case of grass tetany in spring, the rumen contents are characterised by:

—A poor flora of micro-organisms: the species occurring in the summer flora are practically always absent.
—A pH which is definitely alkaline.
—A large ammoniacal residue.

He believes that these conditions are a result of excess ammonia thanks to massive dressings of nitrogen, especially non-protein nitrogen. Due to the insufficient quantities of immediately fermentable glucides (carbohydrates) present, the micro-organisms cannot multiply sufficiently to cope with this sudden enrichment in nitrogen, a large part of which may be non-protein in character.

Breakdown (or catabolism) is far in excess of synthesis (or anabolism). *The production of ammonia is excessive.* We have just seen the consequences of this for the contents of the rumen. Let us now examine the effects on the animal itself.

Toxic effects of excessive ammonia production in the rumen

Latteur (71) writes:

"Resorption of non-protein nitrogen takes place almost entirely in the form of ammonia. This resorption provokes reactions which are not absolutely linked together:

"1. Fixed alkalosis which can be explained by the mere presence of ammonia which does not occur in the blood in a normal state. From time to time when the ammonia content of the blood reaches 6–10 mg. per thousand, *the alkalosis quickly sets in motion a fall in the content of Ca^{++} and Mg^{++}.*

"2. Poisoning of the bulbar respiratory centre. If the poisoning is only slight the attack on the respiratory centre often stops at an initial irritative phase. . . . Now the fixed alkalosis, of little importance by itself, is joined by a gaseous alkalosis, producing the most spectacular symptoms of tetany.

"With severe poisoning the respiratory centre becomes quickly paralysed. It appears that this is not due solely to the alkalosis but to the fact that the ammonia is having effect by its mere presence. In fact, return to normal pH does not relieve the inhibition of the respiratory centre."

First of all, it is evident that alkalosis, due to ammonia excess, can bring about a fall in the blood's content of magnesium; this explains the hypo-magnesemia ascertained in grass tetany. Secondly, ammonia can act by itself on the respiratory centre, and *it would have been possible to reproduce grass tetany experimentally by means of intravenous injections of ammonia.*

Susceptibility of individual animals to ammonia poisoning

Mention has already been made of the fact that some animals are more susceptible to tetany than others. This can be explained by the variation in the capacities of the liver and kidneys of the animal to get rid of a surfeit of ammonia, resulting either from the absorption of excessive doses of non-protein nitrogen or some other imbalance in the ration.

Latteur (71) writes:

"Ammonia in the blood is highly toxic. For this reason it is converted by the liver (at least in the case of mammals) into urea to be conveyed by the circulation to the kidney in a related, but harmless, chemical form. It is immediately evident that disorders which affect the ureopoietic function of the liver (synthesis of urea from ammonia) or affect the kidney's excretion capacity are as good as *predispositions to grass tetany.* To be compared with this are the great lesions of granulo-fatty degeneration regularly observed in the liver of animals which have succumed to tetany.

"It is therefore possible that certain animals whose hepatic functions are impaired, are hit by tetany for smaller quantities of ammonia than those whose ureopoietic function is intact."

One case of grass tetany in twelve years of rational grazing

I could say that I have never had a single case of grass tetany in twelve years of rational grazing. I have had one, however, but that was a very special case. The victim was a heifer which suffered from tetany three times in the course of her life: the first two attacks took place in the *stall*, and the animal was got on to her feet again thanks to injections of magnesium salt. The third relapse took place at grass, and once again the heifer was saved. However, since she was wasting away, I sent her to the "sausage maker" and asked my veterinary adviser to take a look at the slaughtered animal. *The liver of this animal was nothing but a mass of tumours.*

This appears to confirm Latteur's thesis that disorders in the functions of the liver predispose the animal to tetany.

Precautions against grass tetany

All these considerations confirm many of the points already examined in the course of this work, or which will be examined in dealing with problems affecting the working of rotational grazing:

1. It is essential to change over **progressively** from stall feeding to grass by limiting the length of grazing time during the first few days in order to allow the rumen flora to modify and to adapt itself to the new feeding so that it can comfortably digest the nitrogen in the grass. This is by no means confined to rational grazing, but should be observed whatever the grazing method. It is a custom practised by peasants from time immemorial.

2. An intensive grazing system with too short rest periods is to be avoided, because it means that the cow is feeding *exclusively* on very young grass, the composition of which is not well balanced. It will be seen later that *the systems advocated by the pioneers of rotation systematically put extremely young grass at the disposal of the stock. This is what happens also with rationed grazing systems and when one makes the mistake that I shall label "untoward acceleration"* (pp. 204 and 226).

Many accidents have taken place and still will do in the future: things could not and cannot be otherwise. *The phenomenon of grass tetany was, and still is, more accentuated when the proportion of temporary swards increases overmuch (ley farming) in relation to permanent grassland.*

PART THREE

LAWS OF RATIONAL GRAZING [1]

---

[1] The term "rational grazing" (a direct translation from the French) is used to describe the author's particular system of grazing. The author is, however, fully aware that the expression closely resembles "rationed grazing". He hopes that this will not cause confusion.

REQUIREMENTS OF THE GRASS AND OF THE COW

The particular "points of view" of grass and cow have been examined in the preceding chapters. As a result of these studies, and after twelve years of practice in rational grazing, I have been able to establish four laws which I consider to be universal and which must govern all rational grazing, whatever the soil conditions, climate, altitude, latitude or longitude. The first are concerned with the requirements of the grass: the last two with those of the cow.

First Law

Before a sward, sheared with the animal's teeth, can achieve its maximum productivity, sufficient interval must have elapsed between two successive shearings to allow the grass:

(*a*) to accumulate in its roots the reserves necessary for a vigorous spurt of re-growth;
(*b*) to produce its "blaze of growth" (or high daily yield per acre).

Corollary I of the First Law

The rest period between two successive shearings varies with the season, climatic conditions and other environmental factors.

Comments on the First Law and its Corollary

A pasture grass, by nature, is capable of re-growth after several cuts because it is able to accumulate in its roots (and the lower parts of the stems) reserves which allow it to make new growth. The example of the well-known graminaceous plant, wheat, was cited. The young wheat plant is destroyed by grazing as it emerges from the soil because it has no reserves as yet. Similarly, when the ripe wheat is cut, the stubbles produce no re-growth, for all the root reserves have passed into the grain.

Figs. 1 and 2 (pp. 12 and 14) show that the curve of grass re-growth is sigmoid or S-shaped, with the result that grass only reaches its maximum daily re-growth after a sufficiently long rest period. For example, in May–June, with a rest period of 18 days, the daily re-growth of grass (maximum) is 240 lb./acre [*266* kg./ha.] against 71 lb./acre [*80* kg./ha.] with a rest period of 6 days which is more or less equivalent to the "hidden" rest periods granted in continuous grazing.

131

These curves, moreover, reveal that (under average conditions in North-West Europe) the optimum rest period for August–September is almost double that for May–June.

Two analogies may be called to mind here with advantage. If one cuts lucerne ten times instead of thrice (normal practice in our regions) a poor yield is obtained. *The same drop in yield takes place in continuous grazing when the grass is sheared twenty times instead of the six to eight times which are average for rational grazing.* Table 23 (p. 59) (unfortunately based on equal rest periods throughout the year) clearly illustrates the enormous fall in grass production when grass is cut every eight days, the most common rest period where continuous grazing is practised.

One does not allow the same interval between the first and second cuts of lucerne as between the second and third. Similarly, the intervals between successive shearings of the grass by the animal must be varied.

Second Law

The total occupation period on one paddock should be sufficiently short for a grass sheared on the first day (or at the beginning) of occupation not to be cut again by the teeth of these animals before they leave the paddock.

Comments on the Second Law

This law could have been made a corollary of the First, because if one grass plant is sheared twice in the course of one occupation period of a paddock, then the rest period which has elapsed between the two shearings has not by any means been sufficient. This rule concerning periods of occupation, however, is so important that it was thought preferable to grant it the status of a fundamental law.

So as to better understand the implications, let us assume that in May, when the grass is re-growing to a height of 6 in. [*15* cm.] in 14–18 days, 5 days are required for the grass to reach the 2 in. [*5* cm.] necessary for the cow to be able to grasp and graze it. Let us likewise assume that all the animals remain for 8 days in one paddock. It can then be said that a grass which is grazed on the first day of occupation is capable of being grazed by the cow on the sixth day, which means that it has been subjected to two cuts in the course of the 8-day occupation. *Such a grass plant, cut at a time when it has not yet renewed the reserves in its roots,* will have great difficulty in making new growth. It will take a very long time to attain a height of 6 in. [*15* cm.] again; indeed, the lack of reserves in its roots may even be fatal to this very young plant. Cut at the beginning of its re-growth, not only will the plant not achieve its "blaze of growth" but it will also produce only a very low daily re-growth of green matter.

It has already been pointed out, and will be mentioned again in studying rationed grazing (Table 51, p. 236), that prolongation of the periods of occupation leads to a fall in grass production, a fall the more marked as conditions are more unfavourable, particularly if rainfall is slight.

From the practical point of view, to avoid this double shearing during a single grazing passage on one paddock, the period of occupation should not exceed 4 days, at the very maximum 6 days.

In spring, moreover, when re-growth is particularly vigorous, the grass may have re-grown sufficiently at the end of 4–5 days for the cow to be able to grip it. Six days is therefore the maximum for spring; an occupation period of 4 days is certainly preferable if it can be adhered to without giving rise to too many complications.

The first two universal laws concerning the requirements of the grass can be summarised in one sentence:

Just as there is a time when grass is ready for cutting with the blade of the mower, so also is there a time when it is ready for shearing by the tooth of the animal.

Third Law

The animals with the greatest nutritional requirements must be helped to harvest the greatest quantity of grass of the best possible quality.

Corollary I of the Third Law

Grass with an average height of 6 in. [15 cm.] in the case of permanent pastures (and of at least 9 in. [22 cm.] in temporary grazings) allows the cow to harvest maximum quantities of high quality grass.

Corollary II to the Third Law

The less scraping (or finishing off) imposed upon the cow, the more grass she harvests.

Comments on the Third Law and its Corollaries

Table 31, p. 82, showed that a 10-cwt. [500 kg.] cow put out on to a permanent pasture where the grass was 6 in. [15 cm.] high harvested, per day:

(a) 100 lb. [48 kg.] of grass, if she was forced to graze the sward absolutely bare.

(b) 116 lb. [56 kg.] if she had only to harvest half the grass present.

(c) 133 lb. [64 kg.] if she was obliged to harvest only a third of the grass present.

Possible daily yields of milk for a cow of 10 cwt. [*500* kg.] in these instances were respectively (Table 36, p. 91):

(*a*) 2·4 gal. [*11* litres];
(*b*) 3·3 gal. [*15* litres];
(*c*) 4·0 gal. [*18* litres].

The same would be true of gains in live weight or progress in development.

Fourth Law
If a cow is to give regular milk yields she must not stay any longer than three days on the same paddock. Yields will be at their maximum if the cow stays on one paddock for only one day.

Comments on the Fourth Law
When a cow is shifted to a new paddock she reaches her maximum yield after the first day. Thereafter the yield continues to decrease as the period of stay is prolonged. This is a result of the Third Law, for as the sward becomes more bare the cow harvests reduced quantities of grass of an inferior quality.

Different mechanisms that the cow possesses allow the width of these rhythmic variations in her milk production to be reduced. They are ultimately revealed, however, in a steep drop of the lactation curve, less good growth or less rapid fattening.

It is the First Universal Law that determines the colossal increases in yield obtained from rational grazing
By applying the last two laws concerning the requirements of the cow one will probably be successful in increasing the individual yield of the animal by 20 perhaps 30%. I do not believe that more can be hoped for, but we can wish for research to alter this position in our favour.

But when we obey the First Law (and the Second, which, in fact, derives from it) *we will at least double the yield of grass.* Since it is only in obeying this Law that one can apply high rates of nitrogenous fertilisers without endangering either the grass or the animal, it may be said that in observing the First Law we are multiplying the yields of our grazings by three or even more.

One principle governs the four Universal Laws
So we have the four Universal Laws, two for the grass and two for the cow. All in fact derive from one great principle which, in the future, will direct rational grazing.

Up till now it has been thought that grass grows alone, and the cow eats it alone. From now on, our thought must be that *grass does not grow alone, neither does the cow eat it alone.*

The conclusion to be drawn is:

We must help the grass to grow and guide the cow in harvesting it.

PART FOUR

PRINCIPLES FOR THE CONDUCT
OF RATIONAL GRAZING

Chapter 1

DEFINITION OF THE BASIC ELEMENTS

The different names applied to intensive grazing

IT is quite obvious that many different names are applied to intensive grazing —grazed on a rotational basis—rotational grazing—controlled grazing— strip-grazing—close-folding—rationed grazing—on-and-off grazing—break grazing—Warmbold system—Hohenheim system. This multiplicity of names necessarily leads to a great deal of confusion. For many people, even at the present time, all the various forms of rational grazing consist in dividing up the pasture initially to a greater or lesser degree.

Discussion follows discussion; it is declared that one system produces more than another, figures even being calculated with the utmost precision so as to differentiate between them. We may read, for example, that "rationed grazing produces 25% more than rotational grazing" (pp. 216–219). But if the difference in yield is so exactly defined, there is hardly a superfluity of precise definitions of the two systems and the details of how these are put into practice.

Rational grazing

In recent years I have adopted the term *rational* grazing, and it is to this that the present work is devoted. This term is, to my mind, synonymous with good grazing.

Necessity for defining the basic elements

The same confusion which reigns among the terms designating different rotational systems is at work among the basic elements of rational grazing. Before putting a rational system of grassland management into practice and before judging the effectiveness of such a system, it is essential that one is acquainted with the basic elements which must act as guides to both the scientist and the farmer. If misunderstanding is to be avoided, these elements must be clearly defined; and this is what I have endeavoured to do. Basing my efforts on the work of such German pioneers of rotation as Geith, Klapp, Könekamp and others, I created certain French expressions in translation of the German terms (Voisin, 114); and I shall here give the equivalent British terms.

Animal unit

In order to simplify calculations, the animal unit (*Grossvieh-Einheit*) has been defined: this is a bullock (or dry cow) of 10 cwt. [*500* kg.] live weight. Where a herd is composed solely of cattle the weights of the beasts are added up and the total divided by 10 cwt. [*500* kg.] thus giving the equivalent in animal units.

Where a herd is made up as follows:

10 calves of 4 cwt. [*200* kg.]	=	40 cwt.	[*2,000* kg.]
15 heifers of 6 cwt. [*300* kg.]	=	90 cwt.	[*4,500* kg.]
10 bullocks of 9 cwt. [*450* kg.]	=	90 cwt.	[*4,500* kg.]
12 bullocks of 10 cwt. [*500* kg.]	=	120 cwt.	[*6,000* kg.]
20 cows of 12 cwt. [*600* kg.]	=	240 cwt.	[*12,000* kg.]

$$\overline{\phantom{580 \text{ cwt.}}}$$
$$580 \text{ cwt. } [29,000 \text{ kg.}]$$

it is equivalent to

$$\frac{580}{10} \left[\frac{29,000}{500} \right] = 58 \text{ animal units}$$

If other animals are involved in the grazing, conversion factors such as those listed in Table 40 (p. 139) are employed. These coefficients are obviously very relative. All cows, for example, have a coefficient of 1·00 whether they weigh 8 cwt. or 14 cwt.; an 8-month-old foal has the same coefficient as a two-year old horse, etc., etc. But nevertheless, the inexactitude of our units of measurement, as will be seen later, is such that these coefficients can be considered as acceptable for *comparative* estimations.

Take as an example a herd comprising:

20 calves of less than 1 year
5 yearlings (1–2 years)
6 fattening bullocks
10 cows
4 colts of 1 year
20 sheep

This herd is equivalent to:

$$20 \times 0·12 = 2·4$$
$$5 \times 0·70 = 3·5$$
$$6 \times 1·00 = 6·0$$
$$10 \times 1·00 = 10·0$$
$$4 \times 0·75 = 3·0$$
$$20 \times 0·10 = 2·0$$
$$\overline{26·9 \text{ animal units}}$$

TABLE 40

Conversion factors into animal units for different classes of animal

Animal	Conversion factor	Animal	Conversion factor
Horses		*Pigs*	
Less than 3 years old .	0·75	Piglets under 8 weeks .	0·02
Over 3 years old . .	1·35	Fattening, under 6 months .	0·10
		Fattening, over 6 months .	0·25
		Sows and boars . .	0·30
Cattle		*Sheep*	
Dairy cows . . .	1·00	Ewe with lamb . . .	0·10
In-calf heifers over 2 years	1·00	Ram	0·10
Fattening cattle . .	1·00	Fat sheep . . .	0·10
Draught oxen . .	1·20	Lambs	0·05
Bulls	1·40		
Other cattle, 1–2 years .	0·70		
Other cattle under 1 year .	0·12		

From Schlipf (88, p. 454).

Limitations in the exactness of the animal unit

Simple and valuable as the animal unit is for the practical man, is it sufficiently exact to indicate the productivity of a pasture?

If this unit is to have any value, the following approximate assumptions must be made (in the case of cattle only):

(a) An animal of 5 cwt. [250 kg.] consumes half as much grass as an animal of 10 cwt. [500 kg.].

(b) An animal of 15 cwt. [750 kg.] consumes one and a half times as much grass as an animal of 10 cwt. [500 kg.].

(c) A cow yielding 4½ gal. [20 litres] of milk consumes the same amount of grass as a dry cow.

Mention has already been made of our lack of knowledge concerning the saturation point of the cow's appetite and how this makes it difficult for the practical man to apply feed tables. Knowledge of the quantities of grass harvested by animals is even more limited. It is therefore very difficult to judge whether these three hypotheses fundamental to the animal unit are correct. Indeed, even in Table 40 above some of these rules are not observed.

In conclusion, it may be said that the animal unit is certainly inexact, but for want of anything better it will continue to be used for a long time to come, in commo , unfortunately, with so many other units employed in the feeding of livestock, and in sciences in general, which deal with living things.

"Cow-days" (or animal units grazing days)

The production of a sward can be represented in terms of the grazing days of the animal units which are generally called "cow-days".

Assuming, once again, the 580 cwt. [*29,000* kg.] herd referred to above comprises:

10 calves of 4 cwt. [*200* kg.]   = 40 cwt. [*2,000* kg.]
15 heifers of 6 cwt. [*300* kg.]   = 90 cwt. [*6,500* kg.]
10 bullocks of 9 cwt. [*450* kg.] = 90 cwt. [*4,500* kg.]
12 bullocks of 10 cwt. [*500* kg.] = 120 cwt. [*6,000* kg.]
20 cows of 12 cwt. [*600* kg.]   = 240 cwt. [*12,000* kg.]

$$\overline{\qquad\qquad\qquad}$$
580 cwt. [*29,000* kg.]

let us assume that the grazing policy throughout the season was as follows:

1. From April 1 to 30, the grazing stock consisted of:

15 heifers of 6 cwt. [*300* kg.]   = 90 cwt. [*4,500* kg.]
10 bullocks of 9 cwt. [*450* kg.] = 90 cwt. [*4,500* kg.]
12 bullocks of 10 cwt. [*500* kg.] = 120 cwt. [*6,000* kg.]

$$\overline{\qquad\qquad\qquad}$$
300 cwt. [*15,000* kg.]

This means that 30 animal units were grazing for 30 days, which is equivalent to:

$$30 \times 30 = 900 \text{ cow-days.}$$

2. The whole herd, weighing 580 cwt. [*29,000* kg.] (which is equivalent to 58 livestock units) grazes from May 1 until August 20 (112 days).
The pasture, during that period, therefore provides:

$$58 \times 112 = 6496 \text{ cow-days.}$$

3. From August 20 until October 31 only the following animals are left on the pasture:

20 cows of 12 cwt. [*600* kg.]   = 240 cwt. [*12,000* kg.]
10 calves of 4 cwt. [*200* kg.]   = 40 cwt. [*2,000* kg.]
15 heifers of 6 cwt. [*300* kg.] = 90 cwt. [*4,500* kg.]

$$\overline{\qquad\qquad\qquad}$$
370 cwt. [*18,500* kg.]

which is equivalent to 37 animal units.
(The bullocks have been transferred to graze red-clover leys.)
During this 72-day period the production is:

$$37 \times 72 = 2664 \text{ cow-days.}$$

4. From November 1 until December 20 only the bullocks are grazing, which means

10 bullocks of 9 cwt. [*450* kg.] = 90 cwt. [*4,500* kg.]
12 bullocks of 10 cwt. [*500* kg.] = 120 cwt. [*6,000* kg.]

$$\overline{\qquad\qquad\qquad}$$
210 cwt. [*10,500* kg.]

or 21 animal units for 50 days, which gives:

$$21 \times 50 = 1050 \text{ cow-days.}$$

The pasture in question has therefore supplied:

1.   900
2. 6,496
3. 2,664
4. 1,050

   11,110 cow-days

Assuming that the area of the pasture is 99 acres [*40* ha.], the production obtained has been equivalent to:

$$\frac{11,110}{99} = 112 \text{ cow-days per acre.}$$

$$\left[\frac{11,110}{40} = 278 \text{ cow-days per hectare}\right]$$

Livestock carry

This is the number of animal units (or lb. [kg.] of living weight) being supported on the average by 1 acre [*1* ha.] of the total of the pastures being considered.

If a sward of 25 acres [*10* ha.] (whether or not it is divided up into small paddocks) is being grazed by 40 beasts with a total weight of 200 cwt. [*10,000* kg.], then the livestock carry of an acre [hectare] is:

$$\frac{200}{25} = 8 \text{ cwt.} \qquad \left[\frac{10,000}{10} = 1000 \text{ kg.}\right]$$

or $\frac{8}{10} = 0.8$ animal units per acre $\left[\frac{1000}{500} = 2 \text{ animal units per hectare}\right]$

There were, however, $1\frac{3}{5}$ beasts to the acre [*4* beasts to the hectare]; but since their average weight was 5 cwt. [*250* kg.], they represent only 0·8 animal units per acre [*2* livestock units per hectare].

Stocking density

The term "*Besatzdichte*" was fundamental to the ideas of the German pioneers of the *Hohenheim* System. I translated it in French by "*Charge instantanée*", which is "instantaneous livestock carry". I feel that a simpler and less clumsy expression is "stocking density". It means the number of lb. [kg.] of meat (or animal units) supported by 1 acre [*1* ha.] of *the total area of the paddock being grazed simultaneously*. In other terms, the "stocking density" is the livestock-carry of the total area of the paddocks **being grazed simultaneously.**

Assuming that a herd of 600 cwt. [*30,000* kg.] or 60 livestock units is concentrated in *one single group* grazing a paddock of 2·5 acres [*1* ha.], the stocking density of the grazed paddock is:

$$\frac{600}{2 \cdot 5} = 240 \text{ cwt./acre}$$

$$\left[\frac{30,000}{1} = 30,000 \text{ kg./ha.} \right]$$

or

$$\frac{240}{10} = 24 \text{ animal units per acre}$$

$$\left[\frac{30,000}{500} = 60 \text{ animal units per hectare} \right]$$

Assume now that instead of one group, there are three, but the paddocks are still 2·5 acres [*1* ha.] in extent. In this instance the herd is simultaneously grazing an area of:

$$3 \times 2 \cdot 5 = 7 \cdot 5 \text{ acres } [3 \times 1 = 3 \text{ ha.}]$$

This is the same herd of 600 cwt. [*30,000* kg.] or 60 livestock units. The stocking density being supported by the three paddocks being grazed at one time by the three groups will be:

$$\frac{600}{3 \times 2 \cdot 5} = 80 \text{ cwt./acre} \qquad \left[\frac{30,000}{3 \times 1} = 10,000 \text{ kg./ha.} \right]$$

or

$$\frac{80}{10} = 8 \text{ animal units per acre} \qquad \left[\frac{10,000}{500} = 20 \text{ animal units per ha.} \right]$$

The stocking density has therefore been divided by three. But if, with this herd of 600 cwt. [*30,000* kg.] live weight, which has been divided into three groups, small paddocks of 0·83 acre [*33* acres] are used instead of the 2·5 acres [*1* ha.] paddocks, then the stocking density of the three smaller paddocks being grazed simultaneously will be:

$$\frac{600}{3 \times 0 \cdot 83} = 240 \text{ cwt./acre} \qquad \left[\frac{30,000}{3 \times 0 \cdot 83} = 30,000 \text{ kg./ha.} \right]$$

or 24 animal units per acre [*60* livestock units per hectare], which is exactly the same as in the first instance.

In other words, stocking density is inversely proportional to the number of groups used and the area of the paddocks.

Period of *stay* of a group on a paddock

This is the period (in days or hours) during which *one group* grazes a paddock in each grazing *passage* (that is to say in each rotation).

Period of *occupation* of a paddock

This is the total time (in days or hours) during which a paddock is grazed by *all* the groups in each grazing (that is to say, in each rotation). This time is equal to the total of the periods of stay of the individual groups.

If the period of stay is the same for every group, then the period of occupation is equal to the period of stay multiplied by the number of groups. If, for example, the herd is divided into three groups, each of which spends 2 days on each paddock, then in each passage a paddock will be occupied for $3 \times 2 = 6$ days.

But if the first group remains for 2 days, the second for 2 days and the third for 1 day, the period of occupation of this particular paddock will be: $2 + 2 + 1 = 5$ days.

Where the whole herd is concentrated in a single group, the period of occupation is equal to the period of stay.

Rest period

This is the period between two grazing passages during which the grass is allowed to rest without being grazed. The rest period is equal to the *mean period of stay* of a group multiplied by the number of paddocks resting.

Take a rotation with twenty paddocks, where each group (whether one or several) is moved about every *two* days and assume:

I. *One single group*

A. All the paddocks in the rota:

The rest period is $(20 - 1) \times 2 = 38$ days

B. Twelve paddocks in the rota in this particular passage, the other eight being reserved for mowing:

The rest period is $(12 - 1) \times 2 = 22$ days

II. *Three groups*

A. All the paddocks in the rota:

The rest period is $(20 - 3) \times 2 = 34$ days

B. Twelve paddocks in the rota in this particular passage, the other eight being reserved for mowing:

The rest period is $(12 - 3) \times 2 = 18$ days

Under the same conditions, therefore, an increase in the number of groups effects a decrease in the rest period.

For a fixed number of groups with a fixed number of paddocks in the rota *it is the period of stay which determines the rest period.* Suppose that the entire herd is concentrated in one group and that it arrived on a paddock on the day D and departed on the day $D + PS$ ($PS =$ number of days in the Period of Stay). It will later return on the day $D + X$. The rest period is $(D + X) - (D + PS)$. In other words, the date of *departure* subtracted from the date of *arrival* of the herd gives the rest period for a paddock (in the case of a single group).

Where there are several groups, the date of arrival of the *first* group must be subtracted from the date of departure of the *last* group to give the period of occupation, which in turn gives the period of rest. This is a very small detail, but it has sometimes been the cause of mistakes.

Intensity of grazing

This is the most complicated of the basic elements to understand. In 1950 I wrote (Voisin, 114):

"There is a tendency in studies of rotational grazing always to speak of the stocking density which is a *static* figure and of no value unless the total *number of days* that each paddock is grazed in each rotation is taken into consideration at the same time.

"No complicated calculations are necessary to understand that the quantities of grass eaten by a herd representing a stocking density of 80 cwt./acre [*10,000* kg./ha.] will not be the same if that herd (supposedly concentrated in one group) remains on the same paddock for *three* days or for *six* days.

"I have been brought, therefore, to defining 'intensity of grazing' which I measure in cwt./days/acre [kg./days/ha.]. No doubt these cwt./days [kg./days] will make the reader shudder, but they are simpler than words in the long run.

"The intensity of grazing is obtained *by multiplying the stocking density per acre [per ha.] by the period of occupation of the paddocks.*

"Assume that a herd with a total weight of 800 cwt. [*40,000* kg.] is divided into three groups and is therefore grazing three paddocks simultaneously. The area of each paddock is 2·5 acres [*1* ha.], and the period of stay of each group is 2 days.

"The stocking density is:

$$\frac{800}{3 \times 2 \cdot 5} = 107 \text{ cwt./acre} \quad \left[\frac{40,000}{3 \times 1} = 13,333 \text{ kg./ha.}\right]$$

"The period of occupation of a paddock is: $3 \times 2 = 6$ days
"The intensity of grazing is therefore:

$$107 \times 6 = 642 \text{ cwt./days/acre.}$$
$$[13,333 \times 6 = 80,000 \text{ kg./days/ha.}]$$

"Suppose now that it is decided to double the period of stay of each group on a paddock to 4 days but all the other circumstances remain the same. The stocking density remains as it was, but the grazing to which the paddock are subjected is more intensive. In this case the period of occupation has become $3 \times 4 = 12$ days and the intensity of grazing is: $107 \times 12 = 1284$ cwt./days/acre [*13,333 × 12 = 160,000* kg./days/ha.].

It is evident, therefore, that for a constant stocking density the intensity of grazing varies proportionally with the period of occupation.

"Suppose now that the same herd is concentrated in *one* group staying *two* days on a paddock. In this case, the period of *stay* is identical with the period of *occupation*, both being 2 days.

"The stocking density is now:

$$\frac{800}{1 \times 2 \cdot 5} = 320 \text{ cwt./acre} \quad \left[\frac{40,000}{1 \times 1} = 40,000 \text{ kg./ha.}\right]$$

and the intensity of grazing:

$$320 \times 2 = 640 \text{ cwt./days/acre}$$
$$[40,000 \times 2 = 80,000 \text{ kg./days/ha.}]"$$

In other words, *the intensity of grazing is the same whether the herd is divided into three groups or concentrated in one, on condition that the period of stay remains the same*, Or, to phrase it differently: *For a constant period of stay, the intensity of grazing is independent of the number of groups into which the herd is divided.*

This can be demonstrated mathematically (*vide* Voisin, 114).

Professor Klapp's *Besatzleistung*

Professor Klapp (*vide* his letter in Part Six—Chapter 2, p. 201) has a unit analogous to my "intensity of grazing" which takes into consideration both the stocking density and the factor "time". This unit he has named *Besatzleistung*. In fact, it is the *animal units per acre (per hectare) and per* DAY which are bringing pressure on the sward.

The German term is retained here (Table 41), just as Professor Klapp, when using my unit in his work, always had the courtesy to give it its original French title adding "after Voisin".

One cannot be insensitive to correction from this great scientist, the recipient of the Justus Liebig Prize, which is the highest German scientific distinction. But if intensity of grazing, or *Besatzleistung*, provides valuable units for scientific research, they are too complicated for ordinary usage, especially in work destined for the agricultural advisers' service.

Area required to provide the daily grass ration of one livestock unit

Table 31, p. 82, showed that a cow of 10 cwt. [*500* kg.] live weight (that is to say, one livestock unit), forced *to graze down* a sward 6 in. [*15* cm.] high, harvested 100 lb. [*48* kg.] of grass per day.

Assuming that a paddock is offering 4200 lb./acre [*4800* kg./ha.] harvestable fresh grass at the time when the cow is grazing, then there are 42 daily animal unit rations per acre [or *100* per hectare] available in the sward.

Management in this particular instance can be based on a grazing intensity of:

42 × 10 = 420 cwt./days/acre or 42 livestock units per acre

[*100 × 500 = 50,000 kg./days/ha. or 100 animal units per hectare*]

With this particular herbage yield, therefore, 115 sq. yd. [100 m.²] will supply the daily ration of one livestock unit. It could also be said that 115 sq. yd. [*100 m.²*] are capable of withstanding the pressure of grazing exerted *daily* by *one* animal unit.

For the technical man and those who delight in calculations:

Area (square yard or m.²) required to provide a

$$\text{daily ration for one animal unit} = \frac{48{,}400 \text{ sq. yds. } [5{,}000{,}000 \text{ m.}^2]}{\text{Intensity of grazing}}$$

TABLE 41

Relationship between the period of occupation, stocking density, animal units per day (Besatzleistung), intensity of grazing and the area needed to provide the daily ration for one livestock unit

Period of occu-pation	Stocking density		Besatzleistung: (animal units per day)	Intensity of grazing: (Live weight per day and:)	Area for daily ration of one live-stock unit	
	Live weight per unit of area in occupation	Livestock units per unit of area in occupation				
Days	cwt./acre [kg./ha.]	per acre [per ha.]	per acre [per ha.]	cwt./acre [kg./ha.]	sq. yd.	[m.²]
I. Pasture with 8400 lb. of fresh grass utilisable per acre [9600 kg./ha.]						
4	200 [25,000]	20 [50]	80 [200]	800 [100,000]	57·5	[50]
2	400 [50,000]	40 [100]	80 [200]	800 [100,000]	,,	[50]
1	800 [100,000]	80 [200]	80 [200]	800 [100,000]	,,	[50]
0·5	1600 [200,000]	160 [400]	80 [200]	800 [100,000]	,,	[50]
II. Pasture with 4200 lb. of fresh grass utilisable per acre [4800 kg./ha.]						
4	100 [12,500]	10 [25]	40 [100]	400 [50,000]	115	[100]
2	200 [25,000]	20 [50]	40 [100]	400 [50,000]	,,	[100]
1	400 [50,000]	40 [100]	40 [100]	400 [50,000]	,,	[100]
0·5	800 [100,000]	80 [200]	40 [100]	400 [50,000]	,,	[100]
III. Pasture with 2100 lb. of fresh grass utilisable per acre [2400 kg./ha.]						
4	50 [6,250]	5 [12·5]	20 [50]	200 [25,000]	230	[200]
2	100 [12,500]	10 [25]	20 [50]	200 [25,000]	,,	[200]
1	200 [25,000]	20 [50]	20 [50]	200 [25,000]	,,	[200]
0·5	400 [50,000]	40 [100]	20 [50]	200 [25,000]	,,	[200]
IV. Pasture with 1050 lb. of fresh grass utilisable per acre [1200 kg./ha.]						
4	25 [3,125]	2·5 [6·25]	10 [25]	100 [12,500]	460	[400]
2	50 [6,250]	5 [12·5]	10 [25]	100 [12,500]	,,	[400]
1	100 [12,500]	10 [25]	10 [25]	100 [12,500]	,,	[400]
0·5	200 [25,000]	20 [50]	10 [25]	100 [12,500]	,,	[400]

Table 41 shows, for different yields of grass available per acre [per hectare], the intensities of grazing, *Besatzleistung* and the area required to provide a daily ration. It also gives the stocking density for different periods of occupation. This table is included solely to help the technical man: the only observation that will be made here is that stocking densities can *vary* while intensity of grazing or area necessary to provide a daily ration remain constant.

This confirms the opinion expressed above, namely, that so long as one was content to speak of stocking density without paying heed to the period of occupation, that is to say, the factor "time", rational grazing could hardly make much progress.

Chapter 2

DETERMINATION OF THE NUMBER OF PADDOCKS IS THE BASIS OF THE RATIONAL GRAZING PLAN

The basic problem

IN drawing up a rational grazing plan it is not a case of first determining the *area* of the paddocks. The first essential is to fix the *number* of paddocks: from that the areas can be deduced.

It seems to me quite a wise, average rule to try, using the pasture's resources only, to observe the July–August rest period, but this on condition that the latter is not equal to more than three times the minimum rest period for May–June. Otherwise one must, in drawing up one's plan, confine oneself to realising with the pasture's own resources a rest period equal to three times the minimum rest period. (This, however, is not absolute: such decisions depend very much on a host of local, personal and economic factors.)

Period of stay is the principal factor in determining the rest period

It is the period of stay of each group which chiefly influences the rest period (*vide* p. 251). *The period of rest will be equal to the number of paddocks (in the grazing rota)* AT REST *multiplied by the mean number of days of stay.*

Division into groups leads to a small increase in the total number of paddocks. Given the same number of paddocks, however, an increase in the number of groups reduces the rest period, although it does not basically determine the latter.

Rest periods for the same period of stay and a different number of groups

Take the example of a rational grazing system with 20 paddocks, the period of stay being 2 days. On the assumption that there is only one group of cattle, the period of occupation is equal to the period of stay. The rest period is therefore:

$$\underset{(20-1)}{\overset{\text{(number of paddocks resting)}}{}} \times \underset{2}{\overset{\text{(period of stay)}}{}} = 38 \text{ days.}$$

If there are two groups the period of occupation is $2 \times 2 = 4$ days, and the rest period: $(20 - 2) \times 2 = 36$ days. With three groups the rest period is reduced to $(20 - 3) \times 2 = 34$ days.

Assume now that, as in the preceding example, there are 20 paddocks, the period of stay is 2 days and the cattle are concentrated in *one single* group. The rest period, as already stated, will be 38 days. If the period of stay is to be kept the same but with two groups of cattle, 21 paddocks will be necessary instead of 20, for there will now be $21 - 2 = 19$ paddocks resting (the same as with one group) and the rest period will still be $(21 - 2) \times 2 = 38$ days. Similarly, if division into three groups is desired and the same rest period is to be maintained, the total number of paddocks will have to be increased to 22.

Number of paddocks required for a rest period of 36 days

Table 42 shows the number of paddocks required to give a rest period of 36 days with varying periods of stay and number of groups. Details of the calculations are given for one group only. (Compare pp. 160–162.)

TABLE 42

Number of paddocks necessary to obtain a rest period of 36 days

Period of stay for one group (days)	Total number of paddocks for		
	1 group	2 groups	3 groups
1	$\frac{36}{1} + 1 = 37$	38	39
2	$\frac{36}{2} + 1 = 19$	20	21
3	$\frac{36}{3} + 1 = 13$	14	15
4	$\frac{36}{4} + 1 = 10$	11	12
5	$\frac{36}{5} + 1 = 8$	9	10
6	$\frac{36}{6} + 1 = 7$	8	9
7	$\frac{36}{7} + 1 = 6$	7	8

The laws of rational grazing require that period of occupation and period of stay be relatively short

After examining Table 42 it is understandable that the farmer, in drawing up his rational grazing plan, tends to have the minimum number of paddocks. His reasoning is that if it is advisable to have a rest period of 36 days, he will have one group and move round every 6 days; he will have fixed fences and save money on installation; and if he will employ an electric fence, there will be less work to shift it.

By employing a 6-day period of stay, however, he is disobeying the Fourth Law (p. 134), which demands that the maximum period of stay must be 3 days if these fluctuations in nutrition are to be avoided which reveal themselves in greatly reduced performance on the part of the animal. Moreover, an occupation period of 6 days (Second Universal Law, p. 132) is on the limit of safety, there being a risk that the grass will be sheared twice (in May–June) in the course of the same rotation.

The number of paddocks must not be reduced too far

The simple and cheap solution of a small number of paddocks produces only passable yields. It will be shown, moreover, that with but a few paddocks (six, for example) it is almost impossible to conduct a rotation *with the necessary flexibility*: the result will be "untoward acceleration" (pp. 202–205).

The summer rest period must be observed with the shortest possible periods of stay and occupation, and for obtaining this result it will be necessary to have a sufficiently large number of paddocks. Thus the Laws of Rational Grazing will be respected and the grazing will be conducted with the flexibility essential to its success.

Attempt to classify rational grazings

I have made a *very subjective* attempt to classify systems of rational grazing according to the length of the periods of stay and periods of occupation: the results are contained in Table 43.

TABLE 43

Qualities of a "rational" grazing system according to the length of period of stay and period of occupation adopted

	Perfect		Very good		Good		Passable				Poor								
Period of stay of:																			
Group I	1	1	1	2	2	3	4	2	3	3	5	6	4	4	5	5	6	7	8	→	220
Group II	—	1	1	—	2	—	—	2	3	3	—	—	—	4	5	5	6	—	—	→	—
Group III	—	—	1	—	—	—	—	2	—	3	—	—	—	4	—	5	—	—	—	→	—
Period of occupation of a paddock	1	2	3	2	4	3	4	6	6	9	5	6	4	12	10	15	12	7	8	→	220

This system can become objective only when more precise data are available regarding the annual yields obtained, for example, with:

three groups remaining *one* day on a paddock (period of occupation 3 days), on the one hand, and three groups remaining for *two* days (period of occupation 6 days) on the other;

or

two groups remaining *two* days on a paddock (period of occupation 4 days), on the one hand, and
one group remaining for *four* days (period of occupation 4 days), on the other (cf. p. 197).

Classification of old rotational systems

The Hohenheim system was generally worked with 6–9 paddocks and 3 groups. As the rest period observed was usually 20 days, the period of stay was:

$$\frac{20}{(8-3)} = 4 \text{ days}$$

This in itself was suitable, but it gave a period of occupation of:

$$4 \times 3 = 12 \text{ days}$$

which meant two shearings of the same grass in the course of a rotation.
My classification places such a system in the *poor* category.

Optimum rest period is still the main objective

These short periods of stay and occupation, however, are not an end in themselves: their function is to help to satisfy the requirements of the Universal Laws of Rational Grazing, particularly the first, *which is concerned with rest periods for grass.*
Reference will be made later to the Schuppli system (pp. 196 and 197) in which three groups had a stay period of 1 day, which puts the system in the "perfect" category as far as the occupation period is concerned. It led, how-ever, to some disastrous results, due to the fact that the rest periods observed were only 5 days in spring and 15 days at the end of summer instead of 18 and 45 days, which were probably the optimum rest periods at those seasons under the prevailing climatic conditions.

It is a matter of deciding the number of paddocks, and not the livestock carry

The farmer starting out on rational grazing usually first asks the question, "Up to how many cattle-heads can I carry?" The reply is: "I do not know; I cannot know and neither can anyone else. . . ."
The essential is to draw up a plan of the *times*, for it is these that are fundamental to estimates of areas (*vide* pp. 251, 256) and the search for

possible livestock carry. In the latter instance it is not in fact a matter of a *plan* at all, but of a *progressive search* for the stocking which, under the conditions of management applied, will allow the paddocks to be grazed within the desired periods of occupation, observing therefore the optimum rest periods.

This stocking will depend primarily on the degree of refinement of rational grazing envisaged. If, in order to avoid too much expense or too many complications, it is decided to move only every 6 days and to have only 7 paddocks (with one group) (Table 42, p. 148), then, *ceteris paribus*, the total livestock carry will probably be only half what could have been maintained by moving every day, the grazing area being divided into 37 paddocks.

Moreover, if the farmer, for reasons of economy, applies only small dressings of nitrogen, or none at all, the possible stocking rate will be half what could probably have been carried if 108 lb./acre [*120* kg./ha.] nitrogen had been applied (in well-spaced dressings).

All that can be said to the beginner in rational grazing is that, if he applies the system as it should be applied, he will be able to increase considerably the total number of livestock carried by his pastures in the years that follow. Any more detailed statement would be misleading.

Chapter 3

DIVISION OF THE HERD INTO GROUPS

The German pioneers of rotational grazing advised division into groups

THE German pioneers of rotational grazing recommended division into three groups:

Group 1: High yielding dairy cows.
Group 2: Low yielding dairy cows.
Group 3: Dry cows, calves, foals, sheep, etc.

This method had advantages and disadvantages.

Quality and quantity of grass harvested by the animals in the different groups

The quantities of grass harvested by a cow depend on the height and density of the sward. On a dense sward 6 in. [*15* cm.] high a 10-cwt. [*500*-kg.] cow can harvest, initially, 133 lb. [*64* kg.] of grass: the quantities harvested from a sward that has already been grazed are very much lower (*vide* Table 31, p. 82). In addition, the grass harvested at the beginning of the grazing of a paddock has a much higher nutritive value and is much less fibrous than grass that has already been grazed (*vide* Table 25, p. 63).

A cow producing a high yield of milk obviously requires a larger and more nourishing ration than a dry cow or a young bullock. The German pioneers had no actual figures available, but their idea of what these might be was good, and their principle of division into groups basically sound.

After the leader group (comprising the highest yielders), which needs the largest quantity of rich grass, has grazed a paddock for a certain period it is replaced by the second group consisting of cows yielding less milk. The "scraping" is carried out by the third group (dry cows, calves, etc.), which harvests a smaller quantity of poorer-quality herbage.

Table 31 (p. 82) shows the quantities of grass capable of being harvested by 10-cwt. [*500*-kg.] and 12-cwt. [*600*-kg.] cows where the grass is 6 in. [*15* cm.] high and:

(1) the herd is concentrated in one single group (Case No. 1);
(2) the herd is divided into two or three groups (Cases 2 and 3).

Possible milk production by the various groups

Table 36 (p. 91) shows the theoretically possible milk yields in the various cases, demonstrating that division into groups allows better use to be made of the capabilities of high yielders and good grazers. Moreover, it encourages application of the Third Law, which wants the animals with the greatest nutritive requirements to be put in a position to harvest a greater quantity of grass.

Cows in the first group select their grass

The cow, as has already been seen (Part Two, Chapter 4, p. 114), has two methods of selecting grass:

(*a*) progressive defoliation of the plants;
(*b*) "creaming" a part of the sward.

Attention was also drawn to the fact that high grassland productivity demands a sufficiently long rest period which, however, produces a herbage certain parts of which are already too rich in fibre and whose crude protein content is relatively too low. If high-yielding dairy cows were forced to graze such a sward bare, their production would drop.

An old Norman dictum runs:

"If you want your cows to give you milk
Give them good, tender grass to graze."

But by granting the grass a sufficient rest period to allow it to reach a high productivity level, one is putting at the disposal of one's cows a sward which is relatively less tender, although perhaps just as good. Division into groups provides the solution to the difficulty, the first group selecting and harvesting large quantities of the "good, tender" grass in the sward.

Scottish arguments in the eighteenth century for dividing into groups

James Anderson, a Scottish farmer of the eighteenth century (*vide* pp. 191–192), advocated the division of cattle into two groups. He wrote:

"But, as it would be necessary to allow his fatting beasts always to have a full bite, it would not be proper to keep so many of those as would at any time eat any of these fields quite bare. And as the grass that they would thus leave behind them would, in part, run to seed before they could return to the field, while some other parts of it would be withered, or half rotted, the pastures would be thus rendered less sweet and nourishing than they would otherwise have been. And as there would likewise be a smaller quantity of grass produced on the field in this way than if it had been eaten quite bare to the ground, it would be great want of economy in the farmer not to keep another set of young or lean beasts which should regularly succeed the first, and eat up all that they had left, so as to make it quite bare, and put it in a proper condition for vegetating again with vigour."

This is a clear and reasoned explanation of the necessity, from the point of view of both animal and grass, of division into two groups: from the animals' point of view—the fattening cattle constituting the first group's need "a full bite"; from the grass's point of view—re-growth is less good on a sward not grazed sufficiently bare.

Division of a mixed herd into groups

I should like to call to mind in this connection three great French scholars, Tessier, Thouin and Bosc.

Tessier was a French agronomist born at Angerville (*Seine-et-Oise*) in 1741 and who died in Paris in 1837. He became a member of the *Académie des Sciences* in 1783, then Director of the *Établissement Rural de Rambouillet*, where he was particularly concerned with the rearing and acclimatisation of Merino sheep. Thouin, a botanist, was born and died in Paris (1747–1824); he was chief gardener at the *Jardin des Plantes*, and in 1795 became a member of the *Académie des Sciences*. Bosc, a naturalist, was also born and died in Paris (1759–1828): he was the author of various works on agriculture, including a *Rational Dictionary of Agriculture*. His name has also gone down to posterity for having published the memoirs of his friend, Madame Roland, the beautiful and famous revolutionary, who, before being guillotined, confided her manuscripts to him.

Between 1786 and 1816 these three learned men published, in six volumes, a *Methodical Encyclopædia of Agriculture*. In the beautiful classical language of the age they give an exact description of the advantages of dividing a herd into groups where several animal species are involved (107).

> "The way in which an animal grazes varies from species to species and influences the preservation of the pasture or at least the management which should be applied to it. For example, horned animals seize a handful of grass with their tongues and twist it until it breaks: for them, then, the grass must be high and dense. The horse grips the grass with its teeth and cuts it very short in handfuls. The same is true of the sheep, although, like horned animals, it has no teeth in its upper jaw and can progress only little by little.
>
> "Cows and bullocks can therefore be put on to a pasture first, then horses and then sheep. . . ." (Cf. p. 115.)

Division into groups reduces the risk of fighting among the stock

If animals are to be moved every day or even several times a day their concentration on the sward is high and there is always the risk of fighting. If this is to be avoided an area which has already been grazed must be made accessible to them (p. 227).

Division into groups reduces crowding and consequently the chances of fighting.

Division into groups and the herd instinct

As stated above (p. 74), it is always expedient to form homogeneous groups of cattle, for they can then more easily draw up a common programme of activities. There is much less risk of a cow with "long grazing" qualities not being able to fit in her full grazing time because the other cows start to ruminate.

Arguments against dividing into groups

The pioneers of the Hohenheim system, as was said at the beginning of this chapter, recommended dividing up the herd into three groups. To-day objections to such division are often raised by specialists: these have been summarised as follows by the German specialist Pérignon (83):

"1. Let us assume that a corridor affords the animals on the various paddocks access to one and the same watering point. To avoid mixing the different groups, one can only open the gate from one paddock to the corridor at a time. This means that the animals, not having *constant* access to the watering point, will feel less comfortable and their production will not be so good (pp. 258 and 259).

"2. Another great disadvantage of division into groups is the reduction of the stocking density. The result is a lower yield of grass *due to the extension of the period of occupation of the plots.*

"3. The labour requirement increases with the number of groups. This is particularly true where electric fencing is used: the more groups there are, the more wires and posts are to be shifted."

There is obviously foundation for these three arguments, although the third is valid only in the case of electric fencing. In the case of fixed fencing, division into groups does not greatly increase the amount of manual labour involved. But the first two objections remain. The first will be dealt with later in relation to the division of pastures and access to watering points (Part Eight, Chapter 2, p. 258); but a few words will be said here on Pérignon's second objection. This is a clear case of the grazier being forced to compromise between the demands of the grass and those of the cow.

The requirements of cow and grass are opposed when the herd is divided into groups

The period of occupation is the sum of the periods of stay of each individual group. If these are equal (as is generally the case), then the period of occupation is equal to the period of stay multiplied by the number of groups. The shorter the period of occupation, the higher the yield of grass. Beyond 6 days, in May–June, grass grazed on the first day runs the risk of being sheared again before the animals have completed their occupation of the paddock, and this is prejudicial to the sward. As always, the highest yield is obtained by best satisfying the demands of both cow and grass. Unfortunately division into groups, which satisfies the demands of the animal best, is least satisfactory to the demands of the grass.

Three points are indisputable:

1. Division into groups enables the stock to harvest the quantities and quality of grass most in keeping with their requirements.

2. Division allows reduction of the period of stay on a paddock, and consequently diminishes fluctuations in animal's production due to extended periods of stay.

3. Prolongation of the period of occupation due to division into groups lowers the yield of grass.

If division into groups is to be justified, then the increase in the yields of the animals thanks to their division into groups must be greater than the diminution in yield due to the reduced productivity of the grass as a result of the extended period of occupation.

A Nottingham University student, Dancey (16, p. 12), examining my ideas in grassland management in his Final Year thesis, sums up the position very clearly:

"It must be possible to reach a compromise between the divergent interests of grass and stock. A method may possibly be found which will take the well-being of the sward into consideration while at the same time providing the animal with a herbage of excellent quality."

In the absence of the illuminating results of exact research work carried out with a herd comprised exclusively of cattle, I consider it preferable to work with two groups, although the German pioneers of rotation advocated three.

I have changed over from three to two groups

When, thirteen years ago, I started rotational grazing, I divided my herd into three groups:

Group I: High yielders which were milked at midday.
Group II: Low yielders not milked at midday.
Group III: Dry cows and followers.

As each group advanced (theoretically) every 2 days (period of stay), my period of occupation, in principle, was $2 \times 3 = 6$ days.

From the very beginning I had the impression that due to vigorous growth in May, a grass plant sheared on the first day of the period of occupation by *the first group* was sheared again on the *last day of the period of occupation by the third group*.

If abundant and unexpected growth of grass compelled me to extend the period of occupation by 1 or even 2 days, that is, to 7 or 8 days, so that the "scraping" of the paddocks could be completed, I saw quite plainly that all the young grass "sheared" at the beginning of the rotation was defoliated again before the occupation ended. Nevertheless, I continued working with three groups for quite a long time. Since my rational grazing swards are situated near to my farm, I bring the cows into the shed for each milking. I was afraid that if I united the first and second groups into one I would not be able to separate out the cows that had, from those that had not been milked at midday.

1954, however, was my last year with three groups: thereafter I worked with only two groups:

Group I: All the in-milk cows.
Group II: Dry cows and calves.

Despite this fact *I have never had any difficulty in separating the cows milked at midday from those that were not.* Cows are creatures of habit: they very quickly became accustomed to these differences in management. This is an important point, because I know of graziers who, for the same reasons, hesitate to work with two groups instead of three.

As far as improvement in yield was concerned, it was appreciable. But I carried out other improvements at the same time, and the means at my disposal do not allow me to distinguish between what is due to the reduction in the number of groups and what to other changes.

Chapter 4

COMPENSATION FOR SEASONAL FLUCTUATIONS IN GRASS GROWTH

WHEN the productivity of grass drops in summer it is not a good time to sell beasts in order to ease the pressure on the pastures; this all graziers know full well. One must therefore compensate as far as possible for these seasonal fluctuations in grass in order to maintain all one's stock.

Internal and external methods of compensation

(*a*) *Internal* methods of compensation are those based on the pasture itself, without calling upon *external nutritive resources*. These are:

1. "Disengaging" of paddocks (that is, of a certain area) which are mown when grass growth is most rapid. These paddocks will be brought back into the rota in the period of minimum re-growth.

2. Variation, according to the season, in the quantities of nitrogen applied, that is to say reinforcement of grass growth by nitrogen application.

(*b*) *External* methods are those calling upon *nutritive resources other than those in the pasture*. These are:

Feeding with dry foodstuffs (hay, oats, cake, etc.).
Soiling forage gained from the temporary ley.
Grazing temporary leys.

The method of selling beasts could also be included among these external methods.

Internal and external green areas

There is more affinity between the internal and external methods of green feeding than one might think.

In bringing mown paddocks back into the rota we are compensating for the fluctuation in grass growth *by introducing directly into the grazing cycle green areas that had previously been kept for mowing*. This is the same method as

158

was used by our forefathers when they grazed the aftermath of their mown pastures (generally permanent pastures). Similarly, when we compensate with forage crops (temporary pastures), whether they are grazed or mown and fed green, *we are indirectly introducing green grass into the grazing cycle.*

Soil, climate, gradient, altitude, animal health, economic conditions, relative yields of forage crops and permanent pastures, the requirement of hay for the winter, the crop rotation being employed, etc., these factors determine the relative proportions of the green areas, both permanent and temporary, and the extent, moreover, to which the temporary aspect is stressed.

Chapter 5

COMPENSATION FOR SEASONAL FLUCTUATIONS IN GRASSLAND PRODUCTION BY VARYING THE NUMBER OF PADDOCKS IN THE ROTA

Principle of balancing production by disengaging and re-introducing paddocks

ASSUME that on the average under local conditions an optimum rest period of 42 days is observed in August–September and 18 days in May–June. It is desired to compensate for this difference solely by varying the number of paddocks in the rota *without* interfering with the amounts of nitrogen applied. For the sake of simplicity it will further be assumed that the herd is concentrated in *one* group. Depending on the period of occupation (equal in this case to the period of stay) it is decided to employ, a variable number of paddocks will be disengaged from the rota in May–June.

To get a rest period of 42 days in August–September with one group and an occupation period of 1 day (which is equal in this case to the period of stay), a rotation must be worked out with $42 + 1 = 43$ paddocks. A rest period of 18 days can be achieved in May–June with $18 + 1 = 19$ paddocks in the rota. At this time therefore $43 - 19 = 24$ paddocks must be disengaged.

TABLE 44

Number of paddocks disengaged in May–June for a rest period of 42 days in August–September and 18 days in May–June (with one group)

Period of occupation (days)	Included in May–June		Included in August–September		Disengaged in May–June	
1	$18 + 1 =$	19	$42 + 1 =$	43	$43 - 19 =$	24
2	$9 + 1 =$	10	$21 + 1 =$	22	$22 - 10 =$	12
3	$6 + 1 =$	7	$14 + 1 =$	15	$15 - 7 =$	8
4	$4 + 1 =$	5	$10 + 1 =$	11	$11 - 5 =$	6
5	$3 + 1 =$	4	$7 + 1 =$	8	$8 - 4 =$	4

N.B. In the case of two groups having a period of stay equal to the period of occupation given above, the number of paddocks included must be increased by one. The number of paddocks disengaged does not change.

160

Table 44 shows the number of paddocks disengaged in relation to the period of occupation.

This theoretical and simple example is important: before drawing up a programme of rational grazing and arranging fixed or electric fencing the period of stay of each group must be determined, that is to say it must be decided after how many days the herd is to be moved.

According to prevailing conditions and the site of the pasture (pp. 147 and 148), a provisional estimate is made of probable rest periods in August–September and May–June, and from this the number of paddocks to be anticipated is deduced.

Within what limit is it advantageous to balance up seasonal fluctuations in production solely by varying the number of paddocks being grazed?

Some seasonal fluctuations can, with advantage, be balanced up simply by varying the number of paddocks in use. Other fluctuations are too great to be compensated for by the pasture itself. For example, a well-known fluctuation that one does not try to balance by grazing is the *winter* fall in production: other means provide the remedy, generally stall feeding.

While not accompanied by the same complete and prolonged slowing down of plant growth, there are intermediate seasons, when it would be ill-advised to balance up production from the pasture itself, when one preferably has recourse either to soiling or to complementary areas of permanent, but more generally of temporary pasture.

It is wise (apart from exceptional cases) to use the pasture itself for compensating for differences in amounts of growth not exceeding triple limits. But it must not be forgotten that various factors determine a decision of this nature. With a period of stay of 2 instead of 4 days the number of paddocks multiplies very rapidly when fluctuations in production attain certain proportions.

Moreover, there are many other elements in the management applied which determine the methods to be used to equalise production. Farms solely, or mainly devoted to grassland farming will try to harvest as much winter feed (hay or silage) as possible from their swards. Great variation in the number of paddocks in the rota signify that many paddocks have been mown, and therefore large quantities of hay and silage, which will be needed in winter, have been harvested.

Where there is a large proportion of tillage in a farm, however—and here a rational rotation must make provision for temporary swards of legumes— winter feed can be obtained from the arable area on the one hand and on the other they can be grazed by a section of the herd at times when the vigour of grass growth is reduced, especially in periods of drought and at the end of the season. In this case, there will be less demand for hay from the pastures.

Number of paddocks in and out of the rota in three standard examples

Average rest periods in Normandy and Austria were shown in Tables 7 and 8 (p. 32). Here three *theoretical* systems of rest periods will be assumed which are detailed in Table 45 and probably correspond approximately to average rest periods in three different regions.

Hypothesis A corresponds more or less to the average rest period employed in North-West Europe.

Hypothesis B corresponds to certain continental regions in Europe.

Hypothesis C corresponds to certain regions in southern Europe.

Table 45 shows the number of paddocks in a rota at rest where the period of stay is either 2 days or 4 days. To get the number of paddocks being grazed one need only add to this figure the number of groups.

TABLE 45

Number of paddocks being rested in the case of equilibrium of seasonal fluctuations in production grazing alone

Period	A			B			C		
	Period of rest	Number of paddocks being rested for a period of stay of		Period of rest	Number of paddocks being rested for a period of stay of		Period of rest	Number of paddocks being rested for a period of stay of	
	(days)	2 days	4 days	(days)	2 days	4 days)	(days)	2 days	4 days
April–May . .	24	12	6	24	12	6	24	12	6
May–June . .	16	8	4	16	8	4	16	8	4
June–July . .	24	12	6	52	26	13			
July–Aug. . .	28	14	7				80	40	20
Aug.–Sept. . .	32	16	8	60	30	15	48	24	12
Sept.–Oct. . .	48	24	12	52	26	13			
End of season .	64	32	16						

N.B. 1. It is supposed that these are gross variations to obtain an approximately equivalent production of grass WITHOUT *application of nitrogenous fertiliser.*
2. Compare Tables 7 and 8, p. 32.
3. To get the TOTAL number of paddocks to be included, add the number of groups.

Examination of the above table leaves us, on the average, with the following impression:

Case A: The pasture itself can balance up the May–June and August–September productions.

Case B: It is still possible to equalise production in May-June and August–September from the pasture itself, but it seems essential to call in the aid of nitrogenous fertilisers.

Case C: It is apparently difficult to equalise production by the pasture itself, even if nitrogenous fertilisers are used.

PHOTO 8

Voisin Farm:
single unit rack

An iron rod unites the two sections. Care must be taken that the hole for this rod is off-centre so that the cross bars will not be broken when the rack is closed up (Vide Fig. 8)

Photo Voisin

PHOTO 9

Voisin Farm:
heavy rack in two parts

Photo Voisin

PHOTO 10

Voisin Farm: the mown herbage from each paddock set up on a row of racks

(The sward is thus quickly cleared, allowing for rapid re-growth; moreover,

These are only *impressions*; local conditions will determine the actual method to be used.

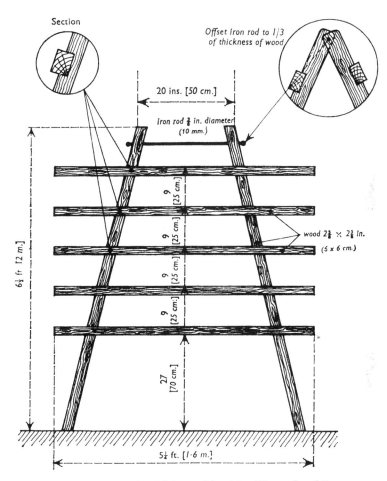

FIG. 8. Sketch of Voisin's quadripod (see Photos 8 and 9).

Difficulties of re-introducing mown paddocks

Paddocks mown at the end of May or beginning of June run the risk sometimes of not being ready for bringing back into the rota at the time when they are required, namely the end of June or during July, when grass productivity is on the decline. One must not hesitate, therefore, to apply all the rapidly acting nitrogenous fertilisers (nitrate of lime) required both before and after cutting so as to speed up the re-introduction of mown paddocks into the rota. But it must be emphasised: *re-introduction of dis-engaged paddocks at the required time is a difficult art of rational grazing*. One must be able to select the paddocks for mowing in time. The soil must be cleared as quickly as possible (crop drying, ensilage) and the nitrate of lime applied without delay. With a little practice all this can be done without any difficulty.

The half-dry meadow

Another art which would be of interest to study is the imitation of what is called "dry meadow grass" in some regions in the centre of France. The system consists in allowing the grass to "bake" *in situ*; in other words, *making hay in situ*. Obviously this hard, rather dry grass will not satisfy the requirements of the cow, but it might be possible to alternate fresh green paddocks with paddocks of "dry meadow grass". Equally the latter might perhaps be reserved for the second or third group.

I think, however, that, in general, only *half-dry meadows* should be aimed at, that is to say, a reserve of paddocks which have baked a little, but are not dry. There might probably even still be sufficient fresh grass to provide an adequate harvest for the *first* group, leaving the harder herbage to the follow-ing group.

In fact, these half-dry meadows are a return to "setting grass aside", something I have often done, and still do, towards the end of June when the grass growth is excellent. This reserve allows me to wait quite safely until, towards the beginning of July, I can bring my mown pastures back into the rota.

If sufficiently long rest periods are observed at the end of the season, a *half-dry meadow*, corresponding to the English foggage, can probably be set aside for the winter. In such a case one has the advantage of a herbage which is only very slightly baked; but, on the one hand, there is the risk of it being spoiled by frost and on the other of being destroyed to a large extent in the field if the sun comes out to melt the snow or heavy rain makes the field muddy.

Chapter 6

COMPENSATION FOR SEASONAL FLUCTUATIONS IN GRASSLAND PRODUCTIVITY BY APPLICATION OF NITROGEN

Influence of nitrogen on grass growth

IT has already been seen (Tables 9, 10 and 11, pp. 35, 36, 37) that considerable increases in the yields of grass or starch equivalent are effected by the judicious apportionment of nitrogenous fertilisers on grass swards. The action of nitrogen, moreover, is very rapid and its effect not as persistent as that of the phospho-potassic fertilisers (Table 10, p. 36), which means that one can aim one's dressings of nitrogen at the target with a certain amount of precision and at the time when their action is required.

Table 10 also showed that at the beginning of the season 1 unit of nitrogen was producing, per day, 5·30–3·40 units fresh grass against 2·03–2·26 units at the end, the mean for the whole season of growth being 3·25 units.

Principle of compensating for fluctuations in grass yield by means of nitrogen

To return to Table 45 (p. 162), it was stated that Example B corresponds approximately to certain continental regions of Europe. In May–June the rest period is 16 days and in July–August 52 days. Assuming that these periods of rest have allowed grass re-growth to the order of 4200 lb./acre [4800 kg./ha.], the mean *daily* growth of grass has therefore been:

$$\frac{4200}{16}\,^1 = 262 \text{ lb./acre} \quad \left[\frac{4800}{16} = 300 \text{ kg./ha.}\right] \text{ in May–June}$$

$$\text{and} \quad \frac{4200}{52} = 80 \text{ lb./acre} \quad \left[\frac{4800}{52} = 92 \text{ kg./ha.}\right] \text{ in July–August}$$

Suppose that the July–August rest period is to be reduced to double that

---

[1] See preliminary note.

in May–June, that is to $16 \times 2 = 32$ days. The daily growth which must be obtained is then:

$$\frac{4200}{32} = 128 \text{ lb./acre} \quad \left[\frac{4800}{32} = 145 \text{ kg./ha.}\right]$$

being an increase over the previous growth obtained of

$$128 - 80 = 48 \text{ lb./acre} \quad [145 - 92 = 53 \text{ kg./ha.}] \quad \text{per day}$$

It will be assumed that, in August, 1 unit nitrogen furnishes the average figure of 3·25 units' fresh grass per day. On this basis:

$$\frac{48}{3·25} = 14·7 \text{ lb./acre} \quad \left[\frac{53}{3·25} = 16·3 \text{ kg./ha.}\right] \quad \text{nitrogen}$$

or 98 lb./acre [*109* kg./ha.] nitrate of lime at 15% must be applied to produce increased grass growth to the order of 48 lb./acre [*53* kg./ha.] per day.

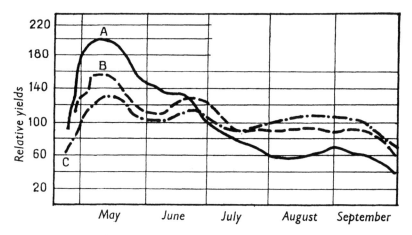

FIG. 9. Seasonal fluctuations in grass productivity for different distributions of the same total quantity of nitrogen.

From Klapp (68), p. 175.

A—Total application of nitrogen before growth starts.
B—Equal applications spread over the whole year.
C—One-sixth of the total when growth starts, one-third in
    summer and one-half before the end of the season.

As has been said, growth of grass is not proportional to the rest period, but follows a sigmoid curve (Fig. 2, p. 14). Moreover, the reciprocal influence of rest period and the amount of daily growth is such that these calculations can be resolved only by means of successive approximations (*vide* p. 31).

In this instance it is obvious that if the mean daily grass growth was 80 lb./acre [*92* kg./ha.] for 52 days it could hardly be more than 45 lb./acre [*50* kg./ha.] for the first 32 days of re-growth. Assuming this figure to be correct, it is essential if the July–August rest period of 32 days is to be attained, that fertiliser nitrogen provide a supplementary daily growth of $128 - 45 = 83$ lb./acre [*145 — 50 = 95 kg./ha.*]. This represents a nitrogen dressing of:

$$\frac{83}{3\cdot25} = 25\cdot4 \text{ lb./acre} \quad \left[\frac{95}{3\cdot25} = 29\cdot3 \text{ kg./ha.}\right]$$

or 169 lb./acre [*195* kg./ha.] nitrate of lime.

This second approximation must be *quite close to reality* and so it can be said:

A dressing of 180 lb./acre [200 kg./ha.] nitrate of lime, under the conditions imposed by System B, will allow the July–August rest period to be reduced from 52 to 32 days.

Whether under local conditions the nitrate of lime is capable of supplying the required 3·25 units of grass per day per unit of nitrogen must likewise be investigated.

Judicious apportionment of nitrogen dressings can produce a more regular grass-productivity curve

When grass growth begins to decline it can be made to increase again by means of nitrogen, which means that, provided the climatic conditions are not excessively adverse (as in winter, for example), reductions in grass growth can be completely or partially compensated for. This is well illustrated by a study made by Klapp.

Fig. 9 (p. 166) shows the curve of *relative* seasonal fluctuations in the productivity of grass where three different methods were employed in apportioning the total dressing of nitrogen. This dressing was extremely high, 216 lb./acre [*240* kg./ha.], but it is large amounts of this nature which most clearly illustrate the effect of different methods of distribution on the curve of grass growth.

The full curve A represents production in the course of a season when all the nitrogen was applied at the commencement of vegetation. This has accentuated the peak of production in May, without, however, reducing the end-of-the-season drop, due to the fact that nitrogen is rapidly absorbed by growing grass and is not persistent in effect. Curves B and C, obtained with more judicious apportionment of the nitrogen, are much less unbalanced.

Apportionment of nitrogen and total yield of grass

Under the direction of Professor Klapp, Schulze continued this study of nitrogen distribution at Rengen. Fig. 10 shows the curves of fresh grass production for the year as a function of the apportionment of the total nitrogen dressing of 216 lb./acre [*240* kg./ha.].

It is quite clear that by concentrating the dressings in the second half of the season and even accentuating them towards the end, a Curve C is obtained, the regularity of which is truly remarkable. This obviously harbours a question which does not spring to mind immediately but which Professor Klapp, that genius of grassland science, has not forgotten to take into consideration.

When nitrogen dressings are distributed in the second half of the season, and especially if they are increased towards the end of that season, *the nitrogen will act less efficaciously*. It has already been seen (Table 10, p. 168) that 1 unit of nitrogen at the beginning of the season was producing 5·30 units of fresh grass, compared with only 2 units of grass at the end of the season. In consequence, when the fluctuations in growth are reduced by concentrating the

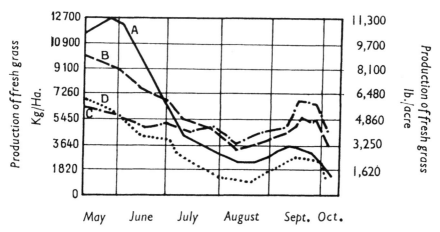

FIG. 10. Influence of the distribution of nitrogen applications on seasonal fluctuations in grass production (Rengen trials).

From Schulze (93).

(It is assumed that the grass contains 22% dry matter.)
A—Total application of nitrogen before growth starts.
B—Equal application spread over the whole year.
C—No application at the beginning of the season; one sixth at the end of spring, one third in summer and *one half* before the end of the season.
D—No application of nitrogen.

nitrogen dressings in the second half of the season and suppressing them at the beginning, the same quantity of nitrogen applied will produce a greatly reduced total quantity of fresh grass.

This is demonstrated by Table 46. In spite of everything, apportionment of the nitrogen dressings over the whole year (Cases B and C) has furnished a total production of grass superior to that obtained by applying the whole dressing at the beginning of the year. But the distribution of the whole amount in four equal dressings (Case B) has furnished approximately 5%

more grass than where the nitrogen dressings were concentrated towards the end of the season.

TABLE 46

Influence of the distribution of nitrogen applications on the total yield of grass

	A	B	C
	216 lb./acre }N [*240* kg./ha.] applied at beginning of season	216 lb./acre }N [*240* kg./ha.] in 4 equal applications	216 lb./acre }N [*240* kg./ha.] applied during 2nd half of season
Total yield of fresh grass, during yield:			
lb./acre	28,300	31,900	30,300
[kg./ha.]	[*31,700*]	[*35,700*]	[*34,000*]
Relative	100·0	113	107

N.B. The figures for yields correspond to the seasonal fluctuation curves shown in Fig. 10.

Calculated from Schulze (93).

The question, therefore, is whether it is better to get the maximum yield of grass or to aim at 5% less grass better distributed over the whole season. The difference is so slight that there can hardly be any doubt as to the answer; but even if the difference had been greater, the answer would no doubt be the same.

To carry matters to extremes, it can be said that if there was a nitrogenous fertiliser available that made grass grow in winter it would be used even if the yield of grass was only one-third per unit of nitrogen of the yield in spring. But even if nitrogenous fertilisers do not make grass grow in winter, they help it to grow at the beginning and at the end of that season, as will be seen later, which compensates for a grass productivity tending to become non-existent.

Extension of the grazing season thanks to nitrogen

Nitrogen also increases the total yield of grass by extending the season at its beginning and at its end, that is to say *by advancing the commencement of grazing and postponing the end of the season.*

It is generally stated that nitrogen allows the grazing season to be extended by 2 weeks at the beginning and at the end. Watson (139, p. 78), however, claims that in Yorkshire nitrogenous fertilisers have enabled the season to be extended by 27 days. This figure seems very high, if not improbable; but what must be remembered is that they are under the control of the method of management (*vide* pp. 181–182).

Management determines the efficiency of nitrogen at the beginning and end of the grazing season

It must be borne in mind that nitrogen can act only if the plant is in a position to utilise it. This principle, which on first sight may appear to be a platitude, is nevertheless of fundamental importance.

If one does not allow sufficient time to elapse between the second-last and last shearings of the grazing year, then the grass will go into the winter with insufficient reserves in its roots. It will suffer more from the cold, but even if the winter has been mild it will start again in the spring with much less vigour. Moreover, with its diminished capacity for re-growth, it will hardly be in a position to utilise efficiently the nitrogen applied to it at the beginning of the year.

In addition, too frequent cutting during the season has a *cumulative, exhausting effect* on the grass which is revealed at the end of the season in greatly reduced re-growth vigour (*vide* Table 5, p. 24). *Such a sward, exhausted throughout the season by over-frequent cutting at too short intervals (as is the case in continuous grazing), will likewise not have the strength to utilise efficiently the nitrogen applied to it at the end of the season.*

A plot without fertilisers is a valuable aid to the farmer

The farmer does not command many exact means of measuring the influence exerted by fertilisers on grass yield. I myself use square plots (at least 16 × 16 ft. [5 × 5 m.]) separated off with four posts and some string; no fertiliser is applied to these areas. One can see the difference between the plot and its surroundings. As a method of measurement it is more than crude, but it is a very valuable aid indeed for the farmer. It was this method that enabled me, little by little, to regulate my dressings of nitrogen and to see at the end of the season up to what date applications of nitrate of lime could efficiently be made in order to prolong the grazing season as long as possible.

Nitrogen applications are only of interest in rational grazing

It must be emphasised once more that nitrogen applications are only of interest where rational grazing is practised. Only rational grazing gives the grass the necessary vigour to utilise nitrogen efficiently at the beginning and at the end of the grazing season.

Theoretical apportionment of nitrogenous fertilisers

Table 47 details some systems of apportioning dressings of nitrate of lime (at 15% nitrogen) which, according to the circumstances and the sacrifices one is willing to make, may be suitable in many cases in North-West Europe. Once again it must be stressed that these figures are merely indicative, serving as guides and examples. They will require in every case to be adapted to local conditions and to the method of management.

TABLE 47

Three possible methods of application of nitrate of lime in North-Western Europe

Rotation No.	Approximate period	Method					
		A		B		C	
		lb./acre	[kg./ha.]	lb./acre	[kg./ha.]	lb./acre	[kg./ha]
1	April–May	90	[100]	90	[100]	135	[150]
2	May–June	0	[0]	0	[0]	90	[100]
3	June–July	0	[0]	0	[0]	0	[0]
4	July–August	90	[100]	90	[100]	90	[100]
5	August–September	90	[100]	135	[150]	135	[150]
6	September–October	0	[0]	90	[100]	180	[200]
7	October–November	90	[100]	135	[150]	180	[200]
	Total nitrate of lime	360	[400]	540	[600]	810	[900]
	Total nitrogen	54	[60]	81	[90]	122	[135]

N.B. The above figures are given only *as an indication*, and should be adapted according to local conditions and system of management.

Chapter 7

EXTERNAL METHODS OF COMPENSATING FOR SEASONAL FLUCTUATIONS

Calling in other green areas

In re-introducing into the rota paddocks that were previously disengaged (which we have considered as an *internal* method of compensation) we are calling upon a green area that is not solely grazed. This is our forefathers' method of grazing the aftermath.

The pasture itself can compensate for quite considerable variations

It is impossible to state precisely within what limits seasonal fluctuations can be compensated for by the mere inclusion or otherwise of paddocks of the pasture itself in the grazing rota. It seems difficult to ask a pasture to compensate, by itself, for fluctuations outside the limits of 1 to 3, that is to say, variations in growth going from simple to triple. This, I believe, is the maximum; it would be better to keep below it.

Temporary pastures in the *Pays de Caux*

In the *Pays de Caux* (Seine-Maritime, Normandy) where the author lives, one-third of the cultivable area is generally in permanent pasture and two-thirds in tillage. A sixth of the area under tillage is devoted to temporary pastures of red clover, which are grazed once (by tethered stock in the first year) and always mown at the beginning of the second year. The aftermath of this second year is either mown or grazed, but most often mown. This means (and this is the case on my own farm) that almost one-third of the cultivated area is permanent pasture and one-ninth temporary pasture (*vide* p. 269).

Tether grazing is therefore practised (this method will be discussed later in detail) either on the second cut of red clover in the second year or/and on the young clover in the first year sown out under oats. After harvesting 20–24 cwt./acre [*2500–3000* kg./ha.] oats, one has the advantage of being able to graze a superb temporary pasture in mid-September. The act of grazing, thanks to compression of the soil by the animal's hooves, reduces attacks by Sclerotinia (*vide* pp. 209, 313).

Soiling

Green areas, whether permanent or temporary, can help to compensate for seasonal fluctuations in two ways:

(*a*) by being grazed (as has just been described);

(*b*) by being mown, the green forage being carried to the stock at grass, in the loafing area or in the stall.

It was stated above (pp. 25, 26) that the Breton peasants of the *Finistère* region definitely prefer soiling to grazing. From their irrigated old permanent pastures the peasants of the Elorn valley produce 48 tons/acre [*120* tons/ha.] grass annually, which is cut and fed directly to the stock.

These methods mainly involve manual labour. It is interesting to consider some of the mechanised American methods which might be adopted in the future in other countries.

Mechanisation of soiling in the U.S.A.

In the hot areas of the U.S.A. stock are kept during the day away from the glare of the sun and are fed cut green forage either in loafing areas or in the stall (*vide* p. 248). The whole operation is mechanised.

The herbage is cut with a mower, loaded by means of a forage elevator on to a trailer behind, transported to the loafing area and thrown by hand into the racks set up there for the purpose (*vide* Photo 12, facing p. 178). But now the operation is mechanised even further. A hay chopper (or field chopper) which cuts, chops and blows at the same time is in common use. The chopped forage is blown into a specially constructed trailer attached behind the hay chopper (Photo 20, facing p. 242). To avoid delay due to breakdowns, the mowing is carried out first and the forage is collected immediately by the hay chopper functioning as a pick-up. The trailers often have a moving bottom, worked either by an electric motor or by power taken from the tractor; they can likewise have a mobile side which pushes the forage mass (Voisin, 117).

The herbage is thus discharged into the trailer by the blower, and removal of this discharged forage may also be mechanised.

It is possible, moreover, that preliminary chopping, by reducing the work of harvesting and masticating for the cow, allows her to eat greater quantities of grass (*vide* my Theory of Satiation (118), and cf. pp. 79–80, 87, 97, 248).

The classical method of compensation in continuous grazing

It is customary in continuous grazing for the grazier to remedy drops in the production of grass by removing from the pasture a certain number of beasts, which he sells, puts on to complementary grazings (red clover, vetches, etc.) or even maintains in the stall. Sometimes he soils the stock on the pasture itself. But, despite the reduction in numbers or the fact that the cows have received a supplementary feed, *they will still continue to graze down the tender young grass plants as soon as they have reached any height at all*. Moreover, the fact that they are looking for food that is scarce in a large pasture increases the damage they cause with their hooves. An old Norman peasant saying tells us that "the cow eats with five jaws: her jaw and her four hooves", meaning that each of the hooves weakens the grass as much as the animal biting it.

It is therefore not merely a case of reducing the stocking in summer or in autumn, or of providing them with complementary soilage; but *these two measures should serve as an aid towards observing the optimum rest periods for the time of year in question*.

The aim in removing stock or practising soiling should be to observe optimum rest periods

The pioneers of rotation certainly recommended reduction of livestock carry or of stocking density when the vigour of grass growth begins to flag; but they made no mention of the rest period. As will be seen, they returned (with reduced stocking, certainly) to a paddock during the period of reduced grass growth after the same, sometimes even after a shorter period of rest.

Let us assume a rest period of 20 days, which is sufficient on the average in spring for the grass to produce a high daily growth. There will be no point in reducing the number of stock or employing soilage. If the animals are put on in summer after a period of 20 days they will show no mercy in mowing down the young grass, which has made only little growth as yet and accumulated only reduced reserves to aid its re-growth. In autumn the same rest period of 20 days will have even more disastrous consequences for the grass.

The last passage in rational grazing in the *Pays de Caux*

I myself get a rest period of about 45 days between the 5th and 6th rotations, that is between the penultimate and last rotations, by using, in my rational grazing, the centuries-old method of the *Pays de Caux*. This consists of removing the beasts in the second group from the permanent pastures towards mid-September and tethering them on young red clover of that year.

PART FIVE
RATIONAL GRAZING IN PRACTICE

Chapter 1

FLEXIBILITY IN MANAGEMENT IS ESSENTIAL

The basic figures are only indicative

IT is often said that a rotation is one in which the group (or groups) is moved every 2 days. Two days, however, is an average time. In practice, it will sometimes be necessary to leave the last group half a day, or a whole day, *longer* to allow the last paddock to be well grazed down. On the other hand, one might be forced to take the last group off before its 2-day period of stay is completed because the sward is already bare.

Variations in the anticipated basic times are warning signals

Assuming that the stocking of beasts is adequate to keep the pasture under control in average normal circumstances, variations in the period of occupation from the basic period anticipated represent valuable *warning signals*.

If the period of occupation has to be extended, growth of grass is greater than what is normally expected. Nitrogen dressings can therefore be reduced, paddocks taken out of the rota, etc.

If, on the other hand, the last group leaves the paddock before the time set, because it has finished the scraping of the sward earlier than was anticipated, this means that growth is slowing down. Nitrogen must be increased, and one must not be over-ready to remove paddocks from the rota. A paddock disengaged for mowing, but not yet mown, may even have to be re-introduced.

The grass commands

Average rest periods in use in Normandy and in Austria have been listed (Tables 7 and 8, p. 32) and three systems of rest periods outlined (Table 45, p. 162) which correspond, with some probability, to three possible regions. It must not be forgotten that these average rest periods are given merely as guides. When May was dry and cold my grass has had to have a rest period of 27 days, whereas if August was humid the rest period could be reduced to 24 days. **It is not a case of rigidly obeying figures: one must follow the grass.**

One has no right to say: so many days after grazing at such and such a

177

time of year, I will start grazing again. One must look for the plots that are ready for grazing, and graze them. **Figures are only guides: in the end it is the eye of the grazier that decides.** The figures, however, must always be kept at the back of one's mind. If, for example, in the month of July, due to unfavourable weather conditions, one is compelled to make several, consecutive reductions in the anticipated period of occupation of the paddocks by a single group, one must not hesitate to stimulate forthwith the failing vigour of grass re-growth by increased dressings of nitrate of lime.

The paddocks are not always grazed in the same order

It is generally believed that the herd is always moved round in the numerical order of the paddocks. Grazing systems very often indicate that the paddocks are to be grazed one after the other. This is not, and should not be, possible in practice; as has just been said, *it is the grass that commands.*

Re-introduction of mown paddocks into the rota is, in itself, sufficient to disrupt the well-ordered sequence of grazing. But even leaving this disturbing element aside, the grass of two paddocks, for various reasons, does not always re-grow with the same vigour. The one, for example, may have been more severely grazed than the other which was grazed subsequently, with the result that the second will grow up again more quickly than the first. Moreover, all paddocks are not equally exposed: one paddock on a gentle slope facing south will grow more quickly in humid weather than another facing north, whereas in hot, dry weather the converse is true. These differences can therefore be the cause of various transpositions in the course of grazing, as is almost always the case in actual fact.

The art of "jumping" paddocks and then going back

The art of carrying out a rational grazing policy consists in being able to "jump" (i.e., leave aside) a paddock

    (a) *that is not sufficiently advanced*, so as to allow it to reach the required height of 6 in. [*15* cm.] and produce its blaze of growth; or
    (b) *that is too advanced*, so that it may mature sufficiently to be mown.

One must equally be able to go back to a paddock that has been "jumped" but has now reached the required height for grazing.

To repeat: **the grass commands; the eye of the grazier follows in its train, ready to receive its orders.**

Alternating mowing and grazing

Flexibility in the conduct of rational grazing necessarily assumes many forms. The beneficial influence on the flora of alternating mowing and grazing will be discussed in detail in my work *Dynamic Ecology of Pastures.*

L'AGRONOME.

DICTIONNAIRE PORTATIF

DU CULTIVATEUR,

CONTENANT

Toutes les Connoiffances néceffaires pour gouverner les Biens de Campagne, & les faire valoir utilement ; pour foutenir fes droits, conferver fa fanté, & rendre gracieufe la vie champêtre.

Comme l'herbe trop mûre durcit & perd beaucoup de fon fuc ; que celle qui n'eft point mûre n'en a point affez, & que les beftiaux vont toujours à la plus tendre, il faut, pour ménager fes Pacages, & afin que toute l'herbe foit pâturée en maturité, & qu'elle repouffe ; il faut, dis - je, féparer les Pâturages en quartiers, grands à proportion du bétail qu'on a à y mettre ; enforte qu'il trouve dans chaque quartier de quoi paître pendant trois ou quatre jours, au bout defquels on le met dans un autre quartier, afin que le premier fructifie, & ainfi fucceffivement. C'eft en relevant les terres, & plantant du bois fur les levées, qu'on fait les féparations, ou avec des hayes & des faules.

✤

A PARIS,

M. DCC. LX.
Avec Approbation, & Privilege du Roi

PHOTO 13

Description of pasture rotation in a French Agricultural Dictionary of 1760

Photo-montage by the Documentation Service of the Centre National de Recherches Agronomiques

It is worth the grazier's while, therefore, to direct his grazing policy in such a way that it is not always the same plots that are taken out of the rota and mown each year. The resulting improvement of the flora will contribute to no little extent in improvement of the yields obtained.

Shifting the first group

I have stressed several times that *judicious apportionment* of nitrogenous fertiliser dressings in rational grazings not only prevented the depression of white clover but even favoured its development. It is possible to have paddocks with 17–26% white clover, that free producer of nitrogen (Table 22, p. 52).

For reasons which have not yet been explained, indigenous white clover very rarely gives rise to bloat. Many farmers will even say that it never causes bloat. Many selected clover strains, whether S.100, Ladino or New Zealand, have unfortunately been shown to cause bloat, the number of disasters which have taken place being beyond counting.

And so one must be very careful in shifting the first groups if they are going on to a new paddock fairly rich in clover. The only recommendation I can make is *never to shift stock in the morning dew or (in hot regions) in the glaring heat of the afternoon sun.* For reasons unknown, these are the two times most favourable to bloat. If the stock are to be shifted several times a day, it is not easy, or, to be more precise, it is *impossible* to observe this rule— one more reason for restricting oneself to one shift each day. This daily shift is itself representative of highly efficient rational grazing, and other reasons will be seen later for not moving the stock more often.

Flexibility in grazing management is possible only with a sufficiently large number of paddocks

Figures are only guides, as has been well emphasised, and one must be able to adapt one's grazing management to unforeseen and unforeseeable variations in climate.

Apart from able handling of nitrogen, the number of paddocks included in and excluded from the rota can usefully be varied. The weather can lead to an excluded paddock being brought back into the rota: this is not a serious matter. *What is very serious, however, is when this cannot be done because the paddock has already been mown and has not yet made sufficient re-growth to be grazed.* In other words, and this happens unfortunately in July, one cannot re-introduce into the rota the mown paddock which is urgently required to extend the periods of rest and to allow the grass to furnish a sufficiently developed daily growth in spite of the slowing down of re-growth due to climatic conditions abnormal for the season.

If there are *twenty* paddocks and re-introduction of one is delayed, then only a *twentieth* of the total area is involved; this is annoying, but it is not very serious. But if there are only *six* paddocks, then a *sixth* of the total area is lost. This is extremely serious: one finds oneself running into "untoward acceleration" and in the end, there is no grass available at all.

As the Swiss worker, G. Heim (38) reminds us:

> "The more paddocks the farmer has, the less he is dependent on atmospheric conditions and the rate of re-growth. On the one hand he has reserves available if he is overtaken by a period of drought; on the other he is able in spring when vegetative power is at its peak, not to graze stock in some of the paddocks but to reserve these for cutting, hay-making or ensilage. Later, as growth diminishes, he can re-introduce these paddocks into his grazing rota. . . ."

Only a large number of paddocks allows one to observe the required rest periods while keeping the periods of stay and occupation sufficiently short. Moreover, this large number of paddocks is essential to flexibility in the conduct of rational grazing.

Chapter 2

PUTTING OUT TO GRASS IN RATIONAL GRAZING

Importance of a good start

IF the putting out to grass is badly done it can throw the whole rational grazing out of gear for the remainder of the season. And so the stock must be introduced on to the pasture under the best possible conditions.

Comparative phenology

Comparative phenology, a science which unfortunately is almost unknown, consists in determining the time for certain farming operations according to the development of wild, natural plants. It is of particular value for fixing dates of sowing which can vary from one year to another by 4, if not by 6 weeks. For myself, I always sow my oats when the wild primroses are in flower, and my beet when the buds of the chestnut-tree are beginning to burst.

It has been said that it is quite a good rule to begin grazing the first of the paddocks in the rotation, having received their 20–30 units of nitrogen, *when the wild cherry-trees have barely started to bud.* If nitrogen is not applied, the start of the rotation should be deferred until the wild cherries are beginning to flower.

What the general validity of such rules is, I know not; they have always seemed good to me in practice.

My advice to those practising rational grazing is to make phenological observations of this nature, whether with wild cherry-trees or some other wild plants, when they start grazing. After a few years the exact phenological rule for their particular case will have become evident.

It would be interesting to see this science of comparative phenology studied more closely by regional research stations; its development could, I believe, put another valuable aid in the hands of farmers.

Differential acceleration by nitrogen of the first spurt of grass growth

It has been seen that nitrogen helps to extend the grazing season both at its beginning and its end (p. 169).

181

Putting out to grass in the case of rational grazing, that is, where pastures are divided, presents a special problem. If one waits until the grass in the first paddocks has grown sufficiently, there is the risk that grass in the last paddocks, grazed in the first rotation, will be too advanced and too fibrous. If, on the other hand, the first paddocks are grazed too soon, the re-growth of the grass may be so retarded that it will not have developed sufficiently by the time the stock return in the next and second rotation.

Nitrogen can help in this connection, in that it can be used to accelerate the initial growth of the paddocks that are to be grazed first. For example, take a rotation with 10 paddocks grazed in numerical order in the first rotation, paddocks 9 and 10 being reserved for mowing.

TABLE 48

Nitrogen application at the beginning of the year to accelerate grass growth to different degrees

Order in which paddocks are first grazed	Nitrogen				Nitrate (15% N) of lime			
	1st plan		2nd plan		1st plan		2nd plan	
	lb./acre	[kg./ha.]	lb./acre	[kg./ha.]	lb./acre	[kg./ha.]	lb./acre	[kg./ha.]
1	20	[22]	33	[37]	134	[150]	223	[250]
2	20	[22]	33	[37]	134	[150]	223	[250]
3	20	[22]	27	[30]	134	[150]	178	[200]
4	13	[15]	27	[30]	89	[100]	178	[200]
5	13	[15]	20	[22]	89	[100]	134	[150]
6	13	[15]	20	[22]	89	[100]	134	[150]
7	0	[0]	13	[15]	0	[0]	89	[100]
8	0	[0]	13	[15]	0	[0]	89	[100]
9} 10}	Kept for mowing. (Nitrogen application according to plan for rotation.)							

Table 48 shows two possible hypotheses:

1. The two last paddocks to be grazed receive no nitrogen at all; in other words, their initial growth is not accelerated. Paddocks 1–3 receive 134 lb./acre [150 kg./ha.] nitrate of lime and paddocks 4–6, 89 lb./acre [100 kg./ha.].

2. The re-growth of all paddocks is accelerated, including paddocks 7 and 8, which are grazed last. The dressings are graded as follows—

223 lb./acre [250 kg./ha.] nitrate of lime for paddocks Nos. 1 and 2, which are grazed first.

178 lb./acre [200 kg./ha.] nitrate of lime for Nos. 3 and 4.

134 lb./acre [150 kg./ha.] nitrate of lime for Nos. 5 and 6.

89 lb./acre [100 kg./ha.] nitrate of lime for Nos. 7 and 8.

Naturally, similar quantities of nitrogen could have been applied using another kind of nitrogenous fertiliser than nitrate of lime, for example, cyanamide, which appears to have the advantage of destroying certain weeds, such as Ranunculus (buttercup) and in addition is an alkaline fertiliser. Other nitrogenous fertilisers are, of course, not excluded, but I personally have always been inclined to prefer alkaline nitrogenous fertilisers for grassland.

Influence of the date of putting out to pasture on the evolution of the flora

The influence of the date of putting stock out to graze on the flora of the pasture is profound. Many examples will be included in my *Dynamic Ecology of Pastures*, which will follow this present work. For the present, reference will be made only to a study carried out by Martin Jones (80) of flora changes brought about by date and method of putting stock out to grass.

This British worker used a newly sown young ley dominant in rye-grass and cocksfoot with a little white clover. He wanted to see how different methods of putting stock out to graze affected the balance between the two grass species. Grazing was carried out by sheep. The rye-grass used began its active growth earlier in the season than the cocksfoot. Both these species are very sensitive to defoliation during their active growth periods, which do not take place simultaneously in the spring. *The result was that, by grazing*

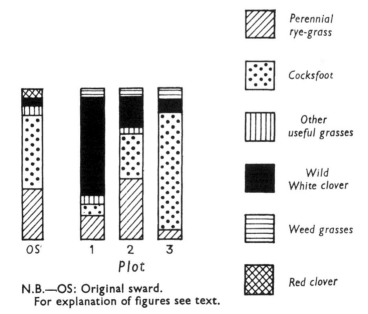

Perennial rye-grass

Cocksfoot

Other useful grasses

Wild White clover

Weed grasses

Red clover

Plot

N.B.—OS: Original sward.
For explanation of figures see text.

Fig. 11. Influence of the date and the method of grazing on the evolution of the flora of a young grassy sward after three years treatment.

From Martin Jones (60).

early, one weakened the rye-grass, a situation of which the cocksfoot took advantage for its own development. If, on the other hand, a sufficient period without grazing was allowed at the beginning of the season, the rye-grass had time to fortify itself. But *when grazing was late in starting the cocksfoot was attacked in its period of active growth and greatly weakened in the process.*

Paddock 1 was heavily and constantly grazed from the beginning of March. Paddock 2 was allowed to rest for the whole of March and the first half of April, that is, during the period when rye-grass is growing rapidly. It was grazed towards mid-April, when cocksfoot was observed to be developing leaves.

Paddock 3 was grazed in March like Paddock 1, but was then left resting until mid-May, that is until the cocksfoot had made strong growth.

All paddocks received high dressings of phosphate.

Fig. 11 shows how the original flora of the young sward had changed after two consecutive years of these three different methods of putting stock out to grass. On Paddock 1 dense and uninterrupted grazing from the beginning of March has weakened both rye-grass and cocksfoot but in the process has allowed the white clover to develop strongly, strengthened by phosphate. Retardation of commencement of grazing on Paddock 2 until mid-April has protected the rye-grass during its period of active growth and allowed it to gain ground. Cocksfoot, on the other hand, has been sheared by the animal at a time when it was very sensitive, with the result that it has been weakened and has retrogressed. Paddock 3 was grazed, like Paddock 1, early in March, reducing the vigour of the rye-grass. But unlike Paddock 2 it was not grazed in mid-April, the grass being left to rest until mid-May after the early grazing. Cocksfoot was therefore left in peace during its period of active growth and was well able to withstand the mid-May cut. Thereafter its development was the more rapid as it became in association with rye-grass weakened by grazing in the beginning of March and white clover suffering the effects of shading by a well-developed stand of cocksfoot.

The three different methods of starting grazing have therefore favoured three different flora:

1. Predominance of white clover.
2. Cocksfoot–rye-grass equilibrium.
3. Predominance of cocksfoot.

These flora evolutions obviously depend on earliness of the species, climate, etc. But Martin Jones' experiment has considerable merit in clearly illustrating the great influence exerted by time of putting out of stock on the flora of almost pure mixtures (not forgetting its effect on herbage yields).

The rota should start on different paddocks each year

In order to avoid these disparities in the flora it is essential that rational grazing should begin each year with a different paddock. If grazing is begun every year in the same order of paddocks, considerable differences between the flora of the various paddocks will become evident after two years at the most, and often even after one year.

Time of putting out to graze in one year is related to the end of grazing in the preceding year

Neither must it be forgotten that the putting out of stock to graze in one year cannot be separated from the end of grazing in the preceding year. Rational grazing definitely offers a notable aid towards complete clearing of the sward before winter sets in, thus avoiding tufted, uneaten areas that hinder early grass growth in the following spring. But it may happen, where rational grazing is practised, that, due to an early winter or some other circumstance, the last "revision" passage cannot be carried out over all paddocks. Early turning out of the stock to grass will help re-growth of the uneaten patches, but, on the other hand, premature shearing of the "tip" of the grass plant retards its subsequent re-growth. Moreover, if scraping is carried too far, the fatigue of the grass makes itself felt in the two ensuing rotations at least. These exhausting effects on the plant of over-short rest periods or exaggerated scraping are both cumulative and persistent in nature.

Animals should be put out to grass gradually

The requirements of the *grass* as regards the beginning of grazing have been examined: what are the requirements of the *cow*?

A general rule in the feeding of livestock is to avoid sudden changes in the quantity or quality of the ration. This rule should be obeyed in particular when animals are changing over from stall feeding to grazing, whether rational or continuous. Animals must be progressively re-accustomed to grass. A Norman expression says that "the cows must cut their teeth on the grass", to which one might add that the micro-organisms of the stomach must become accustomed to grass as a food.

According to the classical method, which is equally applicable to rational grazing, cows are put out for one or two hours on the first day. This number of hours is progressively increased in the days that follow, the amount of stall feeding being proportionately reduced (*vide* Ohms, 81).

Failure to observe these rules of accustoming cows to grass may lead to trouble.

PART SIX

COMMON ERRORS IN SUPPOSEDLY RATIONAL SYSTEMS OF GRAZING

Chapter 1

ROTATION WAS RECOMMENDED BY THE ENCYCLOPÆDISTS OF THE ENLIGHTENMENT

Rational grazing has always been known

RATIONAL grazing has always been known, for shepherds have always possessed the most marvellous of all electric fences, the *living* electric fence known as the *dog*. With the help of their dogs shepherds have traditionally practised what, at a later date, we would call strip grazing, or rationed grazing. These methods were not described on parchment but transmitted orally from generation to generation. As time went on, however, men committed their knowledge to books, and it was the ambition of the Eyclopædists to draw up a complete picture of all this human knowledge, not forgetting agriculture, which feeds mankind. In 1760, therefore, we find an accurate written description of rational grazing and its principles.

To make country life pleasant

The oldest description of the rotation of pastures known to me is found in an anonymous French dictionary dated 1760, which I happen to have in my own private library. The title of the book (Photo 13, facing p. 179) is *The Agronomist, pocket dictionary of the farmer*, which comprises "all the knowledge necessary to deal with and make good use of the resources of the country, maintain its rights, preserve its well-being and *make country life pleasant*". The chapter on "Pastures" gives a description as accurate as it is delightful, of the rotation of pastures:

> "Grass that is too mature becomes hard and loses much of feeding substances. Grass that is not mature does not possess enough of these substances. As beasts always go to the most tender herbage, it is essential when managing grassland, so that all the grass will be grazed at maturity and re-grow, that the pastures be divided up into sections, the size of which is in proportion to the number of beasts they are to carry; the aim being that each section contains sufficient keep for three or four days, after which the stock are put on to another section so that the first can bear fruit. Division is achieved by banking up the soil and planting trees on top, or by hedges and willow trees."

This is the description in 1760, by an anonymous author, of the founda-

189

tions of rational grazing which lead to the green pastures that contribute, in a small measure, to "making country life pleasant".

The *Maison Rustique* of 1768

Eight years later, in 1768, the *Maison Rustique* provided an analogous description of pasture rotation. The words are more or less identical, giving rise to the assumption that it is the same anonymous author in both cases (Photo 14, facing p. 194). In the latter instance, however, the author has added a few observations on the advantages of rotational grazing as he has described it. He writes, for example:

> "All the grass is eaten at once: there is *no trampling, no waste*. The cattle have more grass and better grass, because they move round. The grass grows again more rapidly and more vigorously and one can let it mature as much, or as little, as one wishes. . . ."

There is no better explanation of the Norman proverb that "the cow eats with five jaws: its own and its four hooves".

Abbot Rozier's *Course in Agriculture*

Jean-Francois Rozier (1734–93) was born and died in Lyons, where he entered the seminary and took holy orders. First he was a teacher in the Royal Academy of Lyons and then became prior of Nanteuil-le-Haudouin. An eminent agronomist and botanist, he published in 1785 the first volume of his *Complete Course in Agriculture*, subsequent volumes appearing in the years that followed. This first edition comprised nine volumes in all and was later re-edited in the days of the Empire.

In the seventh volume (1786) (85), in an article on grazing, we find this description of rotational grazing.

> "The intelligent land-owner divides up his acreage into several parts, enclosed by hedges, living or dead, over which the animals pass in succession. The result of these divisions is that while the grass of one section is being grazed, that of the other sections is re-growing, so that the animal is always sure of fresh food and plentiful grass.
>
> "If the area is not divided, the animal eats in one day and destroys with its trampling more grass than it would have consumed in a week. If it is found that it takes too long to produce hedges, they can be replaced by ditches, the soil from which is thrown up on either side and sown with selected seed suitable for meadows. . . .
>
> "Division of pastures is the greatest essential of all when foals and horses are being reared. Without this precaution they attach themselves to the most tender grass, and the more there is of this, the more they scorn the remainder which in the end becomes too tough.
>
> "As soon as the animals have consumed all the grass in one section they are transferred to another. If there are facilities for irrigation this should be done immediately after the stock have left and as often as is necessary. By following this method one is assured of always having excellent grazing available."

Here in this description the principles and advantages of rotational grazing are well illustrated. *Why, then, did this system that was advised by the eighteenth century French Encyclopædists not develop?* The answer is, I believe, to be found in the writings of a great Scottish agriculturalist of the eighteenth century, James Anderson.

A great Scottish agriculturalist

James Anderson, born in Edinburgh in 1739 and died at West Ham (Essex) in 1808, is known as the inventor of a plough generally referred to as the Scottish plough. But he appears to be particularly remembered as the theorist of ground rent, his ideas having been adopted and developed by the great English economist, Ricardo. Anderson's work on economic questions was translated into German in 1893 by Lijo Brentano.

It fell to Professor Johnstone-Wallace (51) to draw attention in 1944 (51) to the fact that James Anderson, in 1791, had described the rotation of grassland. The text that Johnstone-Wallace cites is the *4th edition* of Anderson's work published in 1797. Since my own first quotation in the eighteenth century is from a French book dated 1760, I tried to go farther back into Anderson's work. In the Library of the School of Agriculture, Cambridge, I found the second edition of his book, but despite the efforts of the Cambridge librarian, it was impossible to locate the first. Photo 15 (facing p. 195) is a reproduction of the frontispiece of the second edition of 1777.

Rotation of pastures as seen by James Anderson

Having described accurately and in detail the disadvantages of continuous grazing, Anderson also describes what he looks upon as rational grazing (Photo 16, facing p. 210).

". . . As every kind of animal delights most to feed upon fresh plants that have newly sprung up from a bare surface, in which there is no decayed or rotted stalks of any kind; there can be little doubt but that, if cattle that are intended to be fatted were always supplied with a constant succession of this kind of food, they would be brought forward in flesh as quickly as the nature of that food could in any case do it.

"To obtain this constant supply of fresh grass, let us suppose that a farmer who has any extent of pasture ground should have it divided into fifteen or twenty divisions, nearly of equal value; and that, instead of allowing his beasts to roam indiscriminately through the whole area at once, he collects the whole number of beasts that he intends to feed into one flock, and turns them all at once into one of these divisions; which, being quite fresh, and of a sufficient length for a full bite, would please their palate so much as to induce them to eat it greedily, and fill their bellies before they thought of roaming about, and thus destroying it with their feet. **And if the number of beasts were so great as to consume the best part of the grass of one of these inclosures in one day, they might be allowed to remain there no longer;—giving them a fresh park every morning, so as that the same delicious repast** might be again repeated. And if there were just so many parks as there required days to make the grass of these fields advance to a proper length after being eaten bare down, the first field would be ready to receive them by the time they had gone over all the other; so that they might thus be carried round in a constant rotation. . . ."

Here, then, is a description of rotation of pastures and even the word itself *carried round in a constant rotation.* The reader is well reminded that the grass must be allowed to re-grow to a sufficient height before being grazed again. It is plainly stated that the number of paddocks must be high, namely fifteen to twenty, which, with a herd remaining for one day on each paddock, is equivalent to a rest period of 15–20 days.

Why rational grazing of pastures, which was known at the beginning of the eighteenth century, was not adopted

Both in France and in Britain, therefore, clear and detailed descriptions of pasture rotation were supplied in the eighteenth century, but when one examines the agricultural literature of the nineteenth century, one finds hardly any mention of the subject at all. One must wonder, therefore, why, since the principles of rational grazing were known, the method did not develop and become generally practised. Anderson provides an answer to the question when he says—

"The farmer should have his pasture divided into fifteen or twenty divisions, nearly of equal value. . . . He collects the whole number of beasts that he intends to feed into one flock and turns them all at once into one of these divisions. . . . And if the number of beasts were so great as to consume the best part of the grass of one of these inclosures in one day, they might be allowed to remain there no longer but be given a fresh park every morning. . . .

"*If there were just so many parks as there required days to make the grass of these fields advance to a proper length after being eat bare down,* the first field would be ready to receive them by the time they had gone over all the others; so that they might thus be carried round in a constant rotation. . . ."

This is a very precise description of the general principle of rotation, but at the same time it reveals the vice that prevented the development of the system. In fact, if there are 15–20 paddocks and the single group remains one day on each this means a rest period of 14–19 days, which will suffice in May–June but from July onwards will lead to "untoward acceleration" with all its serious consequences as outlined in Part Six, Chapter 3 (pp. 202–205).

Anderson, moreover, in common with all the authors who have succeeded him up to the present day, *made absolutely no mention of* VARYING *the rest period.* It is now easier to understand why rotation of pastures has developed so little since the eighteenth century, although the majority of its principles were already known at that time: *the fundamental principle of varying the rest periods was completely overlooked and neglected.* The same, as will be seen, was true where the German pioneers of rotation (Umtriebsweide) were concerned.

Chapter 2

PIONEERS OF ROTATION FAILED TO RECOGNISE THE IMPORTANCE OF THE "TIME" FACTOR

It was thought for a long time, and indeed is too often still thought to-day, that rational grazing consists in *dividing* the pasture into a greater or smaller number of paddocks (whether the fence is fixed or movable) and then shifting the herd from one paddock to the next. No thought was given to the "return" and particularly to the *interval which must elapse before this return*, or to the absolute necessity of varying this interval according to the season. The promoters of the rotational method, like their precursors in the eighteenth century, did not appear to attach much importance to varying the rest periods of the grass, on the one hand, or to the need for these rest periods being sufficiently long, on the other. They also overlooked the necessity for the periods of occupation being sufficiently short. In some books and papers on rational management of pastures, indeed, *the "time" factor* is partially or completely neglected.

Falke, the inspirer of "Umtriebsweide"

In 1907 Falke, a professor at the University of Leipzig, published a book entitled *Permanent Pastures* (20), in which he laid the foundations for intensive pasture management. His lectures and his book were to have a great influence on Warmbold, the workers at the Hohenheim Institute, Geith, etc., that is to say, on the instigators of what is known in Germany as "Umtriebsweide" and elsewhere as "rotation" of pastures or the Hohenheim System.

On p. 22 of his book, Falke writes:

> "Useless shifting of the stock must be avoided: this can be achieved by dividing the whole grazing area into a certain number of small paddocks. . . . In determining the number of paddocks, one bases one's calculations on the fact that the grass in a paddock should be eaten within 10–20 days by the number of livestock it is carrying. Moreover, the number of paddocks required derives from *the principle that grass should only be grazed when it has made sufficient re-growth*—in general this is achieved only after 3–7 weeks. Each paddock must remain free from stock for the period required if re-growth is not to be impeded and so that grass will not be grazed at a time when it has not re-grown sufficiently. . . ."

The principle which I summarised in the following words in the course of studying the first two Universal Laws of rational grazing, has therefore been very clearly established by Falke: "Just as there is a time when grass is ready for cutting with the blade of the mower, so there is a time when it is ready for shearing by the teeth of the animal."

Allusion is also made to a rest period of 3–7 weeks, without express mention being made of its *systematic variation*. Later in the text, however, it is stated that the number of paddocks is varied, being increased when the growth of the grass is flagging by means of reserve paddocks (Reservekoppeln). In this way the total grazing area is increased by one third to one quarter, which, in practice, is insufficient without the use of nitrogen (to which there was no reference) to compensate for seasonal fluctuations in the growth of grass.

These premises were sound enough. Unfortunately, a serious mistake becomes evident: a period of occupation (of 10–20 days) of each plot will very often lead to double shearing of the same grass during a single grazing passage (*vide* Second Universal Law). In other words, to use Falke's expression: during this 10–20-day period of occupation "the grass is grazed (a second time) at a time when it has not re-grown sufficiently". This fact seems to have completely escaped this outstanding precursor of rotation, although he had perceived some of the fundamental principles that his pupils and successors were not going to understand as clearly as he had.

The first research workers at the Hohenheim Institute overlooked the importance of rest periods and perpetrated the crime of "untoward acceleration"

Münzinger and Babo (80) in 1931 outlined the basis of the Hohenheim system of grazing, reporting the results and conclusions of experiments carried out around 1925 at the Hohenheim Institute near Stuttgart. They had three groups and employed the periods of stay and rest listed in Table 49.

From the end of June onwards the rest periods were not long enough, as is evident from the very small quantities of grass present at the beginning of each passage as the season advanced. The Hohenheim workers relate that they remedied this situation by employing the two classical emergency aids: green feeding and reducing the size of the herd (a part of which was put to graze other green areas).

It has been stated again and again that reduction in the size of the herd does not remove the obligation of observing the necessary rest periods: on the contrary, it should help to achieve optimum rest periods. It was also pointed out that the pioneers of the Hohenheim system spoke about the stocking density (that is, the stocking per unit area of the paddocks in the course of being grazed) but made no mention at all of how long this stocking density of livestock remained on the paddocks.

LA NOUVELLE
MAISON RUSTIQUE,

OU

ÉCONOMIE GÉNÉRALE

DE TOUS LES BIENS

DE CAMPAGNE;

La maniere de les entretenir & de les multiplier ;

L'expérience nous apprend que l'herbe trop mûre durcit & perd beaucoup de son suc, & que l'herbe tendre & non mûre n'en a pas assez, & ne fait que passer, ensorte que les bestiaux en mangent deux ou trois fois plus qu'ils ne feroient, si elle étoit en maturité, & celle qui est dure, n'est pas assez succulente. Outre cela, les bestiaux aiment à changer de pâture : dans les grands pâturages, la moitié de l'herbe se perd, parce qu'ils vont toujours à la plus tendre, & la plus dure se dessèche, ou ils la foulent aux pieds ; ils pâturent même si près de terre, les endroits délicats, & y tiennent l'herbe si sujette, que de la tendresse dont elle est, elle ne profite presque point. C'est pourquoi, pour bien ménager ces pacages, & afin que toute l'herbe en soit pâturée, qu'elle le soit en maturité, & qu'elle repousse & profite, il faut séparer ces pâturages par quartiers, grands à proportion du bétail qu'on a à y mettre, ensorte qu'il trouve dans chaque quartier, de quoi paitre pendant trois ou quatre jours, au bout desquels on les met dans un autre quartier, afin que le premier repose & fructifie, & ainsi successivement dans tous les quartiers de l'herbage : par-là, tout l'herbage se mange à-la-fois, il n'y a rien de foulé, rien de perdu ; le bétail a plus d'herbe, & il l'a meilleure, parce qu'il change de lieu ; elle repousse plus vite & plus forte, & on la laisse mûrir tant & si peu qu'on veut : deux arpens ainsi ménagés & séparés, en valent plus de trois en commune ordinaire.

A PARIS,

Chez DELALAIN, Libraire, rue Saint Jacques.

M. DCC. LXVIII.

AVEC APPROBATION ET PRIVILEGE DU ROI.

PHOTO 14

The 1768 edition of Maison Rustique describes pasture rotation

Photo by the Documentation Service
of the Centre National de Recherches Agronomiques

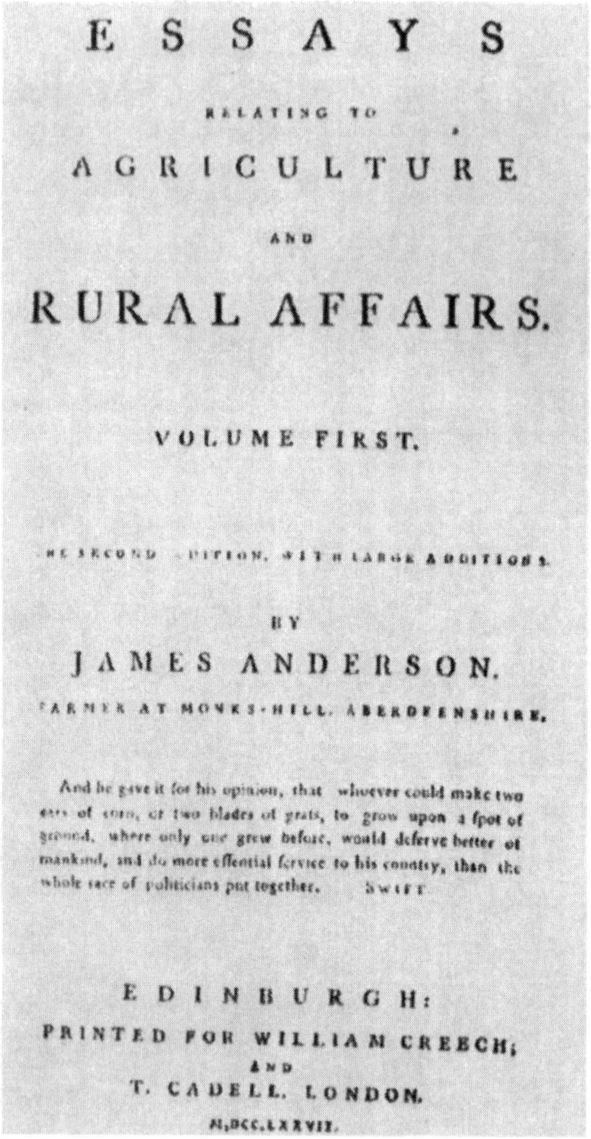

E S S A Y S

RELATING TO

A G R I C U L T U R E

AND

RURAL AFFAIRS.

VOLUME FIRST.

THE SECOND EDITION, WITH LARGE ADDITIONS.

BY

J A M E S A N D E R S O N.

FARMER AT MONKS-HILL, ABERDEENSHIRE.

And he gave it for his opinion, that whoever could make two
ears of corn, or two blades of grass, to grow upon a spot of
ground, where only one grew before, would deserve better of
mankind, and do more essential service to his country, than the
whole race of politicians put together.   SWIFT

E D I N B U R G H:

PRINTED FOR WILLIAM CREECH;

AND

T. CADELL, LONDON.

M,DCC,LXXVII.

PHOTO 15

Frontispiece of the 2nd edition (1777) of Anderson's book

By courtesy of the School of Agriculture, Cambridge
Photocopy by National Institute of Agricultural Botany, Cambridge

TABLE 49

Periods of stay of the groups and rest periods of the grass in the first six years' trials at Hohenheim

Passage number	Period of stay of a group in a paddock (days)	Average rest period of the grass (days)
1	1·60	14·4
2	2·44	22·0
3	2·47	22·2
4	2·04	18·4
5	2·01	18·1
6	2·45	22·1
7	2·12	19·1
8	1·46	13·1
9	0·95	8·6
Average	1·95	17·6

N.B. 1. There were three groups.
     2. The figures are averages for the six years, 1925–30.

From Münzinger and Babo (80).

This table illustrates quite clearly that after the May–June rotations (2nd and 3rd passages), where the rest period was approximately 22 days, rest periods were in fact shorter whereas they should have been extended to meet the requirements of the grass.

Misunderstanding of the principles of the Hohenheim system

I believe that this failure on the part of the Hohenheim pioneers to appreciate rest periods had very serious consequences for the development of the system they recommended. Those who wanted to apply the system, in all countries, based their plans on the elements supplied by its founders, as was only natural and logical. But more often than not they ended up in untoward acceleration of the rotation with all its disastrous consequences.

Without doubt, twenty years ago, as to-day, the conceptions held of rotation, which at that time was generally referred to as the Hohenheim system, were many and various, and sometimes contradictory.

Experiments at Beltsville

When they wanted to study the so-called Hohenheim system in the U.S.A. at Beltsville in 1930–35 (*vide* Voisin 117, vol. II, pp. 425–465) they tried to define exactly what the system was (147). With much frankness the American authors write in 1938:

"United States research workers are not agreed on exact methods characterising the different phases of the Hohenheim system. They are all however agreed that the two fundamental methods of the system are:

"(1) Circulation of the beasts by means of rotation over the various paddocks of a pasture.

"(2) The use of large quantities of fertilisers, especially nitrogen fertilisers.

"The opinion is held in some quarters that the herd must also be divided into two groups having regard to the milk yields of the cows, those giving the most milk forming the leading group and therefore grazing young, intact herbage. Others believe that the harvesting of hay from some of the paddocks at the beginning of the season is a point fundamental to the system. It appears certain that all these various forms of application were in fact employed by Dr. Warmbold at Hohenheim."

It will be appreciated that the basic principles of rational grassland management hardly appear at all among these considerations. Not one word is said about rest periods, and this aspect was completely neglected in the Beltsville experiments. No surprise can therefore be evinced at their results. The American workers found that rotational grazing, or more exactly, the Hohenheim system as they understood it, increased the yields from the pastures by a maximum of 10%.

In my *Diary of Travels in the U.S.A. with the Forage Production Mission*, I have analysed in detail why the method the Americans employed was not in any respect a real rotational system. For twenty years, however, they have lived on these results. How many American professors have inspected rational grazing on my farm and said: "Yes, but it is of no interest to us: according to the Beltsville trials the increase in yield is only 10%; it is not worth while." It was not until 1952 that the experiments of Grundage and Petersen (11) showed that rotational grazing led to *double* the yield of a pasture under continuous grazing management.

This American experience clearly reveals the mistaken ideas which have made the development of rational grazing difficult: *unfortunately these still hold sway to-day*.

The Schuppli rotation

The pioneers of rotation had certainly felt that the shorter the periods of occupation, the greater the tendency of the grass to produce higher yields. And so, to remedy the falls in yield they experienced, they tried reducing the periods of occupation, a fortunate line to follow. One of the advocates of rotation with daily moves was the Swiss Schuppli, who in 1936, gave the following explanation of the system as he conceived it (94):

"In the case of grazing with daily shifting of the animals it is necessary, depending on the quality of the land, to have 12 to 20 paddocks, each with an area per cow of 239 to 479 sq. yds. [*200–400* m.²] (359 sq. yds. [*300* m.²] on the average). For 20 cows, therefore, the area of paddock required is 1½ acres (60 ares). The pasture will be divided into 20 paddocks where some of the paddocks in the rotation are mown in spring. If grazing only is practised, 12 paddocks will suffice. . . .

"The cows that produce the most milk graze the first day on paddock 1, the second day on paddock 2. The cows yielding the least milk graze the second day on paddock 1. The young stock comes on to paddock 1 on the third day and paddock 2 on the fourth day. . . .

"In spring, the rate of re-growth is usually so rapid that a paddock can be *grazed again after 5 days rest.*

"Where there are 18 paddocks half (Nos. 10–18) will be mown in spring. The first group will therefore return on the 9th day to paddock 1, which will have had *6 days rest.* After three rotations, that is to say 24 days grazing, the nine paddocks (Nos. 19–18) previously mown can be grazed, that is to say, *6 days after cutting.* For the rest only a few (four out of nine Nos. 1–4) of the paddocks previously grazed are used, so that the rest period is increased only from 6 to 10 days. The other five (Nos. 5–9) are set aside for mowing.

"The four paddocks which have not yet been mown (Nos. 1–4) are now mown, and half the paddocks are subjected to a second cut, so that, *on the average, each paddock is subject to 15 rotations per year* and is mown three times in two years.

"With this rapid rotation the rest period is continually being lengthened and towards the end of the summer (from the end of July) mowing is discontinued: the rest period of a paddock is then 15 days. . . .

"The day the stock return to a paddock the grass is half a fist high (Halbfausthoch) ($2\frac{1}{4}$–$2\frac{1}{2}$ inches [6–7 cm.]); it is therefore very young and very rich in protein and provides the animals with a diet they eat readily. By using the requisite phospho-potassic fertiliser and 36–54 lb./acre [40–60 kg./ha.] of nitrogen (in the form of nitrate of lime etc.) in summer, one can graze each paddock 12–16 times a year, and mow each 1·5 times. . . ."

Basic elements of the Schuppli system

Schuppli's description has been cited at length for it reveals the errors that brought about the failure of rotation. His system can be summarised as follows:

Three groups.

Period of stay = 1 day.
Period of occupation = 3 days.
Rest period—
> In spring: 5 days.
> End of summer: 15 days.

Compensation for seasonal fluctuations in growth by excluding–including paddocks with the support of nitrogen in summer.
Number of rotations per year: 15.

The periods of stay and occupation place this system in the "Perfect" category according to our classification (Table 43, p. 149). But, as has been said, short periods of stay and occupation are not an end in themselves: their one aim is to allow the observance of the required rest periods while satisfying as far as possible the requirements of grass and cow. In this present instance the *perfect* system has been used only to attain completely *imperfect* rest periods of 5 days in spring and 15 days at the end of summer, when they should no doubt have been 18 and 40 days respectively (*vide* Table 8 from Zürn in Austria, p. 32). Such short rest periods go but little way towards supplying the needs of the grass, which cannot renew its reserves or provide its "blaze of growth". Moreover, the quantities of grass which the cow can harvest from a grass sward $2\frac{1}{4}$–$2\frac{1}{2}$ in. [*6–7* cm.] high are anything but large. But this is not the most serious aspect: there is worse to come where the requirements of the cow are concerned.

A rotational system which does not allow the cows to ruminate

Schuppli himself advises the following procedure:

> "In the case of grazing with daily advancement of the herd the animals must be put on to the new paddocks in the evening, after milking and left there until the next morning. They are milked on the pasture if possible so that they are not in the stalls at that period of the day when grazing is most intensive. . . . Thereafter, throughout the day they remain in the stall where they rest and ruminate. The only interruption to their rest is at mid-day when each cow is fed 33 lb. [*15* kg.] of green grass (young stock 18 lb. [*8* kg.]). (Author's note: we are not told what kind of grass is meant here but I naturally assume that it is mature fairly fibrous grass.) **This green feeding is absolutely necessary for the short and very young grass at the disposal of the animals does not allow them to ruminate sufficiently.**
>
> "Rumination is absolutely essential to the health of the stock. It is noteworthy, moreover, that cows grazing such a sward and being fed no green material tend to eat straw; this demonstrates the need of these animals for food of this kind to enable them to ruminate to their satisfaction. . . ."

One must obviously be a little surprised at the recommendation of a grazing system which does not allow a ruminant to ruminate, especially as the author describes this action as "absolutely essential to their health". But even assuming that the cows had been able to ruminate, I doubt whether they could have successfully escaped bloat or grass tetany for very long with grass $2\frac{1}{4}$–$2\frac{1}{2}$ in. [*6–7* cm.] high. The occurrence of tetany arising from the use of such *very young grass, which was imagined to be very rich in protein when in fact it was only very rich in nitrogen* and very poor in carbohydrates and energy units (starch equivalent) has already been mentioned (Part Two, Chapter 6, p. 123–7).

Unfortunately the Schuppli system is basically only the rotational system that has been universally recommended. Serious accidents ensued, the memory of which still impedes the development of rational grazing.

Professor Caputa's version

The position is restated as follows by that excellent Swiss scientist, Caputa (12):

"It is not advisable to regularly graze a sward that is too young, for at this stage it is less able to withstand the trampling and grazing of the stock. In addition, *the too young sward contains too high a proportion of proteins* undesirable in livestock feeding. With the 'Pacager System' (the French Swiss term for rotation) after the stock have been grazing a limited area for 2–4 days *the sward is left free for a rest period of 3–4 weeks* so that it can grow up again to the requisite height. . . ."

This short review of the ideas expressed by the pioneers of grazing rotation will be concluded with the examination of two manuals published by Geith, which around 1940 were the hand-books of rotational grazing (Umtriebsweide) in Germany. It was these that I myself used when I first started using rotation some thirteen years ago.

Two popular manuals by Geith

In 1943 Geith published two manuals in popular terms on the subject of rational grassland management—*Modern Methods of Grassland Management* (29) and, in collaboration with K. Fuchs, *Grassland Manual* (30). *Nowhere in either of these works is a single word to be found on variation of the rest period,* although in a previous study in 1936 (27) Geith had stated, very vaguely: "Between the various rotations there is a rest period of 14–20 days." (These times correspond approximately with those of the first Hohenheim experiments listed in Table 55, p. 274.)

In his 1943 manual Geith is content with dividing the grazing season into three parts (29), pp. 43–44:

"The first period runs from the beginning of grazing until 1st–10th July. It is characterised by exceptionally abundant and rapid re-growth of the grass which generally furnishes large quantities of nutritive substances.

"The second period extends from the beginning of July until the end of August or beginning of September. The strong rays of the sun and the heat give rise to a great deal of evaporation with the result that even a high rainfall does not allow the grass to grow with the same vigour as during the first period.

"The third period is from the end of August until the end of the grazing season. Vigour of growth is so reduced that even a great deal of heavy rain and heavy rates of fertiliser application cannot prevent a great diminution in the yield of grass."

To compensate for the difference in production during these three periods *Geith makes no mention of varying the rest period but recommends only variations in the stocking rate.* He writes:

"These falls in production are compensated by changing the livestock carry (*Auftriebsgewicht*) in the course of the season, *whether by reducing the number of animals in the herd or by increasing the area being grazed.* For example:

"First period from beginning of April until early July, total livestock carry 12–16 cwt./acre [*1500–2000* kg./ha.].

"Second period from beginning of July until end of August, livestock carry 9–12 cwt./acre [*1100–1500* kg./ha.]

"Third period from 25th August until end of grazing season, livestock carry 6–8 cwt./acre [*750–1000* kg./ha.].

"The area to be grazed can be calculated on the basis of these figures. Assume, for example, that the weight of the herd is 40 cwt. [*2000* kg.]. Applying the above figures, we find that for each of the three periods the following grazing areas are required:

"First period—33–25 acres [*13·2–10* ha.].

"Second period—46–33·2 acres [*18·2–13·3* ha.].

"Third period—67–50 acres [*26·6–20·0* ha.].

"After the end of August, therefore, the grazing area required is almost double that at the beginning of the grazing season. In many regions the herd can be reduced by the sale of beasts when productivity of the grass falls. When this is impossible the grazing areas must be increased by the incorporation of what are called 'secondary pastures' (Nebenweiden): these may be either temporary leys of clover or clover grass mixtures, or meadows which were mown earlier in the season. . . ."

All that Geith says is sensible, and his methods are perfectly applicable. The incorporation of emergency pastures, that is to say the re-introduction of green areas previously dropped from the rota, is certainly mentioned. But in all this there is not one word on the importance of rest periods or the necessity for varying these. It has already been stressed that to bring a herd back, even if it is reduced by half, after about 20 days in August on to a sward which has re-grown only to a height of $2\frac{1}{2}$ in. [*7* cm.] wastes the sward, the plants in which are defoliated by the stock before they have had a chance to renew their reserves; this means a fall in the yield of grass and also deterioration of the flora. Reduction in the number of stock is of no avail.

If can, of course, be said that increasing the total area will allow the rest periods to be extended. But Geith has nothing to say on this matter. It is apparent that he has paid no attention to the "time" factor. Nevertheless, we must be grateful to him for the important work he did and the contribution he made to the progress of grassland science.

Between 1930 and 1945 the idea of the importance of rest periods as the basic principle of rational grazing management was not yet very widely known as Professor Klapp points out.

The most serious mistake made by the pioneers of the Hohenheim system

After having read a paper which I published in the *Bulletin du Herd Book Normand* in 1950 dealing with intensity of grazing and the importance of the "time" factor, Professor Klapp, Director of the Institut für Boden und Pflanzenbaulehre in Bonn, wrote to me in 1951:

"I have read your article on rotational grazing with much pleasure and with great interest: with pleasure, because I have never before read such a clear and logical description of the system (such clarity is difficult to achieve in the German language); with special interest, because I found in it many suggestions for my own work. Since you have gone so deeply into this question, I should like to draw your attention to the following points.

"Geith advised a stocking density of 80 cwt./acre [*10,000* kg./ha.]. There are several weaknesses in this rule. . . . In fact, Geith paid no attention *either to number of groups, period of stay or period of occupation of a paddock by all the groups.* . . . I often discussed this question with Geith but unfortunately he died before a definite conclusion could be reached.

"Since then I have frequently expressed my point of view both orally and in writing but have never succeeded in making myself heard. It is with pleasure therefore, that I see today that I must have been right, for you too have arrived, independently, at the same idea. Moreover, you have successfully translated it into a precise form.

"Almost all the literature on grassland seems to be afraid of tackling these problems, perhaps, as you have said, because 'these cwt./days/acre [kg./days/ ha.] make one shudder'. Whether your formula or another analogous to it is used in the future, there is one point which remains certain and can no longer be overlooked: in calculating the basic elements of a rotation, *consideration must be given to the 'time' factor.* It is evident that:

160 cwt./acre [*20,000* kg./ha.] stocking density where a paddock is grazed for one day are equivalent to:

320 cwt./acre [*40,000* kg./ha.] stocking density where the paddock is grazed for half a day."

This statement by Professor Klapp of the historical evolution of the ideas of pasture rotation shows how difficult it has been to establish certain conceptions essential to the good conduct of rational grazing. It also helps one to understand the slow progress of the system in the past and to-day.

The "time" factor must dominate and rule rational grazing

These few very brief remarks in retrospect reveal that the pioneers of rotation not only neglected but completely failed to recognise the presence of the "time" factor. Even to-day the "time" idea is hardly mentioned in the vast literature on grassland as a basic factor in rational grazing management.

If, as Professor Klapp has rightly said, the "time" factor must be taken into consideration in *calculating the basic elements* of a rotation, it must be given even more consideration in the *practical carrying out of the rotation.* If the attention of the grazier making his first attempt at rotation is not focused on this cardinal point he will almost always end up on a reef that I call "untoward acceleration". This reef, I believe, is the case of almost nine-tenths of the failures experienced with rational grazing.

Chapter 3

UNTOWARD ACCELERATION

The mechanism of untoward acceleration

The ordinary system of grazing, known as continuous grazing, consists in general of putting beasts into a pasture in spring and leaving them there until the end of the season. The farmer usually selects the livestock carry capable of eating the grass in May–June. In consequence, when grass re-growth begins to flag, he has to reduce his livestock carry in some way or other, some of the beasts either being fed by other means or sold.

BRITISH SYSTEM

When a farmer decides to practise rotational grazing he tends to continue in his former habit and try to stock his pasture sufficiently at the beginning of the season to utilise all his grass in May–June. Assume, for example, a rotation with six paddocks each *two* acres in area, and a single group of animals which remains for 4 days on each paddock: the basic rest period is therefore

$$(6 - 1) \times 4 = 20 \text{ days}$$

The farmer is satisfied with this: it corresponds more or less to one of these rare figures quoted sometimes in articles or books on rational management of grassland: "The stock return approximately every three weeks to a given paddock." Vague though it undoubtedly is, this represents the only homage paid to the fundamental "time" factor which must rule any rotational system of management.

A rest period of 20 days will produce in May–June a re-growth of 4200 lb.

METRIC SYSTEM

When a farmer decides to practise rotational grazing with the metric system, he tends to continue in his former habit and try to stock his pasture sufficiently at the beginning of the season to utilise all his grass in May–June. Assume, for example, a rotation with six paddocks each 1 ha. in area and a single group of animals which remains for 4 days on each paddock: the basic rest period is therefore:

$$(6 - 1) \times 4 = 20 \text{ days}$$

The farmer is satisfied with this: it corresponds more or less to one of these rare figures quoted sometimes in articles or books on rational management of grassland: "The stock return approximately every three weeks to a given paddock." Vague though it undoubtedly is, this represents the only homage paid to the fundamental "time" factor which must rule any rational system of management.

A rest period of 20 days will produce in May–June a re-growth of 4800 kg. of

202

consumable fresh grass per acre. This represents 84 daily rations of 100 lb. grass from 2 acres, and allows 21 live-stock-units to remain for 4 days on each paddock 2 acres in area.

But at the beginning of July grass growth has already slowed down and at the end of 20 days only about 3150 lb./acre fresh grass will have re-grown corresponding to only:

$$\frac{3150 \times 2}{100} = 63 \text{ daily rations}$$

This will allow the animals to remain on any paddock for only:

$$\frac{63}{21} = 3 \text{ days}$$

at the end of which time the paddock will be scraped completely bare. It is doubtful, moreover, whether each live-stock unit will be able to harvest 100 lb. of grass from a sward of this height and openness. The result is therefore that the stock must be moved every 3 days, so that towards the end of July the rest period finds itself reduced to:

$$(6 - 1) \times 3 = 15 \text{ days}$$

As the summer advances the rate of growth of the grass slows down, even more. This rest period will have allowed the production of only about 2100 lb./acre fresh grass, which will supply:

$$\frac{2100 \times 2}{100} = 42 \text{ daily rations}$$

The stock will be able to remain for only 1 day on a paddock with the result that, towards the middle of August, the rest period is reduced to:

$$(6 - 1) \times 2 = 10 \text{ days}$$

Assuming that these 10 days will hardly allow the re-growth of 1050 lb./acre fresh grass, representing

$$\frac{1050 \times 2}{100} = 21 \text{ daily rations}$$

consumable fresh grass per hectare equivalent to, 100 daily rations each of 48 kg. which will allow 25 livestock units to remain for 4 days on each paddock 1 ha. in area.

But at the beginning of July grass growth has already slowed down and at the end of 20 days only about 3600 kg./ha. fresh grass will have re-grown corresponding to only:

$$\frac{3600}{48} = 75 \text{ daily rations}$$

This will allow the animals to remain on any one paddock for only:

$$\frac{75}{25} = 3 \text{ days}$$

at the end of which time the paddock will be scraped completely bare. It is doubtful, moreover, whether each live-stock unit will be able to harvest 48 kg. of grass from a sward of this height and openness. The result is therefore that the stock must be moved every 3 days, so that towards the end of July the rest period finds itself reduced to:

$$(6 - 1) \times 3 = 15 \text{ days}$$

As the summer advances the rate of growth of the grass slows down even more. This rest period will have allowed the production of only about 2400 kg./ha. fresh grass, which will supply:

$$\frac{2400}{48} = 50 \text{ daily rations}$$

and allow the 25 beasts to remain on each paddock for only:

$$\frac{50}{25} = 2 \text{ days}$$

At the end of August, therefore, the rest period is reduced to:

$$(6 - 1) \times 2 = 10 \text{ days}$$

Assuming that these 10 days will hardly allow the re-growth of 1200 kg./ha. fresh grass, representing the stock will be able to remain for only 1 day on a paddock with the result that, the middle of August the rest period being allowed to the grass is only:

$$(6 - 1) \times 1 = 5 \text{ days}$$

$$\frac{1200}{48} = 25 \text{ daily rations}$$

the stock will be able to remain for only one day on a paddock with the result that, towards the middle of August, the rest period being allowed to the grass is only:

$$(6 - 1) \times 1 = 5 \text{ days}$$

In other words there will be practically no re-growth when the stock return to a paddock. The rotation is finished and there is no alternative to removing the stock, grazing them elsewhere or feeding them in some other way.

The rotation is being speeded up when it should be slowed down

It will now be understood why I have called this fault in rational pasture management "untoward acceleration". It means, in effect, that the progress of the herd across the paddocks is being *accelerated* when, in fact, it should be *slowed down*. Or, to put it differently, the animals are returning with *greater rapidity* to a paddock when they should be returning more slowly; in brief— *the periods of rest are being reduced at a time when they should be extended.*

The farmer starting out on rotation is taken unawares

With continuous, that is the ordinary normal method of grazing, the farmer is conscious that his grass production is dropping as re-growth becomes less and less. With rotational grazing, the farmer is deceived. He does not see that he is going to run short of grass. The growth slows down, but, by speeding up the rotations, he has sufficient grass to feed his stock until, *at one blow*, he has no grass at all. He has been unaware of the danger and finds himself with paddocks completely denuded and scraped bare to the ground.

The grazier new to rotation is taken unawares because he fails to recognise the danger of *reducing* rest periods when he should be *extending* them. If his attention had been drawn to the fundamental factor "time", and if he had learned the practical methods of varying rest periods, he would immediately have sensed the danger in accelerating passages. He would have taken the necessary measures to remedy the situation before the catastrophe of *total* absence of grass could take place. But farmers, and even grassland research scientists, often let themselves be surprised by the summer reduction in the yield of grass. One often hears it said: "With rotational management I am short of grass in summer." But it is not the grass or the system that is at fault: it is the farmer himself.

Untoward acceleration and stock health

This untoward acceleration will lead to the cow grazing an *extremely young* sward which has not yet produced enough re-growth. The dangers of this have been seen: bloat and grass tetany. Possibly also, as with the Schuppli system, the animals will not even be able to ruminate.

It is also understandable that untoward acceleration has brought about—and is still doing so to-day—the ruin of rotational grazing, by producing very mediocre yields of grass and causing serious set-backs among stock. In this connection, I have memories of my own personal experiences in 1956.

We have all made the same mistakes

In a *département* in the North-East of France where a great effort has been made to establish rational grazing, I was explaining to an audience of experienced farmers this main and basic reason for the failure of rotation, namely "untoward acceleration". I said, "I was the first to make this mistake ten years ago. And so I ask without malicious intention, how many of you have made the same mistake?" Several arms were raised in all honesty, among them that of the president of the Farmers' Federation, who, very readily and with good grace, said: "Chance has it that I have six paddocks in my rotation. Every year I am forced to speed up the movement of the stock from the beginning of July onwards and from more or less the middle of August I have no grass left."

Very recently I asked the same question at a meeting in a *département* in the west. The first hand to be raised was that of the President of the "Association for Grassland Improvement", who said: "For several years I have been trying to explain to myself why, practising rotation, I am short of grass in summer. Now I know: thank you very much for telling me." And note that this is a *département* with one of the highest rainfalls in France, both throughout the year and in the summer.

Unfortunately, untoward acceleration happens even more easily with some of the systems being recommended at present, as will be seen later in dealing with rationed grazing.

PART SEVEN

TETHERING AND RATIONED GRAZING, SPECIAL SYSTEMS OF RATIONAL GRAZING

Chapter 1

TETHER GRAZING

The principle of tethering

TETHER [1] grazing is an age-old method consisting of attaching the animal, by means of a chain or rope, to a stake which is shifted once, or more often, each day, the animal each time being allowed a sufficient grazing area to satisfy its appetite until the next time the stake is moved. To-day, both chain and stake can be replaced by mechanised systems which simplify the work involved.

In England, one can see a tether consisting of a tube turning on an axle which is fixed into the soil by means of slightly curved points. The apparatus can easily be shifted by using a lever to raise up the points. A rod regulates the length of chain and so reduces the number of times the turning spindle requires to be moved.

Tethering rations each cow individually

It may be said that tether grazing is the most refined form of rational grazing, because not only does it limit the area made available to the herd as a whole, but it also measures the area allotted to *each* individual animal. Despite their restricted grazing area, the animals do not fight amongst themselves: their attachment prevents their coming into contact with each other.

Reference has already been made to the tether system of grazing temporary pastures in the *Pays de Caux* (Seine-Maritime), where farmers say that it is becoming difficult to find a herdsman able to conduct tether grazing with any skill. The great *art* is to be able to allot to each beast the area it will be able to graze, no more, no less, before the next time the stake is moved. This area depends both on the animal and on the fresh grass available: for an in-milk cow a skilled herdsman will shift the stake a greater distance than for a yearling. He will shift the stake less when the red clover begins to flower than when it has not even begun to bud, at the beginning of the grazing season (pp. 172 and 211).

---

[1] The word "tether" very probably comes from an old Saxon or Scandinavian word. It is "tière" in French, and "tüdern" in German. The words are analogous and this indicates that this grazing method is ancestral.

209

Watering tethered animals

Watering tethered animals involves much work: even to-day cattlemen can still be seen in my own district passing along the rows of animals once or twice a day with their horses and carts, giving the stock a drink from a bucket.

Where one-year young clover is being grazed, as is by far the most common practice in the *Pays de Caux*, the very young green forage has such a high water content that there is no need to water the stock, especially at the end of the season (September 15–November 1) when the weather is generally cool.

Tethering in Scotland and Ireland one hundred and fifty years ago

This question of tether grazing provides an opportunity to mention the names of two great figures in the agricultural sphere at the beginning of the nineteenth century—the Scotsman John Sinclair and the Frenchman Mathieu de Dombasle.

Sir John Sinclair was an economist who was born at Thurso Castle in 1754 and died in Edinburgh in 1835. In 1793 he set up the first Board of Agriculture of which he himself was president. His *Code of Agriculture*, which was translated into French in 1825 by the famous French agronomist Mathieu de Dombasle (1777–1843), contained several pages on tethering. He described the system thus (97, p. 134):

> "In some districts of Scotland and Ireland, instead of soiling, they tether their stock upon the land.
>
> "In the Agricultural Report of Aberdeenshire, it is stated that there are some cases, where the plan of tethering can be practised with more profit than even soiling. In the neighbourhood of Peterhead, for instance,[1] in a regular and systematic method, moving each tether forward in a straight line, not above one foot at a time, so as to prevent the cows treading on the grass that is to be eaten, care being always taken to move the tether forward like a person cutting clover with a scythe, from one end of the field to the other. In this way a greater number of cows can be kept, on the same quantity of grass, than by any other plan, except where it grows high enough to be cut, and given them green in houses. In one instance the system was carried to great perfection by a gentleman who kept a few sheep upon longer tethers, following the cows. Sometimes also, he tethered horses afterwards upon the same field, which prevented any possible waste, for the tufts of grass produced by the dung of one species of animal will be eaten by those of another kind, without reluctance. This system was peculiarly calculated for the cow-feeders in Peterhead as, from the smallness of their holdings, they could not afford to keep servants to cut, or horses to carry home the grass to their houses, to be consumed in a green state.
>
> "In Ireland, the plan of tethering stock is strongly recommended, in preference to that of the promiscuous pasturage, even though accompanied by a

---

[1] They tether milch-cows in their grass fields.

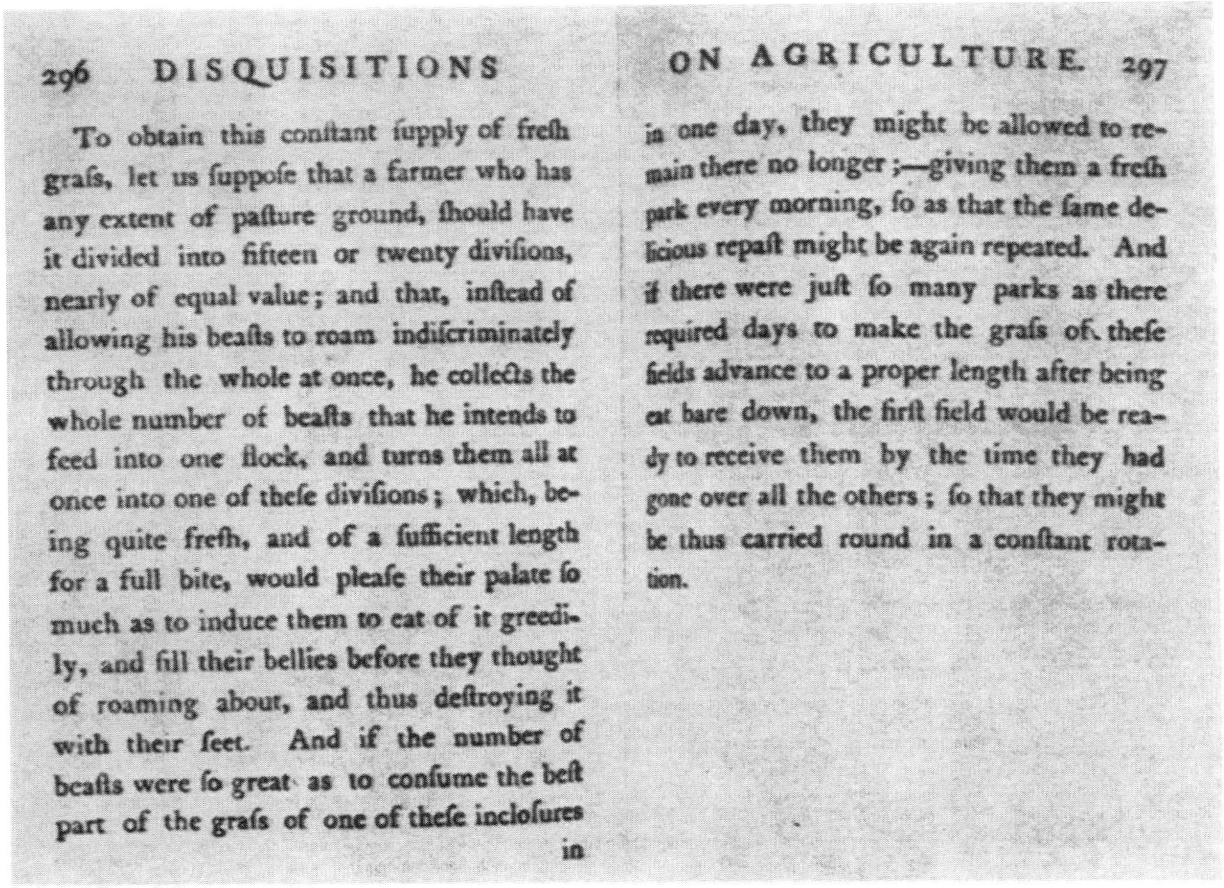

To obtain this constant supply of fresh grass, let us suppose that a farmer who has any extent of pasture ground, should have it divided into fifteen or twenty divisions, nearly of equal value; and that, instead of allowing his beasts to roam indiscriminately through the whole at once, he collects the whole number of beasts that he intends to feed into one flock, and turns them all at once into one of these divisions; which, being quite fresh, and of a sufficient length for a full bite, would please their palate so much as to induce them to eat of it greedily, and fill their bellies before they thought of roaming about, and thus destroying it with their feet. And if the number of beasts were so great as to consume the best part of the grass of one of these inclosures in

in one day, they might be allowed to remain there no longer;—giving them a fresh park every morning, so as that the same delicious repast might be again repeated. And if there were just so many parks as there required days to make the grass of these fields advance to a proper length after being eat bare down, the first field would be ready to receive them by the time they had gone over all the others; so that they might be thus carried round in a constant rotation.

PHOTO 16

Description of pasture rotation in 1777 by the Scottish agriculturalist, Anderson

By courtesy of the School of Agriculture, Cambridge

PHOTO 17

Strip grazing

Farmers Weekly

PHOTO 18

Strip grazing on the Farn
Leicestershire (England)
Period of occupation of t
strip at the back : 16 day.

Photo **Voisin**

herdsman or keeper. It is there observed, that both cattle and sheep must thrive better, and feed faster, when they have a fresh bite of grass regularly given them, than when they are permitted to wander over a whole field. This is effectually done, when they are not allowed to range indiscriminately over pasture lands, destroying more than they consume, but when each animal is secured by a tether, to the spot on which he is allowed to feed. By changing this spot he is enticed to eat, from having a clean and fresh bite, perhaps twice a day given him. He does not acquire rambling habits, which exhausts his strength, and prevents his fattening; but becoming docile, he necessarily thrives much better. *The pasture is also improved, for the young is not bit off prematurely, which checks its progress, but remains untouched, till it is ready for consumption.*

"Some eminent and extensive agriculturists in Ireland have practised this system with success—have produced by it, beef and mutton of the best quality —and their lands have been materially improved, since they followed that plan. In other cases it has been tried with milch-cows, store cattle, sheep and lambs, with all of whom it has completely answered; and by its adoption it has been found that land will improve more in two years, than under indiscriminate pasturage, in five; and that, at least one third more stock may be maintained per acre, under the one system than the other. The reason is obvious, the cattle, being better fed deposit more dung which, falling in a narrow compass, is trod into the ground, by the time the spot of grass, in which they are tethered, is nearly eaten; whereas when the dung is scattered about, the land is not much benefited by it."

Present-day methods of tethering

At the present time tethering, except of bulls, is not much used on pastures. Its use is more or less exclusive to instances, as in the *Pays de Caux*, where there is no "return" (at least no immediate "return") and where, therefore, there is no need to pay attention to the rest period between two successive tetherings.

Tethering systems where no attention need be paid to the rest period

1. *Crimson clover.* After one grazing passage it is ploughed and sometimes fodder beet is then sown. In very exceptional cases two tethering rotations are carried out, only over a very small section.

2. *Red clover* is generally sown in oats and the young clover tether grazed towards the middle of September. There will be no return in that year, for the season is finished and winter is approaching.

3. In principle, the first cut of *red clover* (in the year after sowing) is reserved for forage. Only the *second cut* (and not always) is tether grazed. Thereafter the beasts cannot return to the clover for it is ploughed to be sown with wheat.

These examples show that the farmer practising tether grazing on these temporary pastures of the *Pays de Caux* (as on many others) has no need to trouble himself about when he will be returning to re-graze the plants: that is to say, he need give no thought to the rest period the plant will require to attain the optimum stage of development which will allow it to better withstand another cut.

On normal pastures, temporary or permanent, *the return of stock has always to be reckoned with*: this is what the instigators of rationed grazing appear to have forgotten when they took their pattern from tethering.

Chapter 2

DIFFERENT FORMS OF RATIONED GRAZING

A fashionable word

THE word "rationed grazing" (not to be confused with rational grazing) is very much in fashion; although, as was seen above, many other words are also used to describe the many existing conceptions of rationed grazing. Whereas ten years ago an author would have written "rotational grazing" or "rotation of grassland", to-day he uses the expression "rationed grazing" with the feeling that he is thus showing himself to be very progressive.

One feels a little lost in the face of all the different conceptions expressed in the abundant modern literature on rationed grazing. In most of the studies little information is given to show the reader exactly how this "rationed" grazing is conducted and what it means in practice. Reference has already been made to the confusion surrounding rotation of pastures. That confusion is even greater in the case of "rationed" grazing, so great indeed that some have gone so far as to say that in "rationed" grazing the *electric* fence is used, whereas with rotation only *fixed* fences are employed.

It is not easy, therefore, to determine the particular and different ideas hidden under one and the same name.

Three factors can be "rationed"

I have the impression, without however daring to state categorically, that the various authors have staked their rationing on three factors:

1. Rationing of *fresh* grass.
2. Rationing of grass *already grazed*.
3. Rationing of the animals' grazing *time*.

A word will be said briefly in explanation of each of the three cases.

1. At each shift a *variable* area of **fresh** grass is "allowed".

2. In addition to the fresh grass, a **variable area of grass that has already been grazed** is also made available to the animals.

3. Within the framework of possible methods outlined above the herd is allowed to graze for one part of the day or night only; in other words, *the animals graze for only a* LIMITED *time.*

213

I feel that in the North-West of Europe when one speaks of rationed grazing, one is thinking in particular of "rationing" of the first two factors, except in the exceptional case of very high-yielding cows. "Rationing in time" (Case No. 3 above) is applied particularly in hot regions and becomes more strict, the hotter the climate.

All that has been said is based on the assumption that the herd is concentrated in one single group. I have, however, seen quite a lot of rationed grazing where two groups are involved: this will be dealt with in Chapter 6.

The "time" factor is almost always ignored in rationed grazing

The instigators of the various systems of rotation ignored, and still ignore, the importance of the "time" factor in grazing management (pp. 193–201). In the thirty typewritten sheets where I had listed the different conceptions of rationed grazing, I could hardly find one reference to the importance of the rest period and still less to the period of occupation. This neglect of the principal factor in all rational grazing obviously had to lead to the same difficulties and the same mistakes.

To speak of *rationed grazing* instead of *rotation* does not dispense with the need to obey the Universal Laws of rational grazing; nor does it remedy the faults of the Hohenheim system. Neglect of the laws in both cases has produced failures of a similar nature. Indeed, *grazing rationed "in area" can lead to untoward acceleration even more easily than rotation.*

Rationed grazing has often been the continuation of rotation

I have frequently gained the impression that, for many scientists, the idea of *rotation* is linked up with *fixed fences* and that of *rationed grazing* with *electric fencing*. In fact, however, at the time when they were campaigning for rotation of pastures, or the Hohenheim system, electric fencing was hardly known and division was achieved by fixed fences.

On the other hand, more than six to eight divisions were rarely used, which gave periods of stay of 4–7 days. It has already been said how difficult it is to manage a rotation with such a small number of paddocks. In addition, the importance of the "time" factor was completely overlooked, with the result that untoward acceleration almost always ensued, and when the rate of grass growth was reduced in summer there was a scarcity of grass.

It was, however, felt that too small a number of paddocks was making the conduct of the grazing difficult. Meanwhile the electric fence had made its appearance, and *so the idea was conceived of increasing the number of divisions, that is, of dividing up the fixed paddocks by means of electric fencing.*

They set out with this idea: there is a shortage of grass as soon as summer arrives with the Hohenheim system; therefore, let us use the electric fence and practise rationed grazing. It was, and still is, extremely rare to find precise details of the management of this rationed grazing that was superimposed on the old rotation.

Tethering, the inspiration of rational grazing

It was well known that tethering allowed excellent utilisation of forage crops; and so the idea took shape that the herd could be rationed with the electric fence in imitation of tethering. Unfortunately two fundamental points were forgotten:

1. Hardly any tether grazing systems involve return and so they do not *require* optimum rest periods to be observed.

2. With tethering, the animals are restricted *individually*, whereas with the electric fence the *whole herd* is restricted. It is impossible in the latter instance to "concentrate" the stock as if they were tethered. Moreover, a greedy animal runs the risk of harvesting excessive quantities of certain dangerous foods (bloat), such as very young red clover or a very young temporary sward with a very high content of white clover (especially S.100 or Ladino).

Division of our study of rational grazing

In the following pages a Scottish study will be examined (Chapter 3). Then two forms of rationed grazing with *one* **single** *group* will be studied:

1. Rationed grazing, where a variable area is allotted, but *no* previously grazed herbage made available (Chapter 4).

2. Rationed grazing, with area allotted either fixed or variable, the animals being given access to an area that has already been grazed (Chapter 5).

In Chapter 6 rationed grazing *with* **two** *groups* will be examined and the section ended by a study of grazing rationed in time (Chapter 7).

Chapter 3

DOES RATIONED GRAZING PRODUCE 25% MORE THAN ROTATION?

A popular statement

THE following statement has become generally accepted and recurs in article after article (p. 137): "*Rationed* grazing produces 25% more than *rotation.*" Every time I have read or heard these words I have asked myself the same question: "What rotation are they talking about? For a rotation in which the herd is shifted *every second day* produces double the yield of a rotation where the animals move *every twelve days.*" Then I asked what was meant by rationed grazing.

The answer was (or rather it was provided by the article) that rationed grazing is the system of grazing in use in an experiment at the Hannah Institute in Scotland. Before going on to examine the question of rationed grazing in more detail, it will be advisable to study more closely this Scottish experiment, the summary of whose results quotes a difference in yield between two grazing systems of 25%.

Important contribution of the Hannah Institute to grassland research

The Hannah Dairy Research Institute (Ayr, Scotland) has done extensive and excellent work on questions concerning milk production, not forgetting the grazing aspect. Holmes, Waite, Ferguson and Campbell, working together as a group, have studied the most efficient and most rational methods of grazing. Holmes spent a few days at my farm on his way to an International Congress in Paris, and I remember with pleasure the discussions we had on all these problems which were of such interest to us both.

This group of workers published, in 1950, a paper entitled "A comparison of the production obtained from close folding and rotational grazing of dairy cows" (41) and (136).

The experimental method employed

On the conduct of this experiment, the Scottish authors write:

"Two groups of cattle were used. . . . For various reasons only four cows remained permanently in each group throughout the experiment. Group I was grazed on the *close-folding system* and Group II on *rotational grazing* for approximately 8 weeks from 3rd May until 29th June, after which the groups were interchanged. It was thus possible to compare the influence of the systems of management on both groups of cows.

"After the cows had already been at pasture for about 3 weeks, grazing began on the experimental fields on 3rd May and, *with the exception of 2 weeks in August*, continued on these fields until 10th October. *The cows received nothing but grass throughout this period.*"

Rotational grazing compared with close folding

The two methods are described as follows:

"(a) *Rotational grazing.* The method of rotational grazing adopted waš to allow one group of cows free access to a paddock of about 1 acre in extent. Stocking was therefore at the density of about 6 cows per acre [*15* per hectare] and grazing was continued until it was judged by normal farm-management standards that a change was needed. *These periods varied from 5 to 14 days.* The cows were then transferred to another similar paddock. *Paddocks were rested after each grazing for a period of 3–5 weeks.*

"(b) *Close folding (or rationed grazing).* In close-folding the cows were allowed only a very limited area of grass and *the area was changed each day. The quantity of herbage offered and the area allocated per day varied during the season.* The stocking density varied from 50–80 cows per acre [*125–200* per ha.]."

The daily allocation of grazing area was made with *two* movable electric fences. Each day the forward fence was moved to its new position and the rear fence brought up to the former position of the "forward fence".

Examination of these two grazing methods

In the first case, rotational grazing, the period of stay (which, with one group, is equivalent to the period of occupation) had to be varied *from 5 to 14 days*, that is, from simple to triple proportions. By whatever name it is called, therefore, *this method is but a vague imitation of rational grazing.*

Like all other instances where the principles of rational grazing are not observed, *there was a shortage of grass in August*, with the obviously very grave result that the experiment had to be discontinued. The Scottish research workers were well aware of this fault in the system and write in their discussion of the results obtained. *"Where the rotational grazing periods were shorter, they approached more closely the more efficient conditions of close-folding."* Personally, I would have written: "When the rotational grazing periods were shorter, the rotation became more *rational* and tended to approximate to what has been called *close-folding (rationed grazing), being, in effect, nothing more than rational grazing with an occupation period of one day.*"

In a subsequent paper on this work published two years later (42) the Scottish workers write: "The only differences in 1950 were that the rotational paddocks were slightly reduced in size to *intensify* that system. . . ."

And in the end Holmes (46), that remarkable agronomist, must have seen that rational grazing is something different, for he writes:

> "A more precise definition of rotational grazing is the system where each paddock is grazed for a period of 3–5 days and then rested for 18–28 days before re-grazing or cutting. The rate of stocking [1] is 6–10 cow equivalents to the acre [15–25 cow equivalents per hectare]."

The rationed grazing in these experiments consists in daily shifts with *variable areas*, but without any area already grazed being accessible. In another study (44) Holmes describes as *strip grazing* what he had previously referred to as *close folding*, while in yet another paper (45) he states that this method may equally be described as *rationed grazing*. This proves what was said above, namely that one name can be applied to many systems or many applied to one system.

Results obtained from the two methods of rotational grazing by the Scottish workers

The results of the first two series of trials carried out in two different years are indicated in Table 50.

TABLE 50

Comparison of protein and starch-equivalent yields from two systems of grazing in the Hannah Dairy Institute trials

	Digestible crude protein				Starch-equivalent			
	Rotational grazing		Close folding		Rotational grazing		Close folding	
	lb./ acre	[kg./ ha.]	lb./ acre	[kg./ ha.]	lb./ acre	[kg./ ha.]	lb./ acre	[kg./ ha.]
1st series of trials .	360	[404]	446	[500]	2431	[2725]	2884	[3233]
2nd series of trials.	482	[540]	625	[701]	3350	[3755]	4251	[4765]
Averages: Actual	421	[472]	535	[600]	2891	[3240]	3568	[3999]
Relative	100		127		100		123	

N.B. Calculated by the author from the results of the Hannah Institute.

The conclusion that the experimenters *rightly* reached was that close folding would allow a sward to yield 25% more than rotational grazing: this was correct for the methods of grazing *subjectively* described by these names.

Popular conclusion, its terms badly defined

This conclusion was spread abroad, and from 1950 onwards it has been said and written again and again: "Rationed grazing" allows a 25% higher

---

[1] My equivalent term is "stocking density".

yield to be obtained than "rotation of pastures". But since it was not made definite what in fact these two terms actually meant, much misunderstanding and confusion has arisen, and continues to do so. As has just been demonstrated, what was involved in both instances was a semi-rational grazing system. The only correct and general expression of the conclusion reached appears, therefore, to be that *a more intensive and better conducted rational grazing system has produced a 25% higher yield than the same system when much less intensively and fairly badly managed.*

In my opinion this is the conclusion which best represents the results and at the same time avoids misunderstanding of terms with a multiplicity of possible meanings.

Chapter 4

RATIONAL GRAZING ALLOWING A VARIABLE GRAZING AREA BUT NO AREA OF PASTURE THAT HAS ALREADY BEEN GRAZED (ONE GROUP ONLY)

A simple case

To take a very simple case first of all: one single group is involved and the area allotted at each shift is limited by means of *two movable* electric fencing wires, the front and back wires. The sides are delimited by fixed or semi-mobile fences (*vide* Fig. 12).

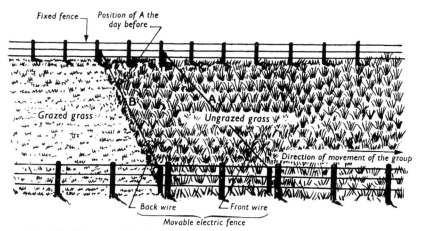

Fig. 12. The area of fresh grass is limited by two electric wires (front and back) which are moved along two fixed fences.

The electric fencing, that is to say the group, can be moved every *x* days or several times a day. The latter is generally recommended in books and journals in imitation of tether grazing practice.

The variable area allotted

Where a herd is of fixed size, the area to be placed at its disposal depends on the height and density of the grass. The general indicated allocation is 60–120 sq. yd. [*50–100* m.²] per day: this figure may obviously vary greatly.

220

Rational grazing where a variable area is allowed leads to untoward acceleration

Assume a large pasture 12·5 acres [*5 ha.*] in area grazed by 25 livestock units. These are moved every day by shifting the two electric wires delimiting an area which provides 25 daily rations, each of 100 lb. [*48 kg.*], representing in all:

$$25 \times 100 = 2500 \text{ lb. } [25 \times 48 = 1200 \text{ kg.}]$$

In May–June the yield of fresh grass available will supposedly be in the region of 4200 lb./acre [*4800 kg./ha.*]. To provide the herd with its daily ration of 2500 lb. [*1200 kg.*] the area that must be allocated to it is:

$$\frac{2500 \times 1}{4200} = 0 \cdot 6 \text{ acre } \left[\frac{1200 \times 100}{4800} = 25 \text{ ares} \right]$$

The area not occupied by grazing is 12·5 — 0·6 = 11·9 acres [*500 — 25 = 475 ares*], which represents

$$\frac{12 \cdot 5 - 0 \cdot 6}{0 \cdot 6} = \frac{11 \cdot 9}{0 \cdot 6} = 19 \text{ "blocks" of } 0 \cdot 6 \text{ acre}$$

$$\left[\frac{500 - 25}{25} = \frac{475}{25} = 19 \text{ "blocks" of 25 ares} \right]$$

Since the "block" of this size is provided for the herd *each day* the rest period in May–June

$$\frac{12 \cdot 5 - 0 \cdot 6}{0 \cdot 6} = 19 \text{ days } \left[\frac{500 - 25}{25} = 19 \text{ days} \right]$$

It will be assumed that under the prevailing climatic conditions, this 19-day rest period will be sufficient in May–June to give a re-growth of 4200 lb./acre [*4800 kg./ha.*] of grass. In this event at the beginning of July, the same rest period of 19 days will supposedly supply only 3550 lb./acre [*4000 kg./ha.*] grass.

In a previous example (Part Six, Chapter 3) a period of 20 days at the beginning of July was assumed to permit the re-growth of grass to the extent of 3150 lb./acre [*3600 kg./ha.*]. But in that instance the period of occupation was 4 days, while in the present case it is only 1 day, which, supposedly, will allow the grass to re-grow more vigorously.

If the amount of grass available is 3550 lb./acre [*4000 kg./ha.*] the daily allocation of land must be:

$$\frac{2500 \times 1}{3550} = 0 \cdot 7 \text{ acre } \left[\frac{1200 \times 100}{4000} = 30 \text{ ares} \right]$$

to supply 25 daily rations.

The average rest period will then be:

$$\frac{12\cdot5 - 0\cdot7}{0\cdot7} = \frac{11\cdot8}{0\cdot8} = 16 \text{ days approx.}$$

$$\left[\frac{500 - 30}{30} = \frac{470}{30} = 16 \text{ days approx.} \right]$$

This reduced rest period, at a time when it should be extended, is going to supply 2400 lb./acre [2857 kg./ha.] of grass towards the end of July and the rationed grazing area will then be:

$$\frac{2500 \times 1}{2400} = 1\cdot04 \text{ acres} \quad \left[\frac{1200 \times 100}{2857} = 42 \text{ ares} \right]$$

This, at the beginning of August, will give a rest period of:

$$\frac{12\cdot5 - 1\cdot04}{1\cdot04} = 11 \text{ days} \quad \left[\frac{500 - 42}{42} = 11 \text{ days} \right]$$

which will assumedly provide fresh grass of the order of 1600 lb./acre [1925 kg./ha.], which will limit the daily area to:

$$\frac{2500 \times 1}{1600} = 1\cdot56 \text{ acres approx.} \quad \left[\frac{1200 \times 100}{1925} = 62 \text{ ares approx.} \right]$$

In mid-August, therefore, the rest period is reduced to:

$$\frac{12\cdot5 - 1\cdot56}{1\cdot56} = 7 \text{ days approx.} \quad \left[\frac{500 - 62}{62} = 7 \text{ days approx.} \right]$$

This means that in August there will be no re-growth of the sward and therefore grazing is finished.

Untoward acceleration is the result of reduced periods of occupation as well as increases in the areas allocated

It was seen (pp. 202–203) that where the area of the paddocks was fixed (whether by barbed wire or electric fencing) reduced grass growth led to a shorter period of occupation (and therefore a shorter period of stay) if the rest periods were not extended as the necessity arose. Indeed, after a shorter period of occupation (and stay) than had been theoretically anticipated, the sward was grazed completely bare and the herd had to be moved on. Since the number of paddocks resting was the same as in the preceding grazing period, *this fixed factor* (number of paddocks resting) *multiplied by a reduced period of occupation gave a shorter rest period at the time when it should, in fact, have been extended.*

Where the area allowed, that is the "block" to be grazed, *varies* in size and every advance of the herd follows an *equal* period of occupation (identical here with period of stay), the rest period is obtained by multiplying this fixed factor (period of stay) by the number of (virtual) blocks resting. This number diminishes, for it is obtained by dividing the area at rest (always *equal* in the present hypothesis) by the area allowed at each shift which, as has been seen, *increases* with each rotation.

It may be said therefore:

    1. With *identical* paddock or block areas:

Rest period $= (Variable$ period of stay$) \times (Fixed$ number of blocks$)$

    2. With *variable* paddock or block areas:

Rest period $= (Fixed$ period of stay$) \times (Variable$ number of blocks$)$

Untoward acceleration is due to:

    1. In the Hohenheim system, reduction of the period of stay.

    2. In rationed grazing, reduction of the number of "blocks" of pasture at rest.

In both cases, for different reasons, untoward acceleration takes place, that is, shortening of the rest period of the grass at the very time when it should be lengthened.

An illustration will make this more easily understood.

The bigger the slice eaten each day, the quicker a tart is finished

A small boy buys a tart at the baker's and decides to eat a slice every day. The bigger the slice eaten daily, the quicker the tart will be finished. This means that the boy will have to go back to the shop and buy a second tart sooner than he would have had to if he had eaten a smaller piece each day.

Width and thickness of the slice of tart

This boy may perhaps eat a wider slice of tart each day, not because he is greedy but simply because he wants to eat an *equal weight* of tart; this is quite legitimate, especially if tart is all he has to eat. In order to eat the same quantity of tart each day, however, he *will always have to be eating a wider slice, for the tart becomes thinner and thinner as he cuts it* due to the fact that it is not properly shaped. This is what is to be understood from Fig. 13, p. 224.

In the case of a well-shaped tart of equal thickness (H) (top of the figure) the boy eats the whole tart in 6 days in 6 slices of equal width and equal weight ($A_1 = A_2 \ldots = A_6$). But with the badly shaped tart (bottom of figure) (height H at the beginning and $\frac{H}{3}$ at the end) if he is to consume a slice equal in weight to those of the well-shaped tart ($B_1 = \ldots = B_4 = A_1 = \ldots A_6$), the boy will have to eat a wider slice every day; with the result that the tart will be consumed in 4 instead of 6 days.

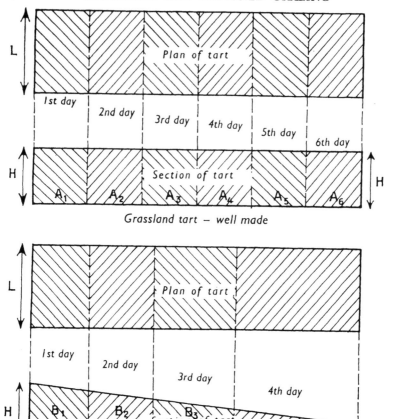

FIG. 13. Grassland tart well made and badly made (the areas
$A_1, A_2, A_3, A_4, A_5, A_6$, and B_1, B_2, B_3, B_4, are equal).

Increased rate of consumption of "blocks" of pasture

An analogous phenomenon takes place in rationed grazing where the allocated area varies, being greater the less grass there is available (that is to say the shorter the grass) as time goes on, just as in the case of the tart at the foot of Fig. 13. The grazier is therefore compelled to cut into a second tart sooner, that is to say, to start the next rotation after a shorter rest period, than he had normally foreseen. In other words, rotational shifts are being speeded up as grass growth declines.

Here, then, once more is the untoward acceleration of the Hohenheim system, but aggravated in this instance because the very principle of rationed grazing provides for a basic rule which automatically creates the situation of untoward acceleration.

Increasing the area allocated results in reducing the stocking density when grass growth is flagging

Increasing the area allocated is no remedy for the chief defect of rotation, untoward acceleration, which has almost always led to a scarcity of grass in summer where the Hohenheim system is applied (pp. 202–205). A different name has been given to the system and the method of shifting the herd has been changed, but in both cases the same mistake has been made: the fundamental factor, *time*, has been overlooked.

The founders of rotation, as has been said, spoke of stocking density, which is equal (assuming one group) to:
$$\frac{\text{Number of animal units}}{\text{Area of paddock}}.$$

No mention was made of period of stay or rest period. When grass growth was flagging the stocking density was reduced by *lowering the numerator* (number of livestock units) but without touching the denominator. In rationed grazing, where the area allowed for grazing is variable, the stocking density is reduced by *increasing the denominator*, that is to say the area of paddock grazed by a fixed stocking.

In both cases the only interest is in altering the stocking density: no attention at all is paid to the "time" factor. The end of such a system is untoward acceleration, which is achieved even more easily with rationed than with rotational grazing.

In all rational grazing, whether the area is fixed or variable, seasonal fluctuations in growth must be compensated for

Allocation of a variable area each time the stock are shifted does not compensate, in itself, for the seasonal fluctuation of grass growth. Whether the area allowed is *fixed or variable*, one of the methods of compensation previously outlined must be employed.

Rationed grazing compels the grass to work with a low productivity

Neither from the plant's nor from the animal's point of view does it come to the same thing to allow a herd:

(*a*) 1·25 acres [*50* ares] of grass 6 in. [*15* cm.] high and supplying a total of 5250 lb. [*2400* kg.] green grass (assuming a yield per acre [per hectare] of 4200 lb. [*4800* kg.] harvestable grass).

(*b*) 3·75 acres [*150* acres] of grass 2½ in. [*6* cm.] high *also* supplying a total of 5250 lb. [*2400* kg.] green grass (which means that the yield of harvestable grass per acre [per ha.] is only 1400 lb. [*1600* kg.]).

From the grass's point of view it has been stated that when it is grazed at this insignificant height it can only partially renew its reserves and certainly cannot produce its "blaze of growth". In the month of August, case A will correspond probably to 36 days rest and case B to 18 days rest. On looking back to Fig. 4, p. 16, it is evident that in compensating for the insufficient re-growth of the grass by increasing the area allowed, one is working with a daily grass production of 80 lb./acre [*89* kg./ha.] against 120 lb./acre [*133* kg./ha.] when the compensatory methods of rational grazing are employed and optimum rest periods observed.

With the method of varying the grazing area allowed, neither the require-ments of the grass nor those of the cow are being satisfied.

Short grass on an increased area does not allow the cow to harvest such a large quantity of grass

In studying the work done by Professor Johnstone-Wallace it was seen that the area grazed could be extended (provided it comprised a similar herbage) without the cow harvesting more grass for, like a good trade-union member, she refuses to work any extra hours (pp. 68 and 78). It is known also that a cow harvests much less grass from a sward $2\frac{1}{2}$–3 in. high [*6–7* cm.] (case B above) than from a sward 6 in. [*15* cm.] high (case A above). **There is no point, therefore, in tripling a sward of this low height: it will not** (or at least only to a slight extent) **increase the quantity of grass the cow harvests,** and certainly the effects on her will be almost imperceptible.

But that is not the most serious feature: the consequences may be disastrous, as will be seen.

Rationed grazing with a variable allocation of grazing area can endanger the health of the cow

If, without waiting for sufficient re-growth, one makes twice, three or four times the area of very young grass *only* available to the animal, one is, in effect, back at what has been called the Schuppli system, by which the cow *always and exclusively* eats grass $2\frac{1}{2}$–3 in. [*6–7* cm.] high. The originator of the system himself admitted that it was impossible for a cow on such a diet to ruminate. Even assuming that this very young grass does not reach the limit of preventing rumination (which is obviously fatal if not remedied im-mediately) it will considerably increase the chances of bloat and tetany.

Ploughing up of pasture, rationed grazing and grass tetany

In studying grass tetany (p. 124) it was seen that the exclusive (or almost exclusive) practice of rationed grazing on temporary pasture led to a con-siderable increase in grass tetany on certain English farms, especially if it was accompanied, as is sometimes the case, by the thoughtless and badly spaced application of nitrogenous fertilisers.

X's farm in Lincolnshire springs to mind as an example in which all three factors were combined: exclusive use of temporary pasture, rationed grazing system with allocation of variable grazing area, bad apportionment of large quantities of nitrogen. I was lunching with him in the month of June, and before we sat down at table Mrs. X, the charming lady of the house, asked her husband a question, the sense of which I did not immediately understand: "The equipment for the intravenous injections is ready. It is hot to-day; and do you remember that the last case happened at lunch-time?"

I asked what all this was about and was told that the equipment was for giving intravenous injections of magnesium salt. Mr. X added: "This is only June and I have already had 21 cases of grass tetany this year which means that a fifth of the herd has already been affected. Luckily, through acting in time, I have only lost two".

This is only one among many cases that I personally encountered in many countries.

If electric fencing is used it must be set up at the same place in every rotation

All these considerations lead one to believe that if the grazing area is delimited by electric fencing, small marking stakes should be used to see that the wires are always put in the same places. Otherwise, involuntarily and in spite of oneself, one will be tempted to allow a varying size of block to compensate for a probable scarcity of grass. This would satisfy the demands of grass and stock less and less, the outcome would be untoward acceleration which, in the end, would lead to total absence of grazing in summer, not to mention the risk of serious disorders among the animals consuming such very young grass *exclusively*.

Where rationed grazing is on a variable area and no area already grazed is allowed, the herd cannot be moved frequently

I have taken the *theoretical* case of a variable grazing area with a wire in front and a wire behind. The herd has access *only* to an area of *fresh* grass without any adjoining area of pasture that has already been grazed.

What was intended in this system, which one often calls *strip grazing*, was to imitate tethering and to try to shift the electric fence (in this case, two electric fences) several times a day. In practice, this is very difficult, for the cows are not individually tethered and restricted and *the concentration of stock is much too great*. If, in fact, it is estimated that 60 sq. yd. [*50* m.²] must be allocated per animal unit per day and the electric fences are moved three times a day, this means 20 sq. yd. [*16* m.²] per shift per beast. Imagine a small paddock 1110 sq. yd. [*900* m.²] in area on which $\frac{1110}{20} = \left[\frac{900}{16}\right] = 56$ beasts are grazing. In all probability a terrific battle will be raging. Obviously the concentration can be higher if the animals are of a peaceful disposition and dehorned. Nevertheless, excessive concentration always presents great difficulties.

And so, in practice, it is wise:

(*a*) not to move the stock oftener than once per day;
(*b*) to divide the herd up into groups (*vide* p. 154).

It should be noted that an adjoining area that has already been grazed is generally incorporated in the system, as will be seen in the following chapter.

Chapter 5

RATIONED GRAZING WHERE THE ANIMALS HAVE ACCESS TO AN AREA THAT HAS ALREADY BEEN GRAZED (ONE GROUP ONLY)

Area of new grass and area already grazed

FIG. 14 illustrates the basic method of shifting the two electric wires to allow the stock simultaneous access to a *fresh* and to a *grazed* area.

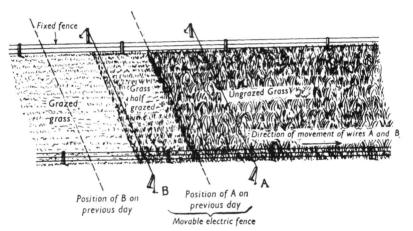

FIG. 14. Paddock with rationed grazing (with a portion already grazed), when the electric fences have just been moved.

The original idea in this system was apparently to:

(*a*) Obtain a more balanced ration, the cow being forced to graze a more fibrous portion (which had already been grazed) at the same time as the new section of fresh grass.

(*b*) Reduce the concentration of the herd.

The second point is perfectly correct: it has already been stated that "de-concentration" of the herd is absolutely essential if the wires are to be *shifted several times a day*. But as far as the first point is concerned, the result obtained is *exactly the same* if the animals are forced to graze to the

229

ground the area they have been allotted. I feel that this has been forgotten to a large extent.

Access to the watering point

In practice, this system of rationed grazing with an adjoining area that has already been grazed has developed so that very large pastures can be divided, and more particularly, so that, where there are insufficient paddocks in a rotational system, these can be split up into smaller sections.

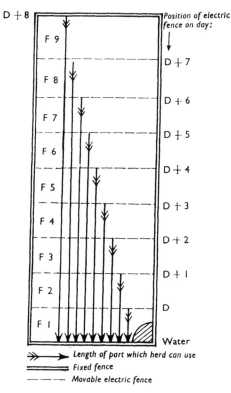

Fig. 15. Method of moving a single *forward* electric fence in a large paddock so that the animals always have access to water.

One of the obstacles arising in a rotational system when division is carried to the extreme is very often access to the watering point (p. 258). For this reason only one electric wire at the front has been used, the animals being able to return to the watering point. This is the case in Fig. 15, which shows a large pasture grazed in 9 days, the wire having been moved on each day.

Periods of occupation and rest are not the same for all sections of the pasture

It is evident from Fig. 15 that the portion F 1 will be occupied for 9 days and F 9 for only 1 day, which is 8 days less. It was thought that wastage, and particularly over-grazing, would be avoided if the animal was allowed only a half-day's (often even a third of a day's) ration. But due to the concentration of stock, they must be allowed access back *to an area which has already been browsed and is terribly over-grazed.*

If the rest period of F 9 is T days, that of F 1 will be $(T - 8)$ days. *Portion F 1 situated near the watering point will therefore suffer the dual disadvantage of* TOO LONG *a period of occupation and too short a period of rest.* In addition, it will be badly damaged by the trampling of the animals going for water. The intervening portions will suffer from the same disadvantages, in proportion as they are nearer to or farther away from the watering point.

A serious difficulty is that it is impossible, due to the necessity of access to the water, to reverse the order of grazing. The result is an increased, cumulative effect on the sections near the watering point which greatly reduces their production, as will be seen later (Table 51, p. 236).

Subdivision of large pastures by means of one electric wire in front is extremely widespread

In view of the failures experienced with untoward acceleration in the Hohenheim system, attempts were made to re-divide the paddocks. This "rationing" was greatly facilitated by the advent of the electric fence. Rationed grazing with *one single wire in front* is what is most frequently encountered.

An excellent photograph from the English farming paper, *The Farmers Weekly*, is reproduced in Photo 17, facing p. 211. All the area that has previously been grazed remains at the disposal of the animals so that they can get water from the watering point near the trees at the back. The variations in shade clearly reveal the strips over which the cattle have advanced.

Photo 18, facing p. 211, is of a rationed grazing on a Leicestershire (England) farm. Only one electric wire in front is used, the animals being able to return to the watering point near the hedge. The wire is moved on once daily, the whole pasture being grazed in 15 days or so. This means that the *period of occupation of the first section near the watering point is 16 days*, the rest period being reduced in consequence. The photo shows the cows going back to graze the grass that has just re-grown as soon as the allocated area has been scraped bare. The young grass has accumulated hardly any reserves, and its daily growth yield is very poor.

Rationed grazing with one wire in front is suitable only if there is no return

This procedure, which has unfortunately become so popular, presents no difficulties if there is no "return" (at least, not very soon), as is the case with the tethering systems already referred to, on crimson clover, for example. The cows can go back to graze the crimson clover that has grown again, a matter of no importance as there is no other rotation to follow and the clover will be ploughed in after grazing.

The method is equally suitable for marrow-stem kale, which is obviously ploughed up after being eaten. Care must be taken, however, to cut down the kale with a bill-hook along the line of the wire each time the latter is shifted so that the wire does not come in contact with the kale, which is much higher than ordinary grass.

Degradation of the flora by rationed grazing using only one wire in front

Although the photographs reproduced of this system are English, many could have been supplied by France.

In the mid-East area of France I saw a trial which was considered to be perfect rationed grazing. It consisted of a large pasture, several acres in extent, with a fixed fence all round. An electric fence had been installed which was shifted forward twice daily, and the *whole* of the area that had already been grazed was left open to the stock to give them access to the watering point. At the time of my visit the first part of this large pasture had been grazed some 20 days previously, and the grass on it was re-growing vigorously. As a result, a large number of animals were grazing this old part while the others were busy on the fresh grass area.

What stood out very clearly was the marked deterioration of the flora on the over-grazed, over-trampled section the period of occupation of which was in the region of 20 days, while the rest period was insignificant. When I remarked on this, I was told that it was of no great importance as the pasture was to be ploughed in next year and re-seeded, the regular practice every four years.

But ploughing will not change the position of the watering point; and if this defective system of management is not amended the same damage will appear after the ploughing, only it will be more severe as the effect is cumulative (*vide* p. 271).

An access corridor to the watering point is essential

So that periods of occupation and rest periods may not be unfavourable to the paddocks situated near the watering point, *a corridor giving access to the latter is absolutely essential. Two electric wires must be used instead of one*, the back wire preventing the animals from returning to the area that has already been grazed.

This is what is shown in Fig. 16, where the corridor is made by fixed fencing and gates, these being considered preferable.

If excessive concentration of the stock is to be avoided, it is difficult to move the electric fences oftener than once a day. The herd could always be divided into two groups, but then, as will be seen in studying pasture division (Part Eight, Chapter 2, p. 258), it would be necessary to have two corridors if one was insistent that the groups have *constant* access to the watering point.

Personally, I prefer corridors with fixed gates, but they may equally well be made with electric fencing, as the following example shows.

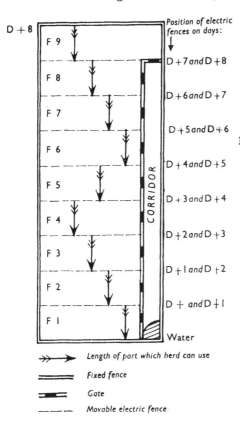

FIG. 16. Method of moving *two* electric fences, front and back, in a large paddock with a fixed corridor giving access to water.

Heine's schemes for rationed grazing with one group

Experiments on rationed grazing have been carried out by Heine at Rengen and Dikopshof (Germany). Recently I had the opportunity to visit these pastures and discuss their management with one of the research workers involved. The schemes and results of these trials (39) described below illustrate some of the variants of rationed grazing with their advantages and disadvantages.

The system employed was rationed grazing (Rationsweide), the area allocated varying with the requirements of the animals in the herd and the quantity of grass available per unit area.

Grazing either continued all day in the normal fashion (Vollweide) or was restricted to part of the day (Teilweide). The animals spent the rest of the day in their stalls, where green feeding was practised.

On the farm at Dikopshof, six grazing passages were achieved during the year, the rest periods being varied. Production at different times of the season was relatively balanced thanks to high rates of nitrogen application (up to 168 lb./acre [*188* kg./ha.] for the year) and to irrigation. In spite of all this, however, the area allocated had to be varied. Here is how the paddocks were divided.

Rotational grazing paddocks (with fixed fencing) were used and divided up by means of two movable electric fences: the front wire, which the experimenters called the "grazing wire" (Fressdraht), and a back wire, called the "rest wire" (Ruhedraht), limiting the area. The paddocks with fixed fences were 4–6 in number, the figure varying with the time of season. The front wire was moved only after 3–4 days so as to avoid overcrowding of the animals and to allow return to the corridor leading to the watering point.

Heine writes:

> "Where the width of a paddock does not exceed 66–88 yd. [*60–80* m.] two movable wires are shifted (*Plan I* of Fig. 17). If the paddock is wider than 88 yd. [*80* m.] the shifting of electric wires would involve too much work, and it is preferable that subdivisions be made with semi-permanent electric fencing."

Plan II of Fig. 17 shows how semi-permanent electric fences A and B form a sub-division. In addition, the semi-permanent electric fence C forms a corridor running to the watering point. Each of the subdivisions of a paddock is grazed by advancing the front wire three times every day. *The back wire is not used.* The result is that the first third is occupied for 3 days, the last for only 1 day.

Plan III of Fig. 17 represents the lay-out I saw at Rengen. The system was being applied on a pasture of 17·5 acres [*7·00* ha.] being grazed by yearlings. The paddock to be grazed is divided in two by the *fixed* fence A. Each of the halves in turn is divided in two by the *semi-permanent* electric fence B. The strips thus obtained are grazed as before, with one *movable* electric fence providing the front wire. There is no back wire, and each of the subdivisions is grazed within 3 days, the first third near the corridor being occupied for 3 days as opposed to 1 day for the last third farthest away from the corridor.

Fall in production on the sections longest occupied

Of particular interest in these experiments of Heine's is the fact that he was able to measure differences in yield due to the prolonged action of the animal's teeth and hooves on one section of the paddocks: the section, moreover, with a shorter rest period. This is shown in Table 51.

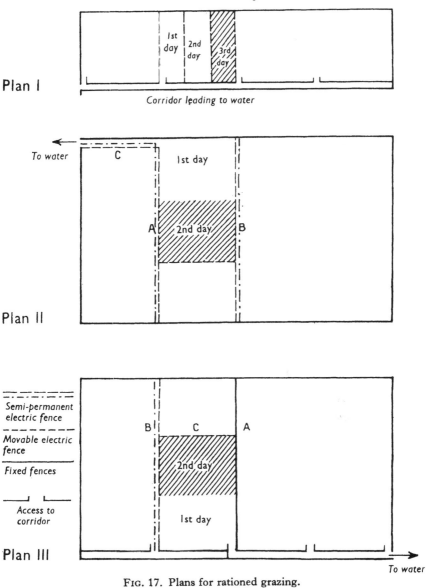

FIG. 17. Plans for rationed grazing.
From Heine (39).

The first half of this table corresponds to the plans in Fig. 17 above, the differences in the occupation and rest periods being only 2 days or so. Despite this slight variation, the effect of which was cumulative on each rotation, the reduction in the yield of the sections at a disadvantage is very marked.

TABLE 51

Yields from rationed grazing of the fractions of the same paddock with the lowest and shortest period of occupation

Strip	Period of occupation of fraction considered	Period of rest	Rainfall		Yield of fresh grass from one rotation	
	(days)	(days)	in.	[mm.]	lb./acre	[kg./ha.]
No. 2 . {	1	38	2·77	[70·5]	5530	[6200]
	3	35	2·77	[70·5]	3390	[3800]
No. 3 . {	1	38	2·77	[70·5]	2850	[3200]
	3	35	2·77	[70·5]	2320	[2600]
H. 3 1st rotation {	9	37	2·72	[69·0]	4120	[4620]
	12	34	2·86	[72·7]	2430	[2720]
H. 3 2nd rotation {	3	30	0·81	[20·7]	920	[1030]
	6	27	0·81	[20·7]	380	[430]

From Heine (39).

Observance of the rules of rational grazing is particularly important in periods of drought

The second part of Table 51 is concerned with another experiment on the same basis, but where the variation in periods of occupation and rest was 3 days. *It is particularly notable that these cumulative, unfavourable effects on the part longest occupied are accentuated when rainfall is very slight.* In fact, when the period of occupation is systematically and repeatedly extended by 3 days (the rest period being shortened by 3 days) production, where the rainfall is about 2·77 in. [70 mm.], falls from 4120 lb./acre [4620 kg./ha.] to 2430 lb./acre [2720 kg./ha.], a drop of 40%. But when the rainfall is 0·8 in. [20 mm.] and the period of occupation is extended by 3 days, production falls from 920 lb./acre [1030 kg./ha.] to 380 lb./acre [430 kg./ha.], a drop of 58%.

As has often been emphasised, therefore, **the rules of rational grazing must be particularly observed in periods of drought**: it is then that they give the greatest return.

An observation by Professor Klapp on rationed grazing

Heine carried out his experiments under the direction of Professor Klapp who has the following observation to make (70), p. 426:

"The conduct of rationed grazing is not unaccompanied by difficulties and disadvantages, especially for the grazing animals: grazing very quickly and often even in haste, poor appetite for the rations low in protein they are fed in the stall, etc.

"Some authors have even reported a reduction in milk yield and in fat content. . . ."

Chapter 6

RATIONED GRAZING WITH TWO GROUPS

Rationed grazing with two groups is quite common

RATIONED grazing with two groups is common. It is noteworthy that although the system may assume many different forms, it is always practised according to the same general principle (for reasons which I have not been able to ascertain). That principle is set out here as it has been explained by Staehler, a very able German grassland specialist, who, in 1951, published a hand-book on the rotation of pastures (99a). This statement of the principle is followed by a typical case such as I have often encountered in France and which applies Staehler's principle in fact.

The most common principle of rationed grazing with two groups

The principle is illustrated by Fig. 18, p. 238. Staehler describes the system as follows:

On the first grazing day under consideration (D) the first group grazes the first third of paddock No. 2. No area already grazed is available as an exercise park. The second group grazes paddock No. 1 for 3–4 days, it having been grazed previously by the first group.

On the second grazing day ($D + 1$) the first group has access to the second third of paddock No. 2 as fresh grass and the first third of the same paddock which was grazed the day before. The second group continues to graze paddock No. 1.

On the third grazing day ($D + 2$), the first group grazes the last third of paddock No. 2, and has still all the rest of the paddock available to it for exercise or to finish scraping. The second group continues to graze paddock No. 1.

On the fourth day of grazing ($D + 3$) the first group grazes the first third of paddock No. 3. Just as at the beginning of grazing paddock No. 2, the first group has no access to a grazed area for exercise purposes.

For 3–4 days the second group grazes paddock No. 2, which had been grazed on the days D, ($D + 1$) and ($D + 2$) by the first group. Paddock No. 1 is rested and nitrogen applied.

It must be borne in mind that the allocation of fresh grass to the first group

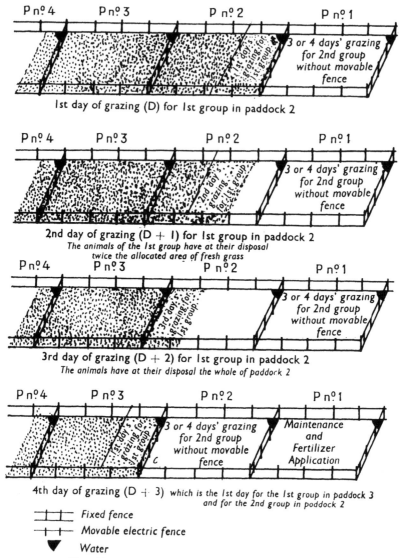

P nᵒ 4   P nᵒ 3   P nᵒ 2   P nᵒ 1

3 or 4 days' grazing
for 2nd group
without movable
fence

1st day of grazing (D) for 1st group in paddock 2

3 or 4 days' grazing
for 2nd group
without movable
fence

2nd day of grazing (D + 1) for 1st group in paddock 2
The animals of the 1st group have at their disposal
twice the allocated area of fresh grass

3 or 4 days' grazing
for 2nd group
without movable
fence

3rd day of grazing (D + 2) for 1st group in paddock 2
The animals have at their disposal the whole of paddock 2

3 or 4 days' grazing
for 2nd group
without movable
fence

Maintenance
and
Fertilizer
Application

4th day of grazing (D + 3) which is the 1st day for the 1st group in paddock 3
and for the 2nd group in paddock 2

‾‾|‾|‾|‾ Fixed fence
—+—+— Movable electric fence
▼ Water

Fig. 18. Rationed grazing in the case of 2 groups.
From H. Staehler (99A).

is not necessarily one-third of a paddock daily: this is rationed grazing with a *variable* allocation of area, the proportion varying according to the vigour of grass re-growth.

As the author has conceived the system of grazing, the first third of paddock No. 2 (and therefore of every paddock) is *occupied* for 6–7 days. It should be noted, moreover, that the last third of paddock No. 2 (and likewise of every paddock) will still be occupied for 4–5 days, which, for a system involving daily shifts is considerable, although less by 2 days than the period of occupation of the first third of each paddock.

The back-run on each paddock is made necessary by the position of the single watering point in each; this will be seen again in an actual example of the system.

The herd advances thrice daily and returns every 32 days in summer

Before a conference I toured the North-East of France accompanied by two agricultural advisers who had read some of my published work and were aware of the significance I attach to the "time" factor. They told me they were going to take me to see a farm on which rationed grazing with *two groups* had been perfected, the stock being *moved twice per day*. They added that the question of rest periods had not been lost sight of, and that, in summer, the herd returned every 32 days, which was quite sufficient in that area of plentiful summer rainfall. Obviously, my curiosity was greatly aroused.

The visit to these pastures was to show me that the fact that the herd returns to a paddock every 32 days did not mean that the grass had a right to approximately 32 days rest or anything like it.

The electric fence returns every 32 days, but the rest periods are only 16 days

Four pastures were involved, all opening on to a river from which the animals derived their water (Fig. 19). The electric fence was moved twice a day, although the farmer complained of the amount of work this entailed. The stock could run back to the river, along the side of which was a race with a gate which was only closed when the animals were being moved but remained open the rest of the time so that they always had access to the watering point.

The first group took 8 days to graze one of the pastures (which will be referred to here as a paddock). As the wire was moved twice daily, 16 portions were grazed in succession. When the first group (comprising milch cows) had finished grazing a paddock the **whole** *area, without restriction, was made available to the second group*, exactly as in the preceding example in Fig. 18.

Fig. 20 shows the position of the two groups on the day $(D + 13)$. By this method of grazing a portion next to the river was occupied for

$$8 \cdot 0 + 8 \cdot 0 = 16 \text{ days}$$

and the portion farthest away from the river for

$$0 \cdot 5 + 8 \cdot 0 = 8 \cdot 5 \text{ days.}$$

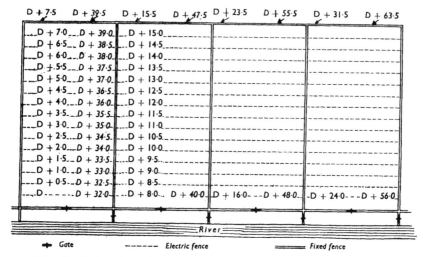

FIG. 19. Rationed grazing with two groups, the first moving twice a day and the second every eight days. (The figures indicate the position of the electric fence *in front of the first* group on the day (D + x).

(*Roman lettering: first rotation. Italic lettering: second rotation.*)

The electric fence certainly returned to the same position after 32 days, but the rest periods were not 32 days. In fact, they were:

(*a*) For a portion next to the river:

$$(D + 32 \cdot 0) - (D + 16) = 16 \text{ days}$$

Day of arrival   Day of departure

(*b*) For the portion furthest from the river:

$$(D + 39 \cdot 5) - (D + 16 \cdot 0) = 23 \cdot 5 \text{ days}$$

Day of arrival   Day of departure

The portion most favoured, therefore, had a rest period of 23·5 days, which is just right, for a good summer; but the portion at the greatest disadvantage rested for only 16 days, which is clearly insufficient except in the spring period of *vigorous* growth.

In fact, in spite of the refinements of advancing his beasts twice per day and returning the electric fence after 32 days, the farmer had fallen foul of untoward acceleration because, instead of grazing a paddock for 8 days with the first group, as intended, he had to reduce this time to 6 days at the end of June, then to 4 days, etc., etc., *at each shift allocating a larger and larger grazing area* (the period of stay in the second group was naturally reduced in the same proportion).

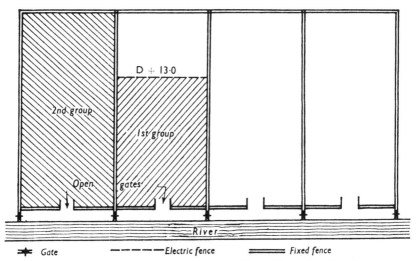

FIG. 20. Position of the two groups on day (D + 13) with rationed grazing where the electric fence is moved twice a day.

To put it briefly, as in all rotations or rationed grazing where the "time"
factor is neglected, the farmer ran short of grass in summer, although his area
is famed for its abundant summer rain. It might be added that the flora of the
portions near the river had been seriously damaged.

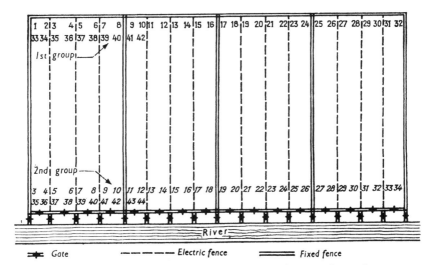

FIG. 21. "Rational" grazing with groups where each moves forward every
two days.

(The figures indicate the dates (D + X) during which each group occupies
each fraction of the pasture.)

The yield is twice as high when the stock are shifted every 2 days as when they advance twice daily

I remarked to the farmer, to his joy, that it would be sufficient for him to use three or four electric wires and shift one or two of them every 2 days, instead of four shifts in 2 days.

In addition to simplifying his task he would get a considerable increase in yield. The only additional expense he would have would be for some additional gates to the corridor. The scheme I outlined for him is reproduced in Fig. 21.

It is always assumed that division within the paddocks is by means of electric fencing. But to allow the two groups permanent access to the river,

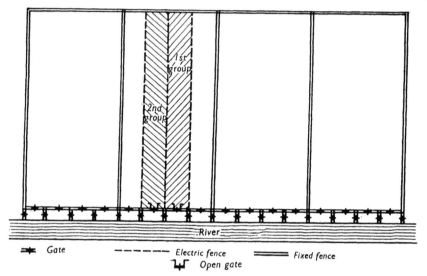

| ⊣⊢ Gate | – – – – – – Electric fence | ══════ Fixed fence |
| | ⊔ Open gate | |

FIG. 22. Position of the two groups on day (D + 13) in a "rationally" grazed pasture where the groups move forward every two days.

gates have had to be erected leading on to the race while other gates divide up the race itself (*vide* Fig. 26, p. 260).

On the day (D + 13) the position of the two groups was as indicated in Fig. 22.

By advancing each group *every two days* in this way, the following results are obtained:

(*a*) Each portion is occupied for *four* days as opposed to 8·5 and 16·0 days for the most favoured and least-favoured paddocks respectively under the old system.

(*b*) The rest period of each portion is:

$$(D + 33) - (D + 5) = 28 \text{ days}$$

Day of arrival   Day of departure

against 23·5 and 16 days respectively before.

PHOTO 21

Stone walls dividing grazing land into strips on Farm B (Derbyshire, England)

Photo Voisin

PHOTO 22

Stone wall dividing pastures on farm B with, in the background, a megalithic stone of pre-historic times

Photo Voisin

Observance of the time factor should more than double the yield of rationed grazing where the occupation periods are unbalanced

It is immediately obvious that this enormous reduction in the period of occupation and extension of the periods of rest will increase the yield considerably while simplifying the work and making it less arduous. Summer grazing will therefore be much easier to manage, and it is improbable that grass will become scarce if the principles of compensating for seasonal fluctuations are observed, particularly in this region, where summer rainfall is plentiful.

The above example demonstrates once again the danger of being hypnotised by words and imagining that one is practising a "greatly improved rationed grazing system" because the electric fence is shifted thrice daily.

As soon as one fixes one's attention on the basic factor in all rational grazing, namely, time, everything appears in a very different light and yields increase rapidly.

Chapter 7

GRAZING RATIONED IN TIME

An old controversy: stall feeding or grazing?

IN all ages the question has been asked whether it is better to graze stock or feed them in the stall (or in a compound) with cut green forage. The same questions were being asked and the same hesitation shown in supplying the answers 150 years ago. I have already mentioned the names of three great French agronomists at the end of the eighteenth century: Tessier, Thouin and Bosc (p. 154). In their *Methodical Encyclopædia of Agriculture* (107, vol. VI, published in 1816) they write:

"Men are not agreed on the question of whether it is better to put stock into the meadows to graze or feed them green inside. Grazing animals harm a pasture:

"1. **By eating the grass before it is mature and consequently retarding its re-growth.**
"2. By tearing up or crushing certain areas and thus killing them.
"3. By making the surface uneven with their trampling.

"These disadvantages are compensated for by the economy of leaving animals out at grass, by *their good state of health*, the better tasting meat they provide, the firmer fat and the milk with a higher butter content, etc. Moreover, these can be diminished by taking precautions which have already been, and will again be discussed.

"The number of farmers, therefore, who mow their grass and feed it green in the stall is very limited, although there is no doubt that this method reduces wastage, speeds up fattening, increases milk yields and the supply of manure, etc. I am, however, of the opinion that green feeding should be employed only in special circumstances, unless one has only a few beasts or the need for manure is imperative.

"**Nevertheless I do not wish animals to be denied grass in their stalls** either in the evening, or **to supplement the insufficient quantity of food they have found in the pastures,** or on wet days, when they are ill, or after parturition, etc. . . ."

The passages in bold type call to mind the Universal Laws of Rational Grazing.

244

Grass Re-growth

The authors write: "Grazing animals harm a pasture **by eating the grass before it is mature and consequently retarding its re-growth.**"

Here, in a nutshell, are all the principles laid down in Part One. But what is of interest from the point of view of this present chapter is the realisation that as they were unaware that this fault in continuous grazing could be rectified by the methods of rational management, the writers had recourse to the mower and to green feeding in the stall.

Quantity harvested and quantity eaten

At the end of the above quotation the authors express in simple terms what measurements, especially those of Johnstone-Wallace, were to translate into figures one hundred and twenty years later (p. 76). It has been said above that if a sward is not very high and is very open, the cow **would harvest** small quantities of grass, in the region of 77–88 lb. [*35–40* kg.] while if the same grass was cut and fed green she would eat 154–176 lb. [*70–80* kg.] perhaps more. Now the writers of this *Encyclopædia* state that soiling may be necessary to "supplement the insufficient quantity of food the animals *have found* in the pastures". They thus confirm that where the grass in a pasture is short and not very plentiful, the animal *harvests* only a small quantity of food, which may with advantage, if not indeed of necessity, be supplemented by soilage.

Health of the grazing animal

Mention is also made of the good state of health of the grazing animal. There is no doubt that life in the open air, the effects of the sun's rays, etc. all contribute thereto. But neither must it be forgotten that grass *under the hoof* possesses many substances (antibiotic, hormones, œstrogens, etc.) which are *very rapidly destroyed* as soon as it is cut. *From the strictly nutritional point of view*, therefore, it does not by any means come to the same thing whether the cow is *fed* cut grass or made to *harvest* that grass in the field.

Probably farmers who feed their beasts in the stall have also felt that grazing is necessary for the animal's good health, as revealed in the appearance of their coats or the percentage losses. The Breton farmers, who are greatly addicted to cattle feeding in the stall, as was seen above, certainly take care to let them roam about a little on grassland situated near the farm buildings. This exercise, small though the amount of grass harvested may be, seems to them to be indispensable for the cow's health (*vide* p. 99).

This is a compromise to solve the problem: the greater part of the ration being fed as soilage, but a small quantity of grass being harvested by the animal on the hoof. Other forms of this compromise, combining grazing and stabling or, to be more exact, grazing and green feeding are about to be seen.

Stall feeding and grazing combined at the beginning and end of the season

Even during the grazing season in regions where it is mainly grazing that is employed, there are periods when stall feeding and grazing are combined: that is, at the beginning and end of the season.

Supplementary feeding for high yielders

The maximum quantities of milk a cow can produce from grazing alone are listed above in Table 36 (p. 91). It is obvious that the yielding capacities of many cows exceed the possibilities afforded them by the grass they harvest. For this reason many farmers have been in the habit of always feeding concentrates to high-yielding dairy cows. This is generally done at milking time, but in the case of champion yielders concentrates may also be fed in the course of the day, as is advocated by Professor Boutflour.

The ideas of Professor Boutflour

Professor Boutflour, lately Principal of the Royal Agricultural College, Cirencester (England), is famed for the originality of his ideas, which he defends with much talent and conviction. One of these is that the cow is a machine destined to transform concentrated foodstuffs into milk, and that the feedingstuffs produced on the farm, such as grass or forage crops, are merely supplements whose duty is to ensure that the concentrates are adequately digested, while at the same time helping to maintain the animal's health.

In this way Professor Boutflour has succeeded in getting cows to yield up to 15 gal. [*68* litres] of milk per day, with lactations of 3000 gal. [*14,000* litres] of milk at 3·45% in 305 days.

There are obviously many aspects to this technique: special composition of the concentrate, "steaming up", prepartum milking and special method for drying the cows. Interesting as the details are they cannot be gone into here: we can only see how the grass for a cow producing 15 gal. [*68* litres] is *rationed*.

Rationing grass for a cow yielding 15 gal. [*68* litres] of milk per day

The grass is rationed by reducing the grazing time. In practice, grazing takes place only once per day, namely at the time of the large morning meal, which, as was seen above (Fig. 7, p. 70), represents the largest quantity of grass harvested in the day. At this time the cow is put out on to a pasture where plenty of grass is available and is thought to harvest an average of 26·4 lb. [*12* kg.]. The rest of the day when she is not in the milking parlour (or in the stall eating concentrates) she is out on an almost completely bare sward which serves as an exercise compound. Her appetite having already been satisfied to a large extent, and in view of the height of the grass, it is considered that the animal does not nibble any more than 6·6 lb. [*3* kg.], making a total daily grass harvest of 33 lb. [*15* kg.].

It is estimated that a cow producing 15 gal. [*68* litres] of milk per day must absorb, in addition to this 33 lb. [*15* kg.] of grass, 60 lb. [*27* kg.] of a special concentrate mixture, the composition of which cannot be studied here. When I visited the Royal College at Cirencester, *Beauty*, the highest milk yielder in England, was giving 15 gal. [*68* litres] per day (Photo 19, facing p. 242) and was therefore eating, in accordance with Professor Boutflour's calculations, 60lb. [*27* kg.] of special concentrate in five equal rations of 12 lb. [*5·4* kg.] each, in addition to the 33 lb. [*15* kg.] of grass she was harvesting. Her daily timetable, described to me by Professor Johnstone-Wallace, is to be found in Table 52 below.

TABLE 52

Mealtimes for a cow at grass giving 15 gal. [68 litres] a day

Time	Milking	No. of meal	Concentrates	Grazing
5 a.m.		1	12 lb. (5·4 kg.)	
5.30 a.m.	Milking			
6 a.m.		2		*Abundant* pasture
7 a.m.		3		Bare pasture
10 a.m.		4	12 lb. (5·4 kg.)	
11 a.m.		5		Bare pasture
1 p.m.		6	12 lb. (5·4 kg.)	
1·30 p.m.	Milking			
2 p.m.		7		Bare pasture
5 p.m.		8	12 lb. (5·4 kg.)	
6 p.m.		9		Bare pasture
8.30 p.m.		10	12 lb. (5·4 kg.)	
9 p.m.	Milking			
9.30 p.m.		11		Bare pasture

From verbal information received from Professor Johnstone-Wallace.

The then Vice-Principal of the College, Kenneth Russell, provided the following illustration of this method of feeding:

The first two meals (1 and 2) represent breakfast, the concentrate being the bacon and eggs, the grass from a lush pasture the porridge. The second concentrate ration (4) is the mid-morning tea, which is never omitted in English colleges; and the third concentrate ration (6) corresponds to lunch. The cow has tea at five o'clock in the afternoon, the fourth concentrate ration (8) and the day ends at 8.30 p.m. with supper, represented by the fifth and last concentrate ration (10).

The bits of grass she nibbles on the bare pastures are similar to the sweets sucked in the course of the day.

It is quite obvious that if the grass was not *strictly rationed* the cow would be incapable of absorbing 60 lb. [*27* kg.] of concentrates. It is equally obvious that without the minimum of grass she is harvesting, the cow, among other things, would be unable to ruminate.

This system certainly gives rise to hundreds of questions which cannot possibly be examined here. Let it only be said that this very strict *rationing* of grazing is indispensable if the colossal yield of 15 gal. [*68* litres] per day is to be achieved. Of course, it remains to be seen what the influence of this feeding system is on the health of the cow.

Grazing restricted to the cool period of the day

The problem of rationing in time has yet another side to it. Observation of the cow at grass has confirmed that she completes all her harvesting within a limited time, which, according to the region, atmospheric conditions, etc., varies between 7 and 9 hours. It has been questioned, therefore, if it would not be preferable to leave the animals on the pasture for just as long as is necessary for them to harvest their grass. The advantage envisaged in this practice is a reduction in the trampling of the sward. Moreover, the cow would be out at grass only during the best periods of the day: morning and evening in hot weather and in the middle of the day when the weather is cold.

I was able to observe the following example in Massachusetts (U.S.A.).

Rationing time on a Massachusetts farm (U.S.A.)

Weiko Holopainen, one of the best graziers I met in the U.S.A. (Voisin 117), vol. I, p. 71, had the following timetable for his cows:

5.30	Milking
7.30 to 8.30	Grazing
8.30 to noon	Rest park
12.30 to 1.30	Grazing
1.30 to 5	Rest park
5	Milking
6 to 7.30	Grazing
night	Rest park

There are three grazing periods representing a total of $4\frac{1}{2}$ hours. The greater part of the cows' time is spent in the rest park, also called the "loafing area", which is an area planted with trees and equipped with a watering point. The cows can rest there in the shade and ruminate (p. 173).

I was to encounter similar systems of rationing in time (with a rest park) in all the hot regions of the U.S.A. The mechanisation of green feeding these cows in their exercise park (or in the stall) was dealt with above (p. 173) and illustrated in Photos 12 and 20.

Economic circumstances

Not only technical and climatic factors, or the question of stock health, determine to what extent grazing time will be rationed. Economic factors are also concerned: namely, the relative prices of concentrates, machinery, manual labour and petrol.

PART EIGHT
DIVISION OF PASTURES

Chapter 1

A GENERAL PROBLEM

What must the area of a paddock be?

THIS is the question generally asked by beginners in rotation and my answer is: "I do not know"; to which should be added that it is not the important point in a rotation.

First of all, one must decide on the number of paddocks. How many paddocks is it possible to have without too many complications? Or, what sacrifices is one willing to make to have the highest possible number of paddocks? These are the first questions to ask oneself.

Number of paddocks is principally a function of period of stay

The problem of determining the number of paddocks was dicussed above in Part Four, Chapter 2 (pp. 147–149), from which the following fundamental point emerged.

Assuming that one wants a rest period, in summer, of 36 days with a period of stay of 1 day, 36 paddocks resting will be required, making a total of 37–39 paddocks, depending on whether 1, 2 or 3 groups are involved (*vide* Table 42, p. 148). If a longer period of stay is considered acceptable, 2 days, for example, it will be sufficient, to obtain a 36-day rest period, to have 18 paddocks resting, making a total of 19–21 paddocks according to the number of groups. Finally, if a period of stay of 4 days is decided on, 9 paddocks resting will be sufficient for a 36-day rest period, making a total of 10–12 paddocks for 1–3 groups.

It must, of course, be remembered that the smaller the number of paddocks, the more difficult it is to manage the rotation comfortably, and especially to balance up fluctuations in production. Moreover, it must not be forgotten that extension of the period of occupation leads to a reduction in grass yield, and extension of the period of stay to a lessening of the animal's output (pp. 18 and 95). The fewer paddocks, therefore, the lower the yields per acre; on the other hand, beyond a period of stay of 2 days, serious difficulties are encountered, which in turn lower the yields.

Having examined the topographical, botanical, animal and financial situations, the farmer will then decide how many paddocks he is going to work with. From the total pasture area to be divided, the average area of the paddocks will be calculated. This, as will be explained, is not the exact area of *each individual* paddock.

Paddock areas are not necessarily equal

It is not a case of having paddocks *equal in area*, but of paddocks producing *equal quantities of grass*.

In practice, if one does not wish to make the conduct of the grazing rotation over-complicated, periods of stay (or occupation) must not vary too much from the basic times anticipated. It was stressed above that flexibility was essential to the management (pp. 177 and 179) and that there must be no hesitation in extending a period of occupation to complete the scraping of a paddock or, on the other hand, of reducing it if a paddock is absolutely stripped before the set basic time is up. There is no need, however, for these variations in time to be either systematic or excessive, otherwise management of the grazing will become very difficult, and there is the additional risk that the Laws of Rational Grazing will be disregarded.

The productive capacity of the paddocks must be equal

Suppose that in the divided pastures there are parts which are not planted and quite open and other paddocks closely planted with apple-trees, the productive capacity of the latter being *half* that of the bare paddocks. If both types of paddock are to furnish similar quantities of grass, the area of the planted paddocks will have to be *double* that of the unplanted. Naturally this does not guarantee equal yields from both types; neither does it guarantee that the grazing stock will receive equal quantities and qualities of grass from both. It is, however, a basic measure which helps to achieve this result.

It is preferable, at the beginning, always to use an electric fence

It is wise, first of all, to draw up a theoretical plan of the divisions having as much regard as possible for differences in the productive capacity of the various divisions visualised. If it is intended to use fixed fences, it can only be recommended that provisional fences be used for the first year or two, during which systematic differences in the times of occupation of certain paddocks will become evident.

A help to the determination of paddock areas with equal productive capacities is the electric fence, about which a few words can be said with advantage.

The value of electric fencing

In view of its cheapness and flexibility, the electric fence allows the number of paddocks to be increased at limited cost. Electric fencing has aided, and is aiding, the development of rational grazing: its advantages are enormous. Unfortunately, however, it is not absolutely certain, and so I think it wise always to make the outside fences (those round the periphery of the pasture) *fixed* and to reserve electric fencing for the divisions. It is not a very serious matter if cows from one group get mixed up with those from another group, or spend the night grazing in a resting paddock. But it is always a serious matter if the cows get into the grain crops, especially if these belong to a neighbour.

After a few years of rational grazing with electric fencing graziers generally instal fixed fences. Even if they are more expensive, they have the advantage (as has just been said) of allowing original paddock areas to be altered and, if necessary, modifying the whole plan of the grazing unit on the basis of the experience acquired. It should be remembered that the electric fence used with sheep is not always a sure method.

Two electric fences must be allowed to each group

When the herd is divided into several groups, it does not mean that one group necessarily follows on behind another. In fact, a characteristic of well-conducted grazing is the frequent transposition of the various groups. It must not be supposed that for 3 groups, for example, it is sufficient to have 4 fences because the 3 groups will follow each other. *Two fences must be provided for each group* which means 6 fences for 3 groups or 4 fences for 2 groups so that the position of the groups can be changed if necessary.

Areas fixed or variable

But if electric fencing has done, and is doing, a great service in assisting the development of rational grazing, it has sometimes discouraged farmers away from the system due to the fact that it led them into error of untoward acceleration of the rotation, as was seen in studying rationed grazing with a variable grazing area (Part Seven, Chapter 4, p. 220): The electric fence itself is not, of course, to blame: the fault lies in an erroneous conception of the conduct of rational grazing. The fact is that, fascinated by the flexibility of the electric fence, graziers try to use it to copy tether grazing and to give their animals fresh grazing areas of varying extent each time they are shifted. This, as has been seen, is perfectly conceivable in a grazing system with *no return* (at least immediately), as in the case of a first-year red clover ley, crimson clover or any other similar, temporary pasture.

Use markers so that the electric fence is always put back in the same position

Where electric fencing is used, therefore, it is essential, as already stressed, **that it should always be put back in the same position, thus**

delimiting the same area. A few pegs are all that is required for this purpose. When new green areas are re-introduced into the rota their area, also, should be marked off with pegs.

What is being studied here are permanent, semi-permanent, temporary swards, etc., where *the return of grazing stock is always regular.*

Various fences

When we speak of movable fencing, we think of the electric wire, but I have seen movable fencing in afforested countries made from *wooden poles* which were shifted as required.

In the same way, when we speak of fixed fences, we think, in this advanced, modern age, of barbed wire. But for our ancestors such division was achieved by embankments, the method advocated in the eighteenth century. These, however, were a waste of space. Another method consists in using stone walls, such as I saw on a Massachusetts farm (U.S.A.). The farmer, a hardy Finnish peasant, Weiko Holopainen, went to the U.S.A. with his father thirty years ago (Voisin, 117, vol. I, pp. 70 and 74). With the aid of a bull-dozer he cleared 75 acres [*30 ha.*] or so of grazing land covered with enormous moraine boulders, saving the smaller of these to make walls for dividing his pasture into fifteen paddocks for rotational grazing. This is a possible solution in a country where space does not count and no one bothers about the amount of land taken up by these wide, dividing walls. The total area they occupy is enormous, but they are obviously very inexpensive, especially with the modern bulldozer and in a country where petrol is cheap.

Skilled labour of the European farmworker

The situation is quite different in Europe, and when dividing walls are built of stone they have to occupy as little width as possible. The stones are therefore broken and built up in a regular pattern, either with or without cement. Among the best examples are those of a Derbyshire farmer, Mr. F. W. Brocklehurst who has half a score of paddocks divided by walls made from broken stones. No mortar or cement is used, but the walls are perfectly straight and even, as is shown in Photos 21 and 22 (facing p. 243). Photo 21 shows paddocks in strips demarcated by these stone walls. In Photo 22 one of these walls is reproduced, together with an enormous megalithic stone set up in olden times on the top of the hill. Here modern grassland management is developing within the magnificent historical framework of the Pennine Chain.

Combination of fixed and electric fencing

If movable (i.e., generally electric) fences are used for the divisions, it is wise to make the surrounding corridors of fixed fencing, if complications are to be avoided (*vide* following chapter).

Gates

If these fixed corridors are to be practical, they must be equipped with gates, sometimes in fair number. A cheap system of providing gates is therefore essential. Photos 23 and 24 (facing p. 258) show the type of gate I myself use and which is generally called a "guillotine gate". Wooden poles or hurdles could, of course, equally be used as gates.

It should be noted that gates indicated in the plans I have drawn up do not remain where they are marked throughout the year. At certain times in the rotation, gates are necessary at the points indicated. The guillotine gate can easily be dismantled and shifted from one point to another.

Shape of paddocks

The shape of the paddocks is determined mainly by the particular conditions prevailing on the pasture in question, examples of which will be given in the two succeeding chapters. Constriction, however, should be avoided, as this gives rise to excessive trampling of the narrow parts.

Divisions corrected by Staehler

Figs. 23 and 24, taken from the work of that excellent German agronomist, Staehler, show how a greater number of less-constricted paddocks can be obtained thanks to a conveniently situated corridor. With elongated and

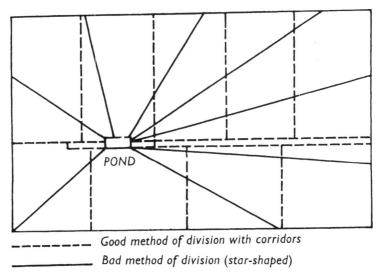

– – – – – – *Good method of division with corridors*

——————— *Bad method of division (star-shaped)*

FIG. 23. Good and bad methods of division of a pasture with water towards the centre (eight paddocks with the star-shaped plan, nine with the corridor plan).

From Staehler (99), p. 15.

constricted paddocks, the number obtained was *eight* (Fig. 23—complete lines). The same length of fencing (dotted lines) inclusive of the double part for the corridor, can produce *nine* paddocks of a better shape.

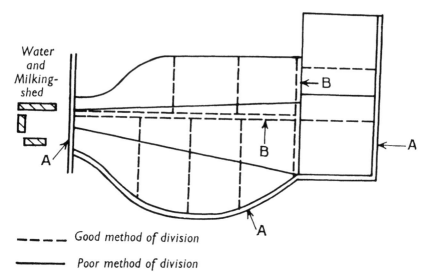

_ _ _ _ Good method of division

———— Poor method of division

FIG. 24. Plan of a pasture divided first into five paddocks of 3 acres [*1·2* Ha.] and then into ten paddocks of 1½ acres [*0·6* Ha.].
From Staehler (99), p. 15.

Fig. 24 shows how, with the requisite corridors, ten rectangular paddocks can be obtained instead of five paddocks, three of which were constricted in shape.

More fencing wire is needed for elongated paddocks

Elongation should also be avoided, as it requires a greater length of fencing wire.

Table 53 shows the lengths of fencing required for five typical possibilities. The square paddocks measuring 100 m. × 100 m. requires 400 m. of fencing, while the paddock 12·50 m. wide will need 1625 m. of fencing. We can also say that a paddock of 1 acre in area can be obtained in the following ways: a square approximately 70 × 70 yd. or a rectangle 40 × 121 yd., or 10 × 484 yd. The square paddock requires 280 yd. of fencing, while the paddock 10 yd. wide will need 988 yd. of fencing.

In other words, this *very elongated* paddock will require *four times* as much fencing as the square paddock of the same area.

These general aspects of the problem of dividing pastures having been examined, a few practical schemes will now be suggested.

TABLE 53

Length of fence required for a 1-acre field of varying dimensions

Dimensions of field		Length of fence, yd.
Breadth, yd.	Length, yd.	
70	69	278
55	88	286
40	121	322
20	242	524
10	484	988

Length of fence required for a 1-hectare field of varying dimensions

Dimensions of field		Length of fence, m.
Breadth, m.	Length, m.	
100	100	400
80	125	410
50	200	500
25	400	850
12·5	800	1625

Chapter 2

WATERING POINTS AND ACCESS CORRIDORS

The question of watering points has sometimes been an obstacle in the way of development of rational grazing

IN most cases large pastures have one watering point, river or pond, which provides the animals with a drink when they want it. When such pastures are divided up, therefore, for the purpose of rational grazing the aim must be to preserve, as far as possible, permanent access from each of the paddocks to the watering points. In the case of *permanent* pastures, new watering points may even be installed, whether in the shape of ponds or water-supply piping.

Where troughs are used for holding water they are either increased in number or mounted on wheels and moved from one paddock to the other. The only change in this instance, therefore, is that the capacity of the troughs must be greater in view of the increased livestock carry employable under rational grazing. In the case of fixed troughs, it is generally economic to provide one trough for two paddocks (Photo 31, between pp. 266–7), or one watering point for two troughs (Photo 27, facing p. 266).

The problem very often becomes extremely complicated where temporary pastures on arable land are concerned. This is the reason for attempts at *rational* grazing on temporary pastures becoming, in effect, systems of *rationed* grazing with *one single* wire in front, allowing the beasts to return at will to the watering point (*vide* Fig. 15, p. 230). The very serious consequences of this method have already been emphasised.

In general, this question of access to watering points is the chief reason for farmers hesitating to undertake rational grazing. Before examining solutions to the problem a preliminary question will be asked.

Must animals have permanent access to watering points?

It is not absolutely necessary for the stock to have permanent access to watering points, be they ponds or troughs; but it is preferable, especially in the case of cows, who, if they are high yielders, will drink more than 22 gal. [*100* litres] of water per day in hot weather (Voisin, 116). There are still quite a number of farms following the old method of continuous grazing without a watering point available in the field being grazed; the stock being driven to a

PHOTO 23

Guillotine gate (half-closed)
on the Voisin pastures

To close the gate completely,
lower the arm into the eyelet
seen on the vertical column and
lock with the pin attached to the
end of the chain

Photo Voisin

PHOTO 24

Guillotine gate (half opened)
on the Voisin pastures

Photo Voisin

pond twice or thrice a day. In the case of in-milk cows the watering point is generally near the milking site (lean-to shed, main farm buildings, etc.). The cows, tethered on second-cut clover in the *Pays de Caux*, are watered by bucket twice daily; this involves a considerable amount of work (p. 210). But these are only makeshift solutions which become less and less acceptable in hot weather.

In very hot regions, however, grazing can be carried out only for a few hours at a time, the animals being taken between grazing periods into a rest park, where they find both water and shade. In such cases, there is generally no need to worry about providing the grazing paddocks with water. Basically the method is equivalent (where watering is concerned) to the old, classical, peasant method of driving the stock to the watering points two or three times a day, at milking time in particular.

Number of groups and number of corridors leading to the watering point

On the assumption that the herd has been divided into several groups (*vide* pp. 152–157) and only one watering point is available for all the paddocks

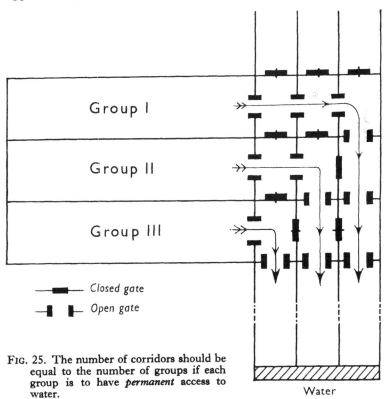

Group I

Group II

Group III

———■——— *Closed gate*

—■ ■— *Open gate*

FIG. 25. The number of corridors should be equal to the number of groups if each group is to have *permanent* access to water.

Water

being grazed, if each group is to have permanent access to the water, then as many corridors leading thereto as groups will be required. This is demonstrated by the simplified scheme in Fig. 25.

If, on the other hand, one accepts the fact that each group goes for water only during part of the day, one single corridor will suffice.

General principle of arranging corridors opening on to the watering point

Assume that the large pasture to be divided has a pond or is situated on the bank of a river. If all the paddocks are to have access to the pond or river, the following basic principle should serve as the rule. A fence must be erected *parallel to the contour of the watering point,* so as to form a corridor or race running along the side of the latter. Where the herd is divided into groups, it is obviously undesirable for these to mix. Not only must a gate be erected between the race and the paddock but gates must also be put up to separate one section of the race, corresponding to a paddock, from the next.

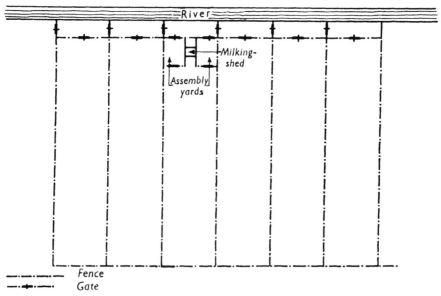

FIG. 26. Plan of paddocks running down to a river.
From Voisin (114).

Like all statements of principle, this appears at first sight to be fruitless, but its sense becomes very plain when one examines Figs. 26 and 27 (Voisin, 114). Fig. 26 represents a pasture running alongside a river, and Fig. 27 the division of a pasture with a pond in its centre. It has been assumed, moreover, that the cows are milked in a milking-shed with assembly yards on either side, the cows passing from one yard, through the milking-shed, where they are milked, and into the other yard. In this way, the milking of a cow cannot be overlooked. The milking-sheds, in both figures, open on to the corridors running along the watering point.

FIG. 27. Plan of paddocks round a pond in the centre of the pasture to be divided.

From Voisin (114), p. 12.

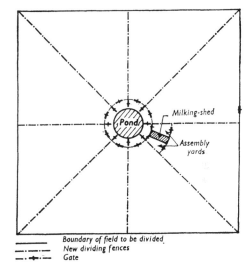

It is recommended, as stated above, that the corridor be achieved by means of fixed fences, the other divisions being made either with electric or fixed fencing. Care must also be taken that the groups do not become mixed at the water itself. Sometimes it may be necessary to put up some means of separation at shallow points where the animals can move around within their own depth.

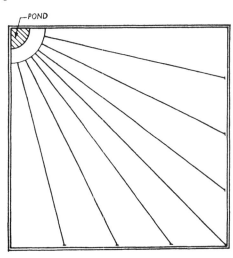

FIG. 28. Plan in the shape of a fan for dividing into paddocks a pasture with a pond in one corner.

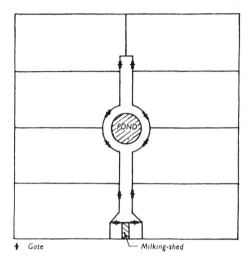

Fig. 29. Division of a pasture with a pond in the centre by means of a single central corridor going round both sides of the pond.

Objections to the general principle

Suppose that the pond, instead of being in the middle of the pasture to be divided, as in Fig. 27, is situated in a corner, as in Fig. 28. If the general principle is to be applied, the paddocks will have to be long and constricted, with all the associated disadvantages:

> excessive trampling of the narrow parts of the paddock leading on to the corridor round the pond;
>
> accumulation of excreta in these same parts to the detriment of the rest of the paddock;

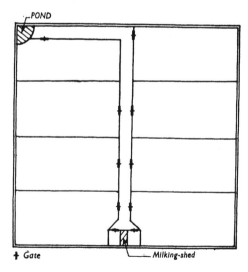

Fig. 30. Division of a pasture with a pond in one corner by means of a central corridor and one down part of one side.

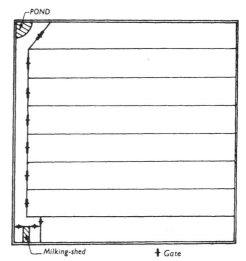

FIG. 31. Division of a pasture with a pond in one corner by means of a single corridor down one side.

distance between the extremity of the paddock and the pond (or milking-shed, or stall);

excessive length of fence to erect and maintain.

A corridor often allows better-shaped paddocks

To avoid these disadvantages it is often preferable to construct a corridor. In this case, Fig. 27 (p. 261) will take the form, for example, of Fig. 29 (p. 262). If the pond (or watering point) is in a corner instead of in the centre, then the schemes set out in Figs. 30 and 31 can be adopted.

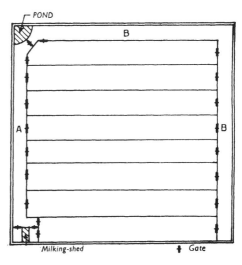

FIG. 32. Division of a pasture with a pond in one corner by means of two corridors allowing two groups to have permanent access to the pond.

By way of example Fig. 32 shows how Fig. 31 can be modified if the herd is to be divided into two groups, each with permanent access to the pond.

These few schemes will be sufficient to convey the general ideas of the principles governing division of pastures so as to give stock access to a single watering point. Some examples will now be given of how pasture division is realised in practice.

Chapter 3

DIVIDED PASTURES

A Bavarian scheme

STAEHLER (99) p. 17, has provided us, among other things, with the plan reproduced in Fig. 33, showing a pasture divided into seventeen paddocks and

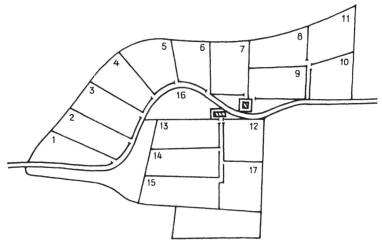

FIG. 33. Plan of a pasture, with water in two places, divided into 17 paddocks (*access to water by existing track and a corridor*).
From Staehler (99), p. 17.

having two watering points. The paddocks are situated on both sides of a road which, with a few corridors, allows for stock shifting and communication.

Rotation practised by M. Jean Fabulet-Laine at Gonneville-sur-Scie (Seine-Maritime)

Photos 28 and 29 (between pp. 266 and 267) are reproductions of aerial photographs (119). Both photographs are of the rotational grazing practised by M. Jean Fabulet-Laine at Gonneville-sur-Scie (Seine-Maritime).

Photo 29 shows the classical plan of paddocks (with parallel fences) opening on to a road leading to the farm, which is opposite the milking site. Some paddocks, as can be seen, are planted with apple-trees; these occupy a greater area than paddocks without trees.

Photo 28 is more unusual, for division here was much more difficult to achieve. All paddocks had to open on to the sunken road, visible on the photo, leading to the farm (situated quite close to the extreme right). Moreover, there was a slope which is not very clear on the aerial photograph and did not allow the classical scheme set out in Fig. 26 (p. 260) to be applied. The result was a system of parallel corridors of increasing length. A certain amount of grass is wasted, but the movement of stock, whether for watering or milking, is greatly facilitated thereby. Corridors and other paddocks open on to a lobby leading to the road. The fences are fixed and there is a common trough for each two paddocks.

M. Bouvier's pasture in the Meurthe-et-Moselle

In 1951 the French Ministry of Agriculture issued a brochure as propaganda for pasture rotation (5). The presentation of this is excellent. It tells the story of rotation on M. Bouvier's farm at Meurthe-et-Moselle and is illustrated by photographs, one of which seemed to me particularly charming and is reproduced in Photo 30 (following). It shows that M. Bouvier's grandson is also interested in rotation. He is making a plan with pieces of straw of his grandfather's divided pastures. From this we see that there are twelve paddocks on either side of a central corridor leading to the watering point and milking-shed, which the boy represents by means of a box of matches.

Division of the Voisin pastures

Photos 31, 32 and 33 present some views of my own pastures.

Photo 31 shows a group of paddocks divided by means of cement posts. In the foreground there is a cement trough astride two paddocks. The trough is fed by running water. Photo 32 (facing p. 267) shows two paddocks planted with apple-trees and divided by an iron fence, while Photo 33 (facing p. 267) shows a group of paddocks in spring when some have been disengaged from the grazing rota for mowing. Three paddocks have just been cut, and the grass can be seen lying drying on the ground.

PHOTO 27

Steelway Farms
(Kentucky, U.S.A.)
One watering point serving
two troughs
and two paddocks

A small shelter has been erected
so that the cows get both shade
and **water**

Photo Voisin

PHOTO 28

Aerial photograph of the second section of M. Jacques Fabulet-Laine's pastures (Seine-Maritime, France)

Vide Voisin [119] · Photo Paris-Normandie

PHOTO 29

Aerial photograph of the first section of M. Jacques Fabulet-Laine's pastures (Seine-Maritime, France)
(part is planted out with apple trees)

Vide Voisin (119) · Photo Paris-Normandie

PHOTO 30

The farmer's grandson is interested in rotation and draws up his plan with pieces of straw

From (15)

PHOTO 31

Voisin pastures divided up by means of cement posts

In the foreground, one trough serving two paddocks

Photo 32

Voisin pastures planted with apple trees and divided with iron posts

Photo Voisin

Photo 33

Voisin pastures: some paddocks have been dropped from the rotation and have just been mown

PART NINE

RATIONAL GRAZING
TRANSFORMS THE FLORA

Chapter 1

EXTREMELY RAPID EVOLUTION OF THE FLORA

Pastures deteriorate because they are badly managed

If a pasture has degenerated it is **because the aggregate of the conditions determining the flora are favourable to the development of a sward of poor quality.** Among the external conditions dependent on Man which determine this poor-quality flora, there are three which must never be overlooked:

1. Bad water conditions usually caused by defective drainage.
2. Bad feeding of the soil with nutritive elements.
3. Defective and non-rational management, the normal and usual form of which is continuous grazing.

It must be emphasised that even if the first two conditions are improved, management still exerts an enormous influence on the flora of a pasture.

Man is to blame, not the grass

Photos 34 and 35 (facing p. 274) were taken on two different continents, America and Europe: in both cases the cows are looking for the little grass that still exists among the weeds that are as high as the animals themselves. It is Man and his methods of management that are to blame for this ingress of thistles and gorse. Ploughing and re-seeding, or, as the Americans call it, "renovation", does not solve the problem. If the pasture is badly managed the flora of the new, re-seeded sward will deteriorate, and very quickly at that.

Confusion of ideas on permanent pastures and temporary leys

Three problems concerning permanent pastures and temporary leys have become confused. The following three distinct questions must in fact be asked:

1. What must be the ratio of tillage to permanent pasture? (*vide* pp. 172 and 173).
2. What proportion and what type of green areas must be included in the crop rotation?
3. Must pasture be ploughed to be improved?

Protracted study must precede any alteration to the crop rotation

The first two questions are specific to each farm, and depend on many pedological, climatic, economic, human factors, etc. These are fundamental questions requiring intensive study of long duration, like all questions concerned with *crop rotation*. It can take a very long time before the general influence of a rotation becomes evident.

For many years, for instance, sometimes twenty to thirty, the rotation can be: wheat–sugar beet–wheat–sugar beet. The agricultural reviews of seventy years ago abound with arguments on this subject. The sugar-beet industry at that time wanted as much beet as possible grown, especially in the areas not too far distant from the factories. It published skilful propaganda in the agricultural Press, and was clever enough to get the support of authorities in agronomy. The minority of scientists who warned against such a rotation did not represent any commercial interest, and were hardly listened to; in fact, their cries were simply stifled. When, after years of repeating this rotation, the nematode appeared, it was a great catastrophe for the farmers, who were the victims of this selfish propaganda. Thereafter, the crop rotations adopted made sure of a greater interval between sugar-beet crops.

A lengthy period is similarly required before exhaustion of the soil by legumes and the fact that such crops must not be grown too frequently on the same soil becomes apparent.

To put it briefly, study over a period of many years is essential before modifications to a crop rotation can be popularised. This is what is being seen in Britain to-day after twenty years of ley farming. In spite of the enormous subsidies offered to encourage ploughing up of pasture and the sale of herbage seeds, British farmers are returning to permanent pastures, the advantages of which they clearly appreciate. Thus the area of permanent pasture is on the increase. The same thing happened also in Germany.

Must pastures be ploughed up to be improved?

This is the third question concerning permanent and temporary pasture, which I answered, in greater detail than is possible here, in two papers on the subject (122) and (124). Although these were written several years ago, there is not much in them that requires alteration. In this present connection two fundamental points should be emphasised.

To know whether a better method of management will improve a degenerate flora one must first be thoroughly acquainted with that method

One sometimes hears it said: "The flora in that pasture has deteriorated. I do not think management alone can improve it. . . . It will have to be ploughed up and re-sown." But it is impossible to know whether the method of management will improve the pasture unless one has, at one's finger ends, the methods of rational management of grassland (*vide* p. 322). One must learn to graze rationally first before one can know what influence such a method will have on the degenerated flora. Common sense and the logic of science require that methods of grazing be carefully studied to know whether or not they are sufficient to improve a flora.

Ploughing up a pasture does not make up for defective management

Even if a pasture is ploughed up, the methods of management must also be improved if high grass productivity is to be achieved and the flora is to be prevented from deteriorating once more. If the flora is degenerate, the management has failed in some respect. Ploughing up the pasture by itself, will not alter these unsatisfactory methods so harmful to the turf.

On p. 232 I described an example encountered in the mid-East of France, where a large pasture was being grazed with the use of one electric wire in front, the cows having free access back to the watering point. As a result, sections near the watering point had a period of occupation of more than 20 days and a rest period reduced by the same length. Flora degeneration of this part was marked, and grass productivity greatly diminished. I was told that this was of no importance: the pasture would soon be ploughed up and re-sown, as was done regularly every four years. *But the defective method of management will remain whether the pasture is ploughed or not, and the new grass will be subject to the same abuse as the old. Whatever happens, with or without ploughing, the decision must be to follow rational grazing.*

The comb and the shears

I often make the following comparison. A father has a son who refuses to use a comb, and his hair is like a bale of straw. Despite his father's scolding, he refuses to keep his hair tidy. In despair, the father takes him to a barber's and has all his hair cut off. But alas! when the boy's hair grows again, it will still be untidy if he continues to refuse to use a brush or comb. Indeed, it will be even worse than before, because his hair, as a result of the close clipping it has received, will be even more wiry. Whatever happens, therefore, the boy will have to learn to use a brush and comb (*vide* p. 322).

Dynamic ecology of pastures

I am working on another book, dealing with the *Dynamic Ecology of Pastures*, that is, the influence of different management factors on the flora of pastures. In this present work, therefore, I will restrict myself to only a few examples illustrating the improvement of flora by rational methods of management.

The opinion of two great ecologists

A few simple examples will be quoted in order to show to what extent a flora, within its particular habitat, is strictly dependent on methods of management. These examples provide a good illustration of statements made by two great ecologists.

Professor Klapp (70) p. 66, writes:

> "The first essential of economic pasture management is to remember that **the flora of a pasture is extremely plastic and varies very rapidly with the management applied.**"

The Austrian ecologist Walter Czerwinka vigorously stresses this fundamental point (15) p. 101:

> "Plant ecology, a young and new science, allows us to understand the *evolution of the plant associations in our pastures. This evolution,* whether in the sense of impoverishment or improvement of the flora, takes place sometimes *with astonishing rapidity.* . . . **Thanks to the requisite measures being taken, completely degenerate pastures are transformed and often within a short time, into pastures with flora of quality.** . . ."

Chapter 2

SIMPLIFIED EXAMPLES OF FLORA EVOLUTION

Influence of the number of annual cuts on the evolution of the flora

TABLE 54 shows the relative drop in yield of white clover and certain grass species when the frequency of cutting is increased. It is quite obvious, for example, that an increase in the number of annual cuts reduces the yield of meadow fescue much more than that of white clover.

TABLE 54

Influence of the frequency of cutting on the yield of some herbage plants

English name	Latin name	Number of cuts per year		
		2–3	4–6	7–13
		Relative yields		
White clover	Trifolium repens	100	64	58
Smooth-stalked meadow grass	Poa pratensis	100	96	35
Perennial rye-grass	Lolium perenne	100	68	31
Cocksfoot	Dactylis glomerata	100	67	31
Red fescue	Festuca rubra	100	65	25
Meadow fescue	Festuca pratensis	100	57	18
Marsh poa	Poa palustris	100	19	8

From Klapp (65).

Table 55 (p. 274) has been borrowed from the Swiss worker, Geering (26): it shows the percentage of different grass species and weeds present in a sward relative to the frequency of cutting. 12, 6, 4 and 3 annual cuts were taken (unfortunately at *fixed* intervals, it may be noted in passing). Moreover, due to certain prior conditions of management (use of liquid organic fertilisers) there was very little white clover present in these swards.

The table shows that increased frequency of cutting has greatly increased the percentage of fiorin and rough-stalked meadow grass, and has decreased the percentage of cocksfoot. The maximum percentage of rye-grass is obtained with 6 cuts and of smooth-stalked meadow grass with 4 cuts. Red fescue percentage contribution varies but little between 3 and 6 cuts, but is greatly reduced by 12 cuts.

TABLE 55

Influence of the number of cuts on certain grasses and weeds

English name	Latin name	Percentage in sward			
		Number of cuts annually			
		12	6	4	3
Fiorin	Agrostis alba	57·4	20·9	5·0	4·4
Cocksfoot	Dactylis glomerata	1·8	16·7	11·0	16·9
Smooth-stalked meadow grass	Poa pratensis	3·8	5·5	9·6	4·2
Rough-stalked meadow grass	Poa trivialis	12·4	11·0	5·8	4·5
Meadow fescue	Festuca pratensis	0·9	0·3	0·7	1·7
Red fescue	Festuca rubra	1·4	14·1	11·2	18·8
Perennial rye-grass	Lolium perenne	10·4	15·6	15·0	9·9
Other grasses	....	—	0·9	1·0	2·0
White clover	Trifolium repens	0·2	0·7	1·5	0·5
Sorrel	Rumex acetosa	—	3·6	0·4	0·3
Meadow buttercup	Ranunculus acer	0·1	0·2	3·2	6·5
Creeping buttercup	Rununculus repens	—	—	0·9	2·3
Ladysmock	Cardamine pratensis	4·1	3·1	1·0	0·6
Cow parsnip	Heracleum sphyondylium	—	0·5	0·6	2·8
Dandelion	Taraxacum officinale	2·9	4·1	30·8	22·5

From Geering (26).

As for weeds, diminished frequency of cutting favours buttercup and dandelion but discourages ladysmock.

Both these tables from Klapp and Geering show how particularly sensitive the various pasture plants are to frequency of cutting.

Influence of different grazing methods on agrostis and white clover

More than twenty years ago the Welsh Plant Breeding Station at Aberystwyth carried out an experiment (143) far removed from actual management conditions generally encountered. Moreover, the rest periods were *the same* throughout the season for all the grazing systems. The five systems chosen, however, are very different, and thus provide striking proof of the extent to which grazing method influences the flora of the re-sown pasture. This is particularly true in the case of bent grass and white clover.

The five methods of management were as follows:

1. *Intensive grazing throughout the season.* Paddocks closely grazed by sheep once every week, from beginning of April until end of October.

2. *Grazing intensive in spring, then moderate.* Sheep grazed every week from beginning of April until mid-June; every month from mid-June until end of October.

3. *Grazing moderate at first, intensive at end.* Sheep grazed every month from beginning of April until mid-August; every week from mid-August until end of October.

PHOTO 36

(Herefordshire)
Professor Martin Jones
explaining the evolution
of the flora as a function
of the methods of
management

Photo Voisin

PHOTO 37

Morgan Farm
(North Carolina U.S.A.)
Green pastures in North
Carolina with the prosaic
view of farm buildings
in the foreground

Photo Voisin

4. *Moderate grazing throughout the season.* Sheep grazed every month from beginning of April until end of October.

5. *Insufficient grazing throughout the season.* Sheep grazed every two months from beginning of April until end of October.

Table 56 shows the influence of each of these grazing systems on the evolution of recently sown bent grass and white clover.

TABLE 56

Influence of different methods of grazing on bent grass and white clover

	Bent grass			White clover		
Method of grazing	Number of tillers per					
	6 in. × 6 in.	m.²	Relative variation	6 in. × 6 in.	m.²	Relative variation
1. Weekly grazing .	503	21,700	400	26	1120	200
2. Weekly in spring .	222	9,560	174	31	1350	240
3. Weekly in autumn .	220	9,480	173	22	950	170
4. Monthly grazing .	173	7,460	136	15	650	116
5. Two-monthly grazing	127	5,460	100	13	560	100

N.B. See text for details of the method of grazing.

From Welsh Plant Breeding Station (143).

With the grazing method most favourable to it, bent grass is four times as abundant as with the method most unfavourable to it. Moreover, white clover is 2·4 times as plentiful when it is favoured as when it is at a disadvantage.

All these plots had the same flora to start with, and the experiment lasted for one year only. This shows the speed with which a certain method of grazing can alter, for better or for worse, the flora of a re-seeded pasture.

On closer examination of the table, it becomes evident (as the authors pointed out) that frequent grazing (at weekly intervals) is most favourable to bent grass, whereas white clover, while encouraged by grazing of this nature early in the year, prefers a more moderate defoliation practice (monthly) towards the end of the season. It is well known, moreover, that bent grass quickly invades swards that are too often tightly grazed and abused.

Influence of the date and method of putting out to grass on the evolution of the flora

It was pointed out in Part Five, Chapter 2 (p. 181), that the date and method of putting out to grass can completely upset the botanical composition, as is clearly illustrated in the diagrams of Fig. 11 (p. 183). Here, then, is the marked influence of one of the elements of method of management, namely putting out to grass, on the evolution of the flora.

Evolution of a simple mixture of smooth-stalked meadow grass and white clover

At Cornell (U.S.A.) Professor Johnstone-Wallace studied the effects of cutting, repeated at different, fixed intervals of time on the evolution of the flora. The sward was a pure stand of smooth-stalked meadow grass (Kentucky Blue Grass) and white clover (52). In every case cutting was to a height of ½ in. [*12 mm.*] above the soil.

When this cut was repeated every week, white clover dominated the pasture and contributed 80% of the flora. Where the cut was repeated only every 4 weeks, a balance was achieved between grass and clover, each representing 50%. If the interval between cuts was made even wider, that is, increased to 8 weeks, the white clover, stifled by the grass, contributed only 10% to the total sward. Finally, if the cuts took place every 12 weeks, which corresponds to two cuts per annum, the white clover disappeared almost completely, barely more than 1% remaining.

The flora of a temporary pasture depends far more on the method of management than on the mixture sown

A mixture containing a much higher proportion of white clover might have been sown, but this would not have prevented the disappearance of the legume when 12-weekly cuts were employed.

On the other hand, the proportion of smooth-stalked meadow grass might have been increased, and this would not have prevented the white clover from dominating the sward when it was cut every week.

It will be repeated once more that the flora of a pasture depends, above all, on the conditions of management. In the present instance of a temporary pasture sown with a simple mixture, comprising one grass and one legume, it is evident that after two years **for the same mixture sown, there can be either 80% or 1% clover, according to the management employed.** In investigational and research work, therefore, priority must be given to methods of management, although in general all eyes are fixed on the seeds mixture to be used. One finds books devoting page after page to formulæ for seeds mixtures, but with not the slightest detail concerning the methods of management which are dismissed in a few vague and subjective sentences.

This does not mean, of course, that we should not try to introduce new species or improved strains into our pastures—which can be done without ploughing up the sward. **But it must constantly be borne in mind that selection must be carried out as a function of methods of management and that fresh species or strains introduced will only succeed in establishing themselves if conditions of management so allow** (whether the pastures be old or new).

These are points which have too often been overlooked, if not completely forgotten.

Chapter 3

COMMON GRAZINGS AT RENGEN (GERMANY)

The Rengen Estates, Eifel (Germany)

EIFEL is a wooden plateau in the Rhineland lying to the North of the Moselle and extending the high upland "Fagnes" (swamps) of the Ardennes from the Luxembourg frontier to the Rhine. Extinct volcanoes tower above the strata of shale, sandstone and limestone. Agriculturally, it is a very poor region, particularly above 1300 ft. [*400* m.].

The Rengen estates are situated in Upper Eifel, and consisted, until 1930, of heathland and marshes. Their altitude varies from 1390 to 1655 ft. [*425–505* m.]. The annual rainfall is approximately 32 in. [*800* mm.], but there is generally a severe period of drought in summer. The soil is very poor and gley-like (see footnote, p. 109) in character, and moisture has difficulty in percolating through it. When the water has evaporated or run off, the soil, in dry summer weather, is like concrete.

The analysis of these soils prior to improvement was as follows:

pH—3·9–5·6 with a mean of 4·3 (*vide* p. 41).
Phosphoric acid—0·96 mg./100 mg. soil.
Potash—9·19 mg./100 g. soil.

The phosphoric acid and potash contents were assessed by the Neubauer method.

Until 1930–34 all this area was used for common grazing, the estimated carrying capacity of 50 acres [*20* ha.] in the period of vigorous grass growth being ten to fifteen head of cattle. After 1934 the land was acquired by the Agricultural Faculty of Bonn University and attempts made to improve the completely degenerate and worn-out grazings after these tens of years of careless, extensive management. This work was carried out under the direction of Professor Klapp (Photo 4).

Restoration of the Rengen grazings

Work commenced in 1934. One of the fundamental measures was drainage, and this presented special difficulties due to the thin arable covering and lack of depth of the impermeable layer. When, however, drainage was finally

277

achieved, the task remained of studying and applying methods which would markedly and permanently improve these greatly deteriorated pastures. This experimental area was to provide remarkable information, some of which has already been referred to (pp. 40 and 44). The influence of different methods of management on the flora will now be examined.

Improvement of pastures by the mower alone

Mowing brought about some improvement, but this was not really perceptible until after fertilisers had been applied.

Table 57 (p. 279) shows that, in the absence of fertilisers, mowing has reduced wavy hair-grass (7% against 16%) and heather (8·3% against 19·3%) while heath-grass (2·4% instead of 3%) and hairy greenweed (3·2% instead of 4·8%) have hardly changed. Nardus (mat-grass), on the other hand, the real curse of this type of pasture tended rather to increase, rising from 11·3 to 15·8%. The application of fertilisers, however, immediately depresses mat-grass, which now represents little more than 4% of the flora. Other weeds are similarly reduced by fertiliser application, except wavy hair-grass which remains more or less the same, showing, if anything, a slight tendency to increase.

Mat-grass always presents a very serious problem in pastures of this kind, it has been thought advisable to include Table 58 also, which shows that, even after five years of mowing, in the absence of a complete basic fertiliser dressing (Ca, P, K), the mat-grass contribution was the same; whereas with the support of a fertiliser dressing, mowing immediately became lethal to this grass, which finally represents, on the average, hardly 2% of the herbage. Very acid soils are known to favour mat-grass, and it has a liking for soils poor in potash. It is probable, therefore, that the application of lime and potash aided its suppression.

Table 59 (p. 280) shows the percentages of the main species (good and bad) after three years of cutting, with or without fertiliser dressings. Where no fertiliser was applied the worst weeds (last five lines of Table 59) successfully resisted mowing, whereas the application of fertiliser reduced them to a very low percentage.

It is particularly interesting to note that the total of the three least bad grass (red fescue, common bent-grass and sheep's fescue) represented 36·6% of the flora in the absence of fertiliser, and 67·5% where fertiliser was applied.

The mower alone can hardly improve a meadow

An important lesson emerging from these trials at Rengen (Tables 57, 58 and 59) is that the mower itself achieves no improvement without the support of basic fertiliser.

In combination with fertiliser a definite improvement is obtained, but this is by no means great (Table 59) in the face of the transformation of these

TABLE 57

Influence of lime and phospho-potassic fertilisers on weeds in a deteriorated pasture when it is cut for hay only

English name	Latin name	Percentage in flora					
		Without fertiliser			*With* fertiliser		
		Year			Year		
		1	2	3	1	2	3
Mat-grass	Nardus stricta	11·3	10·8	15·8	3·0	0·7	4·3
Wavy hair-grass	Deschampsia flexuosa	16·0	14·5	7·0	29·2	16·4	7·4
Heath-grass	Sieglingia decumbens	3·0	0·8	2·4	2·2	0·7	1·9
Hairy greenweed	Genista pilosa	4·8	5·0	3·2	2·0	0·9	1·5
Heather	Calluna vulgaris	19·3	14·7	8·3	5·4	1·1	2·8

N.B. 1. Fertilisers applied in two years (per acre or per hectare): 16 cwt. [*2000* kg.] burnt lime; 2½ cwt. [*300* kg.] P_2O_5; 3 cwt. [*400* kg.] K_2O; followed by normal applications.

2. The grass was mown only, although the number of cuts per year is not indicated.

From Klapp (66).

TABLE 58

Resistance of mat-grass to mowing when the necessary fertiliser applications are not made

	Before start of improvements	Number of years of trials				
		1	2	3	4	5
		Percentage in flora				
Without fertiliser	19·8	18·0	18·0	18·0	25·0	18·0
With fertiliser	19·8	4·1	1·6	1·6	1·3	2·5

N.B. 1. Fertilisers are the same as those in Table 57.

2. The grass was cut for hay only.

From Klapp (66).

low-grade pastures to quality swards achieved by grazing, the effect being the more marked as grazing is more rational.

It is certain, therefore, that where meadows are *mown only* and no, or little, fertiliser applied, it is almost impossible to improve the flora. Ploughing is then the only solution, and this is what is usually done to degenerated hay meadows or lucerne stands invaded by grasses to such an extent that they are no more than dilapidated hay swards.

TABLE 59

Influence of lime and phospho-potassic fertilisers on the general improvement of the flora of a deteriorated pasture when it is cut for hay only (after three years)

English name	Latin name	Without fertiliser	With fertiliser
Red fescue	Festuca rubra	2·0	16·2
Common bent-grass	Agrostis vulgaris	14·3	25·9
Sheep's fescue	Festuca ovina	22·3	25·4
Wavy hair-grass	Deschampsia flexuosa	9·5	9·9
Creeping soft-grass	Holcus mollis	6·4	5·1
Heath bedstraw	Galium Saxatile	0·5	0·4
Common tormentil	Potentilla erecta	2·5	2·0
Manyheaded woodrush	Luzula multiflora	1·7	1·1
Heath grass	Sieglingia decumbens	1·3	0·6
Pill-headed sedge	Carex pilulifera	4·0	1·7
Mat-grass	Nardus stricta	12·9	2·6
Hairy greenweed	Genista pilosa	4·7	0·6
Heather	Calluna vulgaris	15·5	0·9

N.B.   1. The figures show the composition of the flora (in percentage) after three years of mowing, with and without fertilisers.
      2. One cut of hay per year.

From Klapp (66).

Olivier de Serres [1] and hay meadows

This is probably what made Olivier de Serres say many years ago that if a meadow was no longer productive it must be reconverted to ploughed land. From the text it appears that he was speaking of hay meadows and not of pastures: "It would be vain to sing the praises of the meadow. . . . *Hay* grows there with so little care and such facility. . . . The meadow is always ready to be of service."

Olivier de Serres had already felt that the only hope of preventing the deterioration of hay meadows was fertiliser, and he recommended the application of those available in his day: farmyard manure and washed ashes.

Improvement of degenerate flora by rational grazing

Of particular interest in the Rengen trials is the emergence of the fact that the most efficient instrument of improvement of the flora was, in the long run, rational management by grazing, although the *degree* of rotation was not very intensive.

Table 60 shows the initial botanical composition of the paddocks to be grazed. Apart from a little sheep's fescue and bent-grass, there were hardly any suitable grasses in these pastures. Indeed, it is difficult to imagine more run-down swards.

[1] Olivier de Serres, French agronome and author (1539–1619), contributed much to the progress of agriculture in his time. His most important work is *Théatre d'Agriculture des Champs*.

TABLE 60

Influence of method of grazing on the regeneration of the flora of deteriorated grassland (Rengen trials)

English name	Latin name	Initial state	Continuous grazing	Rotation with		Rotation with small paddocks after		
				Large paddock	Average paddock	1 year	2 years	3 years
Heather	Calluna vulgaris	51·4	12·1	3·3	0·1	—	—	—
Mat-grass	Nardus stricta	10·8	2·3	0·2	—	—	—	—
Brooms (various)	Genista (various)	11·0	1·9	0·5	—	—	—	—
Sheep's fescue	Festuca ovina	13·7	48·3	15·6	1·4	0·5	—	—
Wavy hair-grass	Deschampsia flexuosa	3·2	14·6	—	—	—	—	—
Red clover	Trifolium pratense	—	0·5	9·2	4·9	3·3	6·6	1·3
Sweet vernal grass	Anthoxanthum odor-antum	—	0·4	3·2	0·7	2·8	2·2	—
Lesser yellow trefoil	Trifolium dubium	—	0·4	3·1	2·4	0·2	0·4	—
Common bent-grass	Agrostis vulgaris	1·6	5·4	5·3	7·5	5·2	5·5	2·1
Yorkshire fog	Holcus lanatus	—	0·3	3·8	6·0	2·6	2·1	5·3
Creeping soft-grass	Holcus mollis	—	3·0	0·6	3·9	3·5	1·3	—
Rough-stalked meadow grass	Poa trivialis	—	0·2	0·2	6·6	12·6	8·2	8·0
Crested dogstail	Cynosurus cristatus	—	—	5·3	3·0	12·3	11·2	5·0
Red fescue	Festuca rubra	0·1	1·1	22·1	18·1	25·4	26·0	16·2
Smooth-stalked meadow grass	Poa pratensis	—	1·4	1·8	9·6	7·0	6·3	18·5
Perennial rye-grass	Lolium perenne	—	—	0·2	0·5	1·8	1·0	10·8
White clover	Trifolium repens	—	1·1	5·1	13·4	10·0	13·2	14·0
Cocksfoot	Dactylis glomerata	—	—	0·2	0·6	1·0	3·5	7·9
Timothy	Phleum pratense	—	—	—	—	1·3	0·5	1·6

Compare Fig. 34.
From Klapp (66, p. 374).

At the outset grazing animals were left permanently on the pasture, weed clumps were clipped by hand and complete fertiliser was applied (*vide* Table 57, p. 279). Then rotational grazing was practised, first with large, then with smaller paddocks, but even the latter were fairly big. The herbage was cut annually as it became available, and complete fertiliser continued to be applied. Thereafter a suitable rotation with small 2·5-acre [*1*-ha.] paddocks was brought into being, one annual cut taken and the requisite amount of fertiliser applied as before.

The flora created under continuous grazing was not even tolerable. It was only under the semi-rotation that weeds and poor-quality grasses began to be clearly depressed, thus allowing a few better grasses to gain a foothold. But it was reserved for the most intensive and best-managed rotation to achieve definite and decisive improvement:

complete disappearance of heather, mat-grass and brooms;
similarly complete disappearance of sheep's fescue, wavy hair-grass and sweet vernal grass;
suitable percentages of Yorkshire fog, crested dogstail and cocksfoot;
vigorous development of red fescue, smooth-stalked meadow grass, perennial rye-grass and white clover (the rye-grass having taken longer to reach a suitable contribution).

All the species which were not in the original pasture and developed subsequently, *appeared spontaneously.*

Diagram of the evolution of flora under grazing

Fig. 34, below, shows the average evolution of the flora during one of the Rengen trials, and corresponds approximately to the figures in Table 60. Diagrams such as this have the merit of revealing at a glance the influence of each of the methods of management on flora evolution. Fig. 34 illustrates very obviously how rational methods of grazing (and the application of fertilisers) improved the ruined grazings of Rengen.

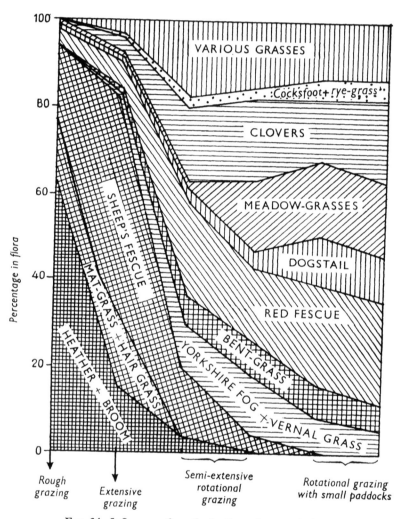

FIG. 34. Influence of method of grazing on the improvement of deteriorated flora.

From Klapp (70), Fig. 74, p. 221.

Yields increase as the flora improves

Yields are also considerably improved by the combined use of fertilisers and rational grazing. These common grazings were producing the miserable quantity of 446 lb./acre [*500* kg./ha.] starch equivalent. With fertilisers and rational grazing the following yields of starch equivalent were obtained:

First year—1963 lb./acre [*2201* kg./ha.]
Second year—2284 lb./acre [*2559* kg./ha.]
Third year—2314 lb./acre [*2594* kg./ha.]
Fourth year—2567 lb./acre [*2878* kg./ha.]

Lessons from the Rengen trials

These tables and diagram show how, in a few years, poor grazings have been changed into pastures of quality thanks to rational methods of management. The extent to which such methods exercise a *positive* effect even on the flora of seriously run-down pastures is obvious. Equally apparent is the fact that these indifferent pastures have become good grazings within a few years, their character improved for all time—based on the special soil structure which constitutes the wealth of permanent pastures.

It has sometimes been said: "In all respects, the effects of applying a good grassland management technique are only negative, consisting in eliminating undesirable species, but not in replacing these by more productive ones. At best, these choice species, if they are present at all in the initial population, are capable of a little expansion. . . ."

The example of Rengen, like so many others, illustrates how far removed from reality this conception is. Not only is the flora improved, but productive species which were not present at all initially, appear spontaneously, brought by the wind, excrement or some other means. To use the American expression, these species are *volunteers*.

All the tables reproduced above illustrate the value of the improvement work carried out at Rengen. But this does not take the place of seeing them personally, as I have done several times. Part of the old mat-grass land has been kept as evidence, and quite close to these poor grazings one can see excellent pastures that many a fertile region would envy. I should like to take this opportunity of thanking the enthusiastic young workers who conducted me round their project. When one walks through those beautiful swards and looks around at the desolate Eifel countryside one is consoled by the thought that men have been able, with their mining agricultural policies, to ravage whole countries, but they are still capable, on occasion, of regenerating the grazings they have plundered. This example has perhaps particular significance in Britain where open-cast mining has laid waste large areas of agricultural land which must be reclaimed when mining is exhausted.

Chapter 4

A PERSONAL EXPERIENCE AND SOME ENGLISH EXPERIMENTS

Pastures ruined by war have been transformed by rational grazing into pastures of quality

TABLE 22 (p. 52) contains the analysis of the flora of my grazings, some of them old pasture and some sown in 1947. The old pastures, probably more than one hundred years old, had suffered all the damage of war: artillery batteries, mines, caterpillar tracks from tanks, etc. One would have searched in vain for a white clover plant; but thanks to rational grazing, combined with applications of basic fertiliser, they were rapidly transformed without being re-seeded. White clover returned spontaneously, and rye-grass, which was rare, has developed vigorously.

Rational grazing has thus completely transformed these run-down grazings into good pastures, the flora of which is superior to those sown down in 1947.

Improvement of wild pastures on the abandoned hill lands of Wales

Martin Jones, one of whose trials has already been cited (Fig. 11, p. 37), has shown in the course of some very fine experiments how grazings on abandoned farms in the hills of Wales could be improved by suitable methods of grazing, methods, however, somewhat different from rational management.

The pastures in question were very poor and comprised what had been arable land until the end of the First World War. Left to their own devices, they re-seeded spontaneously with wind-blown seed. These wild pastures were looked upon as completely inferior ("tumble-down pastures"), and it was thought that the only way to establish a suitable flora was to use the plough. The turf consisted mainly of Yorkshire fog (*Holcus lanatus*) and meadow buttercup (*Ranunculus acris*).

One area (A) was subjected to controlled, semi-rational grazing and received the three fertiliser elements, nitrogen, phosphate and potash. Area (B) received no fertiliser and was grazed continuously without control. Area (C) was treated like Area (B) but, in addition, was mown annually in June. Fig. 35 shows the modifications to the herbage after *two* years of one of the three treatments.

284

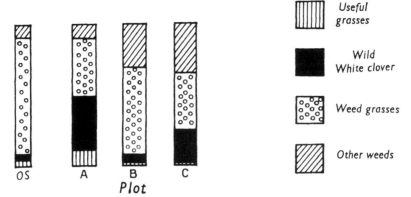

N.B.—OS : Original Sward.
For explanation of figures see text.

Fig. 35. Influence of method of grazing together with manuring on the botanical composition of a "tumble-down" sward after two years' treatment.

From Martin Jones (60).

The combination of fertiliser dressings and controlled grazing (A) has created a greatly improved sward within two years. In the absence of fertilisers, continuous, uncontrolled grazing (B) has achieved little improve-while one cut (C) has assisted the spread of white clover.

Martin Jones concludes:

"The fact remains that the differences in the nature of the various main pasture types of the country can be satisfactorily explained as resulting primarily from the method of grazing management adopted; and it is suggested that this factor, more than any other, is responsible for the differences in botanical composition. **Within wide limits differences of soil, climate, and manurial treatment are of quite secondary importance and should never be adduced to account for the condition of a pasture unless and until it is found that it cannot be explained by a consideration of the grazing management throughout the year.**

"To sum up: There is more than one species in every pasture; these species grow in competition with one another; and **it is largely within the power of the farmer, by appropriate management of the grazing, to decide which shall predominate, and which shall be suppressed.**"

Improvement of an old pasture at Jealott's Hill

From among the many experiments carried out by Martin Jones at Jealott's Hill (56–59), one that is particularly characteristic has been selected. The methods of grazing were as follows:

Paddock 1: high rate of sheep stocking throughout the grazing season (but not in winter).

Paddock 2: not grazed before mid-April and then moderately.

Paddock 3: received an intermediate treatment. Not grazed until mid-April, but thereafter grazed bare and the grass left resting for a month (this might be said to be the beginnings of rotation).

Paddock 4: grazed according to a method in common use in the area. The stocking rate of sheep hardly varied despite seasonal fluctuations in grass growth. Although grass production in May was probably ten times greater than in January, the stocking rate in spring and summer was little more than double that in winter.

The aim of the experiment was to see what improvements could be achieved on a poor, old pasture by use of different methods of grazing combined with application of the fertilisers required. The pasture consisted

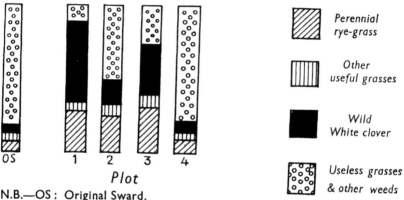

Perennial rye-grass

Other useful grasses

Wild White clover

Useless grasses & other weeds

Plot

N.B.—OS : Original Sward.
For explanation of figures see text.

Fig. 36. Influence of the method of grazing on the botanical composition of an old permanent sward after two years' treatment.
From Martin Jones (60).

mainly of bent-grass with a high proportion of Yorkshire fog and sheep's fescue. Fig. 36 shows the effects on the flora of the different methods of management applied over two years.

Intensive grazing on Paddock 1 brought about a considerable reduction of weeds, which were replaced by white clover in particular and also by rye-grass. Paddock 2, which was not grazed until late spring, had its percentage of weeds or weed grasses reduced to a very much lower extent. On Paddock 3 late defoliation in spring followed by close grazing and rest periods of one month led to a great reduction in weeds and to very vigorous development of rye-grass. The flora could be placed almost half-way between that of Paddocks 1 and 2. Paddock 4, grazed according to the usual method in that part of the country in winter and throughout the remainder of year, retained almost the same bad flora as the original sward, without any great modification.

With two years of suitable management, therefore, the flora of an old and indifferent pasture has been considerably improved: the benefits being not only lasting in character but also progressive, if the required methods of management are maintained.

A striking illustration from Martin Jones

At the Fourth International Grassland Congress, Martin Jones addressed the audience thus (61):

> "Up and down the country you have seen and will see different types of grassland, varying with altitude and soil fertility. Among them you may notice nevertheless how some fields carry very much poorer pastures than others, though on soils which appear quite similar, i.e. they show varying stages in the deterioration of pastures on one and the same type of soil. This is largely due to differences in the botanical compositions of those swards.
>
> "It used to be considered that all such botanical differences were due to the seeds mixture sown, even though the field may have been down to grass for twenty years or more. Experiments have shown, however, that **the way in which grassland is used and particularly the manner in which it is grazed decides very largely the fate of any species included in the seeds mixture.**"

Professor Martin Jones went on to show a photograph (Photo 38, facing p. 290) of a pasture sown under the relatively dry conditions of this part of England, at Jealott's Hill in Berkshire. The re-seeded pasture had been divided into several paddocks, each being submitted to a different system of grazing. On the right of the figure it is evident that suitable management has maintained a good balance between grass and clover. The section on the left of Fig. 35, on the other hand, which was less skilfully managed, deteriorated and became overrun with weeds.

Professor Martin Jones carried out an analogous experiment on a very old pasture (more than eighty years old) in the same area. Photo 39 (facing p. 290) shows the invasion of the pasture by weeds when badly managed (top and back of photograph) and the appearance of the same sward following suitable management (bottom and front).

Thoughts on this subject

The following observations may be made regarding this experiment.

1. A recently re-seeded but badly managed pasture comprises after six years a very poor flora, despite the quality of the mixture sown.

2. *Appropriate management has transformed a very old, degenerate pasture* into one of excellent botanical composition.

This is a very good and very instructive example of dynamic ecology, and readers would do well to meditate on the observations made by the experimenter himself. Of the improvements he has achieved on old pastures as a result of suitable management Martin Jones writes (60): (*vide* Photo 36, facing p. 275):

> "The striking point of this experiment has been the extreme rapidity with which the sward has changed under the influence of the grazing animal, and that despite the fact that the sward had remained for such a long time previously in stable equilibrium. The rapidity with which the wild white clover and the perennial rye-grass increased on this land, which would ordinarily be regarded as naturally inclined towards Bent grass (and indeed consisted of bent, fog and fine-leaved fescue grasses), would suggest that, in order to improve swards of this type, **it would be more economical to invest the available capital, not on ploughing out, cultivating and seeding,** but in such a way as to enable the stocking and its timing to be adjusted for a year or two in such a way as to give paramount consideration to the benefit of the sward."

One cannot do better than add to this British opinion that of a German research worker, Weise (141), who, in 1954, expressed what he considers to be *the* great lesson of pasture ecology:

> "Of all the knowledge afforded us by pasture ecology, one lesson is of funda-mental importance: namely that, **thanks solely to methods of management, it is possible to alter profoundly the flora of permanent pastures, a goal which was long thought to be attainable only by ploughing up and reseeding.** The application of ecology has certainly contributed in no small measure to our better understanding of this problem."

PART TEN

SALIENT POINTS OF RATIONAL GRAZING

PHOTO 38

*Influence of method of management on the evolution of the flora
in a reseeded pasture, photographed six years after sowing*

From Martin Jones (60)

PHOTO 39

*Influence of method of management on the regeneration of the flora
in an 80-year old pasture*

From Martin Jones (60)

PHOTO 40

Farmers visiting the Voisin pastures

One visit is not enough to learn rational grazing

PHOTO 41

Chapel of Harcourt Castle (Normandy) the property of the Académie d'Agriculture de France (French Academy of Agriculture)

Photo Voisin

SYNOPSIS OF THE GENERAL PRINCIPLES

Now that the theoretical and practical aspects of rational grazing have been studied it is worth-while, from the point of view of simplifying the farmer's task, to give a short synopsis of the main points of the system.

What is rational grazing?

Rational grazing affords maximum satisfaction to the requirements of both grass and animal. It should be remembered that **grazing is the meeting of animal and grass,** and these two elements should be kept constantly in mind by the grazier. Rational grazing does not result from the mere division of a pasture or use of an electric fence. Division, whether by means of fixed or mobile fences, is not an end in itself but a means of helping to satisfy, by judicious compromise, the requirements of animal and grass.

Fundamental importance of the rest period

It is essential to allow the grass sufficient periods of rest between two grazing rotations if it is to attain an average height of 6 in. [*15* cm.] and more by the time it is made available to stock. Rest periods can be reduced to 16 days in May–June and may be as long as 100–150 days in winter. It is expedient to reckon on 36–40 days rest in summer in the North-West of Europe, this period being extended as one proceeds southwards.

Observation of these rest periods allows the grass to reconstitute its reserves and thus re-grow vigorously and produce a high rate of daily growth (that is, its "blaze of growth"). Under continuous grazing some grasses are sheared *twenty* times by the teeth of the animal in the course of a year. With rational grazing the grass is sheared only *six* times approximately, the figure varying according to the regional conditions. If lucerne were cut *ten* instead of *three* times in the course of the year, the yields one obtained would be miserable. The same is true of a pasture sheared *twenty* instead of *six* times (Table 23, p. 59).

Lucerne cut ten times in a year deteriorates and is invaded by other plants. The flora of a pasture sheared twenty times in a year similarly deteriorates; and a bad flora is just as incapable of producing results as a worn-out motor-car engine.

291

Periods of stay and occupation must be short

If the period of *occupation* is too long a grass plant will be *sheared* **twice** *in the course of the same rotation*. Its reserves will not have been replenished, and it will therefore be unable to achieve its "blaze of growth".

If the period of *stay* is too long, milk yields will be subject to periodic fluctuations, which will make the lactation curve fall more steeply.

It is recommended not to exceed 6 days for the period of occupation and 3 days for the period of stay. If these times can be reduced to 2 days and 1 day respectively, production will be considerably increased.

Division into groups

It is not absolutely essential to divide the herd into groups. With in-milk cows, however, it is advisable to have at least two groups: this allows the first group to be selective and harvest plentiful quantities of better-quality herbage. The second group, with lesser requirements, will be satisfied with reduced quantities of grass of a lower quality.

The number of paddocks is decisive in the drawing up of a rotation plan

The system of division adopted should, if possible, allow the July–August rest period to be observed. It will then be seen to what extent it is possible to realise the number of paddocks necessary to achieve this summer rest period with the shortest possible period of stay on a paddock. If, for example, with one group, one is aiming at a rest period of 36 days and a period of stay of 1 day, then $\frac{36}{1} + 1 = 37$ paddocks are required. If, on the other hand, a period of stay of two days is desired, $\frac{36}{2} + 1 = 19$ paddocks will be required.

With a 3-day period of stay $\frac{36}{3} + 1 = 13$ paddocks and with a 4-day period of stay $\frac{36}{4} + 1 = 10$ paddocks will be sufficient.

If the herd is divided into two or three groups, then one or two paddocks respectively must be added to the above figures.

It must always be remembered that it is not a case of having a certain area of paddock for a certain number of beasts. It is a question of the number of paddocks necessary for observing the requisite rest periods with the periods of stay envisaged.

The shorter the periods of stay and occupation, the better the requirements of animal and grass will be satisfied. But the number of paddocks will have to be increased.

In practice, it seems difficult to reduce the period of stay to less than 1 day. Moreover, where the period of stay is 1 day it will *be found obligatory* in most cases to divide the herd into two groups in order to avoid over-crowding.

Area and lay-out of paddocks

Having decided on the number of paddocks one wants, one must see that the paddocks are *virtually* all of the same area, that is to say, that their grass-production capacity is more or less equal. If bare paddocks have an area of 2·5 acres [*100* ares], then paddocks planted with apple-trees or situated on a poorer soil will have to measure 5·0 acre [*200* ares] in area.

Where divided pastures have watering points it is good to arrange for corridors allowing the stock permanent access to these points without too long a journey. Permanent access is not indispensable (especially where there are no milk cows), but it is desirable. It must also be remembered that one corridor only gives *one* group permanent access to a watering point, although obviously all the groups could take turns at using it. In cases where electric fencing is being used, it is preferable to make the races with fixed fencing.

With electric fencing, one fence in front and one behind for *each* group should be allowed, since groups do not always follow each other.

In general, moreover, the stock should not be allowed to graze sections that have already been grazed; this avoids excessively long periods of occupation of those sections to which the animals are returning (which can have serious consequences for the grass).

Livestock carry cannot be arranged in advance

It is impossible to establish in advance what livestock carry per acre [per ha.] a rotation will allow. Rational grazing with a 1-day period of stay allows more animals to be carried than 3-day periods of stay. Nitrogen dressings, *which only rational grazing makes allowance for*, will also bring about an increase in livestock carry. It is impossible to foresee, *a priori*, the influence of these various factors. But the grazier who adopts rational grazing will be able to look forward to a considerable increase in stocking from the very first year, that increase continuing in subsequent seasons.

A higher stocking rate raises questions of finance and problems of winter feeding, bedding and housing. It also signifies, however, increased production of excrement and farmyard manure, the foundations of soil fertility, not to mention the direct benefits accruing from improved production by the grass itself.

Balancing seasonal fluctuations in grass yield by "internal" methods

The pasture itself can hardly be expected to do more than balance up the difference in production between summer and spring. This equalisation on the part of the pasture itself is achieved by:

the system of "disengaging" and "re-introducing" paddocks into the rota;

suitable and well-proportioned dressings of nitrogenous fertilisers to stimulate re-growth when it begins to flag.

The number of paddocks is varied by dropping a certain number, which are mown, from the grazing rota in the months of May–June. These paddocks must be properly managed and the necessary nitrogen applied to make them ready for re-introduction into the grazing cycle within a sufficiently short time. This "disengaging" and "re-introduction" of mown paddocks is one of the most delicate problems of rational grazing. It often, indeed very often, happens that they cannot be used for grazing again until much too late (August, for example). This means that there is a risk of the grass supply running out in summer.

It is advisable to alternate the mown paddocks each year.

Careful use of nitrogen should not only permit earlier and later grazing: it should allow fluctuations in grass growth to be reduced. Until fuller information is available it seems preferable to use alkaline nitrogenous fertilisers, especially nitrate of lime with its ultra-rapid effect, which act even in periods of very low rainfall. If nitrogen is wisely used and the rotation well managed, the resulting pastures will be very rich in white clover and their seasonal fluctuations in growth much reduced.

Balancing seasonal fluctuations in grass yield by "external" means

Balancing by internal means, that is to say, relying on the pasture itself (variation in the number of paddocks in the rota, apportionment of nitrogen dressings) is possible in practice only if the May–June production (*without* application of nitrogen) is not more than two and a half (three, as the maximum) times that in July–August (*with* nitrogen application); there are, however, no absolute rules on this point.

If the variation is greater, it appears to be advantageous to make use, at the same time, of the classical means of achieving equilibrium:

green feeding of the animals at grass;
removal of part of the herd for selling or feeding by some other means (stall feeding, direct grazing of arable forage crops, etc.).

When arable crops (temporary leys) are grazed, this is equivalent to "re-introducing" green areas, exactly as when mown paddocks are taken back into the grazing rota.

It is essential to remember that these **"external" measures must help to observe the necessary rest periods; by no means do they dispense with this requirement.** If the herd is halved and a rest period of 16 days in August is made use of when in reality 40 days are required, the grass yield is considerably reduced, the grazing season shortened and the flora deteriorates.

In the last rotations of the season it is preferable to make use of these complementary balancing methods. There is generally no advantage in trying to balance up from the pasture itself.

The final rotation is a "revision" passage, with no imminent return. The periods of stay do not influence the rest period any more. It is necessary, however, to observe long rest periods at the end of the season: the grass is thus enabled to accumulate the necessary reserves to resist cold and, in particular, to produce vigorous re-growth next spring.

Commencing grazing at the beginning of the year

The start of grazing is extremely important because, if it is badly done, it can make management very difficult for the rest of the season. It is advisable to grade the initial growth of the different paddocks by means of graduated dressings of nitrogenous fertiliser spread out over the period February–March. Grazing of the first paddocks as soon as possible so that the last paddocks in the first rotation will not be over-mature when grazed.

The commencement of grazing each year on the same paddocks and in the same order is to be avoided.

The great danger of rational grazing: "untoward acceleration"

Of all the dangers encountered by rational grazing, the most serious is *"untoward acceleration"*.

When the vigour of grass growth begins to wane, the grazier tends to reduce the period of stay on the paddocks because there is less grass available. This reduction has the effect of *shortening* the rest period at the time when it should be *lengthened*, which in turn diminishes the volume of grass on the paddocks at the next passage. The period of stay will then have to be compressed even further, and so it goes on, the animals rotating ever more quickly over the paddocks and being faced with hardly any grass when they return. The grazier blames the summer for this failure; he should blame himself.

Untoward acceleration also takes place when rationed grazing by area is practised. Increase in the allotted area leads to the "portions" of pasture being "consumed" more quickly when in fact their consumption should be slowed down. With normal grazing the stock *do return*; this does not happen with tether grazing of crimson clover or where marrow-stem kale is grazed by means of the electric fence.

In order to avoid untoward acceleration where electric fencing is employed, it is wise to use marking pegs so that the fence is always put back in the same place.

The management of rational grazing must be flexible

The management of rational grazing must be flexible: it is very rare, if not exceptional, for paddocks always to be capable of being grazed in the same order.

Where several groups are involved, it is uncommon for the groups to follow each other in numerical order over the paddocks. The art of the rational grazier consists of being able to jump a paddock:

that is not sufficiently advanced so that the grass can attain the required height (6 in. [*15* cm.]) and thus provide its "blaze of growth";

that is too advanced to allow the grass to mature sufficiently to be mown.

The grazier must equally be able to return at the right time to a paddock he has previously "jumped". He must watch for variations in the vigour of growth so as to:

increase dressings of nitrogen if the vigour is reduced;

reduce, or even withhold, nitrogen if vigour is increased.

Following upon unforeseen variations in growth, the grazier will likewise vary the number of plots in or out of the rota.

Generally, the greater the number of paddocks the pasture is divided into, whether by means of fixed or movable fences, the easier it is to introduce flexibility into one's management.

Increases in yield will be considerable

If rational grazing is well managed (large number of paddocks, fairly short periods of stay, judicious use of nitrogen and flexibility in handling) the grazier will quickly obtain:

very obvious improvement of the flora;

doubled (or even greater) yields per acre [per hectare].

PART ELEVEN
WEALTH OF OUR PASTURES

Chapter 1

SIMPLIFIED METHOD OF CALCULATING PASTURE PRODUCTIVITY

Calculating the production of a pasture

IN studying the basic elements of rational grazing, it was seen that production of a pasture could be calculated in terms of animal unit grazing days, or more simply, cow-days (*vide* Table 40, p. 139). Another method, perhaps better known, consists in measuring the performance of the grazing animal (maintenance, milk, meat, etc.) and calculating the starch equivalent value therefrom.

Production of a pasture in starch equivalent

Table 61 lists the necessary starch equivalent for maintenance, milk production and growth. Expenditure of 2·2 lb. [*1* kg.] starch equivalent has been allowed for the harvesting of grass. The approximate nature of this allowance has already been referred to (pp. 87 and 88).

Assume that a paddock 5 acres [*2* ha.] in area has provided in the course of the grazing season:

    1. 700 animal unit grazing days (or cow-days) (10 cwt. [*500* kg.] live weight).

    2. The production of 11,000 lb. [*5000* kg.] of milk with 3·5% fat.

    3. A live-weight gain of 880 lb. [*400* kg.] with beasts weighing less than 10 cwt. [*500* kg.].

    4. The birth of six calves.

    5. 40 cwt. [*2100* kg.] of hay (cut from plots dropped from the rota).

On the basis of such performance the production of the paddock will be (*vide* Table 61):

	English Units		*Metric System*
(1)	$700 \times 770 = 5390$		$700 \times 3\cdot50 = 2450$
(2)	$11{,}000 \times 0\cdot25 = 2750$		$5000 \times 0\cdot25 = 1250$
(3)	$880 \times 2\cdot50 = 2200$		$400 \times 2\cdot50 = 1000$
(4)	$6 \times 165 = 990$		$6 \times 75 = 450$
(5)	$40 \times \dfrac{77}{2} = 1540$		$20 \times 35 = 700$

                  12,870 lb.   5850 kg.

This paddock produced:

$$\text{per acre: } \frac{12,870}{5} = 2574 \text{ lb.}$$
$$\left[\text{per hectare: } \frac{5850}{2} = 2925 \text{ kg.}\right]$$ starch equivalent

TABLE 61

Starch equivalent required for maintenance and production by grazing cows

	Per unit of live weight	
	lb. starch equivalent per 2 cwt.	[kg. starch equivalent per *100* kg.]
Maintenance (including energy used up harvesting grass): Bullocks and cows—		
4–8 cwt. [*200–400* kg.] live weight . .	1·76	[*0·80*]
8–12 cwt. [*400–600* kg.] live weight . .	1·54	[*0·70*]
More than 12 cwt. [*600* kg.] live weight . .	1·32	[*0·60*]
	lb. starch equivalent per lb.	[kg. starch equivalent per kg.]
Production: Per lb. [or kg.] of milk—		
3·0–3·5%	0·24	[*0·24*]
3·5–4·0%	0·25	[*0·25*]
Over 4%	0·28	[*0·28*]
		0·30	[*0·30*]
Growth: Per lb. [or kg.] of live-weight gain—		
Up to 10 cwt. [*500* kg.] live weight . .	2·50	[*2·50*]
Over 10 cwt. [*500* kg.] live weight . .	3·50	[*3·50*]
	Per unit considered	
	Starch equivalent	
	lb.	[kg.]
Calf born outside	165	[*75*]
1 animal unit (10 cwt. [*500* kg.] live weight) . .	7·7	[*3·5*]
Production of—		
2 cwt. [*100* kg.] of hay . . .	77	[*35*]
2 cwt. [*100* kg.] of fresh grass . . .	31	[*14*]

N.B. It is assumed that 2·2 lb. (*1* kg.) of starch equivalent is expended by 10–12 cwt. [*500–600* kg.] animal on harvesting grass (see p. 87).

From Geith (28, 29, 89 and 90), Klapp (70, p. 435) and Schmidt (89).

Supplementary feeding complicates calculation

Animals can be fed with various other feedingstuffs to supplement the grass they are harvesting. Even in regions where grazing is more or less the animal's *only* source of food for several months of the year, there is an intermediate period at the beginning and end of the grazing season when feeding supplements supply a very large part of the total ration.

Assume that in the course of the grazing season a 25-acre [*10* ha.] pasture has provided 77,000 lb. [*35,000* kg.] starch equivalent and that at the same time the animals received supplementary feeding to the extent of:

1. 3,300 lb. [*1,500* kg.] oats.
2. 2,200 lb. [*1,000* kg.] linseed cake.
3. 8,800 lb. [*4,000* kg.] hay.
4. 11,000 lb. [*5,000* kg.] wheat straw.
5. 44,000 lb. [*20,000* kg.] ensiled sugar beet tops.

On the basis of the starch-equivalent values of the above feedingstuffs provided by feed tables, it can be calculated that, over the grazing season, supplementary feeding amounted to:

1. $3,300 \times 0.63 = 2,070$ [*1,500 × 0·63 = 945*]
2. $2,200 \times 0.69 = 1,518$ [*1,000 × 0·69 = 690*]
3. $8,800 \times 0.35 = 3,080$ [*4,000 × 0·35 = 1,400*]
4. $11,000 \times 0.18 = 1,980$ [*5,000 × 0·18 = 900*]
5. $44,000 \times 0.09 = 3,960$ [*20,000 × 0·09 = 1,800*]

$$\overline{12,608 \text{ lb}} \qquad [\textit{5,735 kg.}] \text{ starch equivalent}$$

The starch equivalent of the supplementary feeding is deducted from the total production from the 25 acres [*10* ha.] of pasture, giving an actual starch equivalent production from grazing alone of:

$$77,000 - 12,600 = 64,392 \text{ lb.} \quad [\textit{35,000} - \textit{5735} = \textit{29,265 kg.}]$$

This is equivalent to a production per acre [ha.] of pasture of:

$$\frac{64,392}{25} = 2580 \text{ lb.} \quad \left[\frac{\textit{29,625}}{\textit{10}} = \textit{2962 kg.}\right] \text{ starch equivalent.}$$

Production due to pasture alone therefore represents:

$$\frac{64,392 \times 100}{77,000} = 85\% \quad \left[\frac{\textit{29,265} \times \textit{100}}{\textit{35,000}} = \textit{85\%}\right]$$

of the total production of the animals at grass.

Actual and effective livestock carry

On the assumption that the production of 77,000 lb. [*35,000* kg.] starch equivalent in the above example was furnished by a livestock carry of 20 animal units, the *effective* carry exerting pressure on the pasture was

$$20 \times 0.85 = 17 \text{ animal units}$$

since 15% of the feeding was in the form of various supplements.

If the milk produced by this herd was equal to 6562 gal. [*30,000* litres], then the pasture produced really:

$$6{,}562 \times 0{\cdot}85 = 5{,}578 \text{ gal. } [30{,}000 \times 0{\cdot}85 = 24{,}500 \text{ litres}]$$

I worked out this simple method of calculation to follow the production of my pastures. I believe that it makes the latter clearer, and so I have used it in the following pages to calculate the production of my swards in 1954.

The reader is reminded once more that the starch-equivalent method of calculation does not take into consideration protein production, that valuable foodstuff which is often in short supply. And, finally, satisfying 15% of the starch equivalent produced by means of supplementary feeding does not mean that 15% of the cow's *appetite* is being satiated.

Chapter 2

PRODUCTION OF THE VOISIN PASTURES IN 1954

Why I chose 1954 production

THE year 1954 was a very peculiar one. The weather was extremely cold and dry in April and May, so much so that for the first time in eight years of rational grazing I was unable to take a grass cut in the spring. But even worse was to come: we almost ran out of grass for grazing in May. In June the grass grew vigorously, but there was never what one could call a sudden "nervous" spurt of growth. Management was therefore extremely difficult.

The Director of Agricultural Services and the Department of Agriculture had sent me a young agricultural advisor who, at the same time as learning rational grazing, was carefully recording stock movements, yields obtained, changes in live weight, etc. Unfortunately he could not stay with me until the end of the season, but the surveys he made, independently of me, have a certain stamp of official guarantee.

It so happened that this was my last year of working with *three* groups: since then I have used only two. Also, I applied a relatively moderate quantity of nitrogen, namely 57 lb./acre [*63* kg./ha.]. My 1955 production was higher than that of 1954, but in that year I used 100 lb./acre [*110* kg./ha.] of nitrogen, and I have the idea that reducing the period of occupation of the paddocks from 6 days to 4 days (by reducing the number of groups from three to two) also improved the output.

The danger of quoting production figures

I hesitated to publish the results given below, because they are exceptional figures, achieved at the cost of eighteen years of farming thought, twelve years of rational grazing practice, many mistakes and very great personal sacrifices.

Basic elements of Voisin grazing in 1954

The total grazing area was 37·5 acres [*15·2* ha.] divided into 19 paddocks of say 2 acres [*80* ares]. Paddocks 1–8 comprised only very old pasture while one-tenth of paddocks 9–17 were very old pasture and the other nine-tenths a pasture sown down in 1947 (*vide* Table 22, p. 52). Paddocks 18–19 consisted solely of pasture sown in 1947.

TABLE 62

Livestock carry on Voisin's "rationally" grazed pastures in 1954

Period	No. of days	Groups I and II						Group III						Total effective cow-days
		Live weight		% fed by grass alone	Live weight fed by grass alone		Effective animal units	Effective cow-days	Live weight		% fed by grass alone	Live weight fed on grass alone		Effective animal units	Effective cow-days	
		lb.	[kg.]		lb.	[kg.]			lb.	[kg.]		lb.	[kg.]			
					A	A	B	C				D	D	E	K	(C + K)
1. Apr. 3–May 10	38	52,646	[23,880]	80	42,130	[19,110]	38·22	1452	32,769	[14,864]	95	31,131	[14,121]	28·24	1073	2,525
2. May 11–July 31	82	53,550	[24,290]	95	50,927	[23,100]	46·20	3788	40,389	[18,320]	100	40,389	[18,320]	36·64	3004	6,792
3. Aug. 1–Sept. 22	53	54,454	[24,700]	90	48,568	[22,030]	44·06	2335	46,297	[21,000]	100	46,297	[21,000]	42·00	2226	4,561
4. Sept. 23–Oct. 10	18	55,997	[25,400]	85	47,620	[21,600]	43·20	777	6,614	[3,000]	100	6,614	[3,000]	6·00	100	877
5. Oct. 11–Oct. 20	10	55,997	[25,400]	70	39,198	[17,780]	35·56	356	6,614	[3,000]	100	6,614	[3,000]	6·00	60	416
6. Oct. 21–Nov. 23	34	57,320	[26,000]	30	17,196	[7,800]	15·60	530	49,825	[22,600]	100	49,825	[22,600]	45·20	1587	2,067
7. Nov. 24–Dec. 16	23	—	—	—	—	—	—	—	50,706	[23,000]	70	35,494	[16,100]	32·20	741	741
Total	258							9238							8741	17,979
													Average{per acre / [per ha.]		473	[1,170]

Effective livestock carry								Effective stocking density				Grazing intensity				Area required for the daily ration of 1 effective animal unit	
Net			Per unit of area							Animal units				Animal unit days (Besatzleistung)			
						Animal units		lb./acre	[kg./ha.]			lb./acre days	[kg./ha.] days				
lb.	[kg.]	Animal units	lb./acre	[kg./ha.]	per acre	per hectare				per acre	per hectare			per acre	per ha.	sq. yd.	sq. m.[1]
(A + D)	(A + D)	(B + E)															
73,261	[33,231]	66	2043	[2290]	1·9	[4·6]		12,310	[13,800]	11	[27]	73,900	[82,800]	67	[166]	72	[60·0]
91,316	[41,420]	83	2427	[2780]	2·2	[5·4]		15,350	[17,200]	14	[34]	92,100	[103,200]	83	[206]	58	[48·5]
94,865	[43,030]	86	2525	[2830]	2·3	[5·7]		15,970	[17,900]	15	[36]	95,800	[107,400]	87	[215]	56	[46·5]
54,234	[24,600]	49	1440	[1615]	1·3	[3·2]		13,740	[15,400]	13	[31]	82,400	[92,400]	75	[185]	65	[54·0]
45,812	[20,780]	41	1219	[1367]	1·1	[2·7]		11,600	[13,000]	11	[26]	69,600	[78,000]	63	[156]	77	[64·0]
67,021	[30,400]	61	1784	[2000]	1·6	[4·0]		11,240	[12,600]	10	[25]	67,500	[75,600]	61	[151]	79	[66·0]
35,494	[16,100]	32	944	[1058]	0·8	[2·1]		17,840	[20,000]	16	[40]	35,700	[40,000]	32	[80]	149	[125·0]

N.B. 1. The 3rd group was not separated on the pastures during the 4th and 5th periods but was mixed with the 2nd group. The 3rd group is shown separately in the Table for these two periods since Groups I and II, but not III, received supplementary feed during this time.

2. The effective stocking density was calculated for 3 paddocks of 2 acres [0·80 ha.] each, total of 6 acres [2·40 ha.], for periods 1, 2, 3, 6: for 2 paddocks of 2 acres [0·80 ha.] each, total of 4 acres [1·60 ha.], for periods 4 and 5; and for 1 paddock of 2 acres [0·80 ha.] for period 7.

3. The figures for grazing intensity were calculated for 6 days of occupation during periods 1, 2, 3, 4, 5, 6, and for 2 days during period 7. These periods of occupation are the theoretical base round which the actual times of occupation are situated.

For the last four years the whole area had received 800 lb./acre [*900* kg./ha.] 16% basic slag and prior to that 450 lb./acre [*500* kg./ha.] of potassium slag 12–12. 4500 lb./acre [*5000* kg./ha.] ground marl were applied in 1947 and again in 1954. 180 lb./acre [*200* kg./ha.] potassium sulphate were applied irregularly in two annual dressings to the parts farthest away from the gate and watering point.

The stock began to go out to grass on April 3 and 4, the transition from stall to grazing being gradual. Groups I (cows milked thrice daily) and II (cows milked twice daily) stayed overnight in the shed until May 10. On May 11 part of the third group was put on to wooded pastures at some distance from the farm and replaced by young stock that had never previously been out of the farmyard itself. From May 11 until September 23 (2nd and 3rd periods) all groups were entirely on grass. Only a few very high yielders were fed a supplement of concentrates in the stall (before one milking).

Throughout the three grazing periods the period of stay of each group on a paddock was 2 days, which meant a *theoretical* period of occupation of 6 days.

On September 23 part of the third group started tethering young red clover that had been sown in that year under oats (harvested at the beginning of August). The tethered animals represented 400 cwt. live weight [*20,000* kg.] or 40 livestock units. In periods 4 and 5 only two groups were in use, each remaining in principle for 3 days on a paddock, making a theoretical occupation period of 6 days. From October 11 onwards the two groups spent the night in the stall and received supplementary feeding.

On October 21 the tethered stock came back on to the rotational pasture. The cows now were only out at grass for a few hours during the warmest part of the day (6th period).

From November 24 the cows no longer left the shed, and only one group remained, the third, which carried out a "cleaning-up" of the paddocks that still had a little grass left. This group received partial supplementary feeding.

Effective stocking of beasts fed from pasture alone

During the transition periods at the beginning and end of grazing, the animals were living partly in the stall and partly on the pastures. Moreover, during the grazing season itself, high-yielding cows were fed a ration of concentrates. In the 1954 rotation I fed ensiled (wet) pulp from August 1 onwards and increased the ration gradually, especially from October 10, when the cows were no longer out at night.

In the Research Stations daily recordings are made of milk production, individual rations fed, gains or losses in live weight calves born, etc. But for a farmer like myself it is impossible to weigh feeds every day. I therefore had to be content with taking sample weights over a few days and calculating from these the *percentage* of the maintenance and production rations that could be *credited* to the pasture itself, according to my own method as outlined above.

Table 62 (pp. 304 and 305) shows the *effective* and actual stockings carried by my pastures at different times during the 1954 season.

Some basic results of the Voisin rational grazing in 1954

The following observations can be made with regard to Table 62:

1. The number of effective animal units grazing days (or cow-days) was 473 per acre [*1170* per ha.] for the season.

2. The total effective livestock carry during the main part of the grazing season (May 10–September 23) was approximately 2·2 livestock units per acre [*5·5* per ha.], although it is the exception in this region to stock up to 0·8 beasts per acre [*2* beasts per ha.]. More often the rate is 0·6 beasts per acre [*1·5* per ha.], which was my own normal stocking rate before I started rational grazing.

3. Except for the last period (final passage for cleaning-up), the effective stocking density varied between 11,600 and 15,300 lb./acre [*13,000* and *17,000* kg./ha.], and the intensity of grazing between 67,000 and 96,000 lb./acre/days [*75,000* and *107,000* kg./ha./days] or a *Besatzleistung* of 61–87 livestock grazing days per acre [*150–215* per ha.].

Area required to supply the daily ration of one livestock unit

The last column on the right of Table 62 shows the area required at different periods of the year to supply the daily grass ration of one livestock unit. Leaving out the last period, this varies between 56 and 79 sq. yd.

TABLE 63

Quantity of harvestable grass available from Voisin's "rationally" grazed pastures in 1954

Period	Area necessary for daily ration of one animal unit		Quantity of harvestable grass available	
	sq. yd.	[m.²]	lb./acre	[kg./ha.]
1. April 3–May 10 .	72	[*60·0*]	6,720	[*8,000*]
2. May 11–July 31 .	58	[*48·5*]	8,340	[*9,820*]
3. Aug. 1–Sept. 22 .	56	[*46·5*]	8,640	[*10,320*]
4. Sept. 23–Oct. 10 .	65	[*54·0*]	7,440	[*8,880*]
5. Oct. 11–Oct. 20 .	77	[*64·0*]	6,280	[*7,500*)
6. Oct. 21–Nov. 23 .	79	[*66·0*]	6,120	[*7,270*]
7. Nov. 24–Dec. 16 .	149	[*125·0*]	3,250	[*3,840*]

N.B. 1. Figures taken from Table 62, pp. 304 and 305.
    2. It was supposed that an animal unit (with 10 cwt. [*500* kg.] live weight) harvests per day 100 lb. [*48* kg.] of fresh grass (see preliminary note).
    The calculation formula of the quantity of harvestable grass available is:
    (*a*) British system

$$\frac{100 \times 4840}{\text{Area necessary for one daily ration}}$$

    (*b*) Metric system

$$\frac{48 \times 10,000}{\text{Area necessary for one daily ration}}$$

[*46·5* and *66* m.²]. Assuming the average daily ration of a livestock unit (scraping the sward bare) to be 100 lb. [*48* kg.], this means that the amounts of *harvestable* grass available have varied (approximately) between:

$$8640 \text{ lb./acre } [10,320 \text{ kg./ha.}]$$

and

$$6120 \text{ lb./acre } [7270 \text{ kg./ha.}].$$

The quantities of harvestable grass *theoretically* available at the beginning of each rotation are listed in Table 63. It was impossible for me to verify these quantities exactly with the means at my disposal.

Meat and milk production

Table 64, below, records the live-weight gains of twenty young beasts in the *third* group between the commencement of grazing and September 23, the date when they went on to tether-grazing young red clover. The objection has often been raised (as already stated) that there is the risk of insufficient keep being available for the last group. It is therefore interesting (and re-assuring) to see that the average live-weight gain of twenty animals in the third group was almost 1·43 lb. [650 gm.] per day.

Taken as a whole, the first two groups (in-milk cows), in the course of the season, showed a live-weight gain of 1650 lb. [750 kg.] and the third group a gain of 18,960 lb. [8600 kg.], making a total gain by the herd of 20,610 lb. [9350 kg.]. These are actual gains. No account has been taken of calves born.

Milk production was checked once per month. The figures are contained in Table 65 (p. 309). The percentage method of "crediting" the pasture with effective production due to grass alone is used. This shows a total

TABLE 64

Live-weight gain on twenty animals in the third group in Voisin's "rationally" grazed pastures, 1954

No. of animal	Date of Start of grazing	Date of Departure to tethered grazing	Number of grazing days	Live weight at Start of grazing lb.	[kg.]	Live weight at Departure to tethered grazing lb.	[kg.]	Live-weight gain Total lb.	[kg.]	Live-weight gain Per day lb.	[kg.]
17	April 3	Sept. 23	173	719	[326]	992	[450]	273	[124]	1·58	[716]
18	,,	,,	,,	632	[287]	899	[408]	267	[121]	1·54	[698]
19	,,	,,	,,	569	[258]	884	[401]	315	[143]	1·82	[824]
20	,,	,,	,,	549	[249]	820	[372]	271	[123]	1·57	[710]
21	,,	,,	,,	445	[202]	686	[311]	241	[109]	1·38	[628]
401	,,	,,	,,	553	[251]	780	[354]	227	[103]	1·31	[594]
387	,,	,,	,,	809	[367]	1074	[487]	265	[120]	1·53	[692]
392	,,	,,	,,	639	[290]	862	[391]	223	[101]	1·29	[583]
393	,,	,,	,,	681	[309]	888	[403]	207	[94]	1·19	[542]
396	,,	,,	,,	527	[239]	747	[339]	220	[100]	1·27	[577]
397	,,	,,	,,	597	[271]	855	[388]	258	[117]	1·49	[675]
399	May 10	,,	136	582	[264]	800	[363]	218	[99]	1·61	[730]
400	,,	,,	,,	597	[271]	776	[352]	179	[81]	1·31	[596]
402	,,	,,	,,	710	[322]	919	[417]	209	[95]	1·54	[700]
403	,,	,,	,,	741	[336]	917	[416]	176	[80]	1·30	[590]
404	,,	,,	,,	633	[287]	842	[382]	209	[95]	1·54	[700]
405	,,	,,	,,	725	[329]	882	[400]	157	[71]	1·15	[522]
406	,,	,,	,,	560	[254]	745	[338]	185	[84]	1·36	[618]
408	,,	,,	,,	582	[264]	741	[336]	159	[72]	1·17	[530]
409	,,	,,	,,	505	[229]	728	[330]	223	[101]	1·64	[744]
			Total	12,355	[5605]	16,837	[7638]	4482	[2033] Average:	1·43	[650]

production from grass of 173,610 lb. [*78,750* kg.] milk, equal to **4620 lb. milk per acre [*5180* kg./ha.] or 205 lb. butter per acre [*230* kg./ha.] for the grazing season.** (The average fat content of the milk was 3·76%.)

TABLE 65

Production of milk and butter with Voisin's "rational" grazing system in 1954

Month	Average daily production of				Days of month when cows were on this grass	% of milk or butter produced from grass *alone*	Net production of grassland alone of			
	Milk		% of butter-fat	Butter			Milk		Butter	
April	lb. 1120	[kg.] [*508*]	3·7	lb. 47·6	[kg.] [*21·6*]	26	80	lb. 23,300	[kg.] [*10,570*]	lb. 1014	[kg.] [*460*]
May	1071	[*460*]	3·5	44·1	[*20·0*]	10 21	80 95	8,580 21,300	[*3,890*] [*9,660*]	355 880	[*161*] [*399*]
June	1036	[*470*]	3·7	44·1	[*20·0*]	30	95	29,540	[*13,400*]	1288	[*584*]
July	988	[*488*]	3·7	41·9	[*19·0*]	31	95	29,100	[*13,200*]	1265	[*574*]
Aug.	888	[*403*]	3·9	41·0	[*18·6*]	31	90	24,800	[*11,040*]	1140	[*517*]
Sept.	750	[*340*]	3·8	33·7	[*15·3*]	23 6	90 85	15,520 3,810	[*7,040*] [*1,730*]	690 170	[*313*] [*77*]
Oct.	635	[*288*]	3·9	29·1	[*13·2*]	10 11 10	85 70 30	5,400 4,890 1,900	[*2,450*] [*2,220*] [*860*]	247 225 88	[*112*] [*102*] [*40*]
Nov.	791	[*359*]	3·8	35·5	[*16·1*]	23	30	5,470	[*2,480*]	243	[*110*]
					Total	232		173,610	[*78,750*]	7605	[*3449*]
					Production per acre Production per hectare			4,620	[*5,180*]	205	[*230*]

TABLE 66

Production of starch-equivalent from Voisin's "rationally" grazed pastures in 1954

It has been assumed that there were:

7·70 lb. [*3·50* kg.] S.E. per animal unit day.
2·50 S.E. per lb. [per kg.] live-weight gain.
0·28 S.E. per lb. [per kg.] milk with 3·7–3·8% butter-fat.

This gives:

Effective animal unit days—

Groups I and II } 9,238   *9,238*
Group III   8,741   *8,741*

$$17,979 \times 7.70 = 138,438 \text{ lb. S.E. or } 17,979 \times 3.50 = 62,926 \text{ kg. S.E.}$$

Effective live-weight gain—

Groups I and II } 1,650 lb.   *750* kg.
Group III 18,960 lb.   *8,600* kg.

$$20,610 \text{ lb. } \times 2.50 = 51,525 \text{ S.E. or } 9,350 \text{ kg. } \times 2.50 = 23,375 \text{ kg. S.E.}$$

Effective quantity of milk— 173,610 lb. × 0·28 = 48,611 lb. S.E. or *78,750* kg. × *0·28* = *22,050* kg. S.E.

Total   238,574 S.E.   *108,351* kg. S.E.

Production per unit of area during the grazing season:

$$\frac{238,574}{37 \cdot 5} = 6,360 \text{ lb. S.E. per acre}$$

$$\frac{108,351}{15 \cdot 2} = 7,130 \text{ kg. S.E. per hectare}$$

Total production from the Voisin rational grazing in 1954

Table 66 shows the total production of the Voisin pastures in 1954. This can be divided up as follows:

> *Maintenance:* 17,979 animal unit grazing days (or effective grazing days).
> 20,610 lb. [*9350* kg.] live-weight gain (effective).
> 173,610 lb. [*78,750* kg.] milk (effective).

This is a total production of 238,574 lb. [*108,351* kg.] starch equivalent, *which is equal to a production per acre [per ha.] for the grazing season of:*

> 6360 lb. [*7130* kg.] starch equivalent.

Analysis of this production

The production of these pastures could also be expressed in the following manner:

1. The herd has produced per acre [per ha.]

$$\frac{20,613 \text{ lb.}}{37 \cdot 5} = 549 \text{ lb.} \quad \left[\frac{9350}{15 \cdot 2} = 616 \text{ kg.} \right] \text{ meat}$$

and

$$\frac{173,610 \text{ lb.}}{37 \cdot 5} = 4620 \text{ lb.}$$

$$\left[\frac{78,750 \text{ kg.}}{15 \cdot 2} = 5180 \text{ kg.} \right] \text{ milk or 205 lb. [230 kg.] butter.}$$

2. The starch equivalent values represented by live-weight gain and milk production are almost equal: 51,525 lb. [*23,375* kg.] against 48,611 lb. [*22,050* kg.] (Table 66). One or other of these productions could therefore be doubled if grazing was devoted *exclusively* to one type of production. It can thus be said that in 1954 1 acre [*1* ha.] of grass produced:

$$549 \times 2 = 1098 \text{ lb. [or } 616 \times 2 = 1232 \text{ kg.] of meat,}$$
or $$4620 \times 2 = 9244 \text{ lb. [or } 5180 \times 2 = 10,360 \text{ kg.] of milk.}$$

These figures certainly go to dizzy heights, and should be considered exceptional.

Voisin pastures under continuous and under rational grazing: a comparison of productions

Part of the old pasture included in the 37·5 acres [*15·2* ha.] under rational grazing was previously devoted to continuous grazing. This pasture received a little phospho-potassic fertiliser at irregular intervals and no nitrogen at all. The livestock carry was slightly less than 0·8 livestock units per acre [*2* livestock units per ha.].

From a succession of calculations I have found that the actual production to be credited to grazing varied between 1800 and 2200 lb. [*2000* and *2500* kg.] starch equivalent per acre [per ha.] per annum. It can therefore be estimated that *rational* grazing, with the assistance of fairly large fertiliser dressings has multiplied the production obtained under *continuous* grazing by:

$$\frac{6360}{2000} = 3\cdot2 \quad \left[\frac{7170}{2250} = 3\cdot2\right]$$

In other words **rational grazing has tripled the yield of continuous grazing.**

The present comparison is not completely fair to rational grazing, for the continuous grazing refers only to old permanent pasture, while under the rational grazing system, almost half the total comprised pasture sown down in 1947. The seedings were very successful: there were two very productive years and then as always, the grass yield diminished. The "hungry years" (*vide* Voisin, 122) were very marked up until 1952, when the yields began to increase again. The young section of my pasture in 1954 was therefore far from having emerged from its "hungry years", and the production it had to offer was clearly inferior to that of the old pastures.

Chapter 3

COMPARISON OF YIELDS OBTAINED FROM ARABLE CROPS AND PERMANENT PASTURES

Production of different German and British pastures

BEFORE comparing the yields of arable crops and permanent pasture, it is of interest to examine the yields quoted for various types of pasture in Germany and Britain (Tables 67 and 68). According to Table 67, my 1954 production of 6360 lb. [*7130* kg.] of starch equivalent (Table 66, p. 309) is far from representing a maximum.

TABLE 67

Annual yield from different types of pasture in Germany

Type of pasture	Variation limits in starch equivalent		Average starch equivalent	
	lb./acre	[kg./ha.]	lb./acre	[kg./ha.]
Rough grazing . . .	350–700	[*400–800*]	540	[*600*]
Continuous grazing in average pastures . . .	700–1400	[*800–1600*]	540	[*600*]
Continuous grazing on good pastures or Poor rotational grazing on average pastures .	900–1800	[*1000–2000*]	1400	[*1600*]
Average rotational grazing .	1400–2100	[*1600–2400*]	1800	[*2000*]
Good rotational grazing .	2200–3100	[*2500–3500*]	2700	[*3000*]
Very good rotational grazing .	3100–4000	[*3500–4500*]	3600	[*4000*]
Exceptional production .	4000–7000 and more	[*4500–8000*] and more	7000	[*8000*]

From Klapp (70, p. 437).
See also Geith (29, p. 11) and Schützhold (95).

Comparison of arable and pasture yields in the *Pays de Caux*

Table 69 reproduces yields obtained in the *Pays de Caux* (my own region). These should be considered as maximum rather than as average yields.

It is immediately evident that the yield of my pastures under *continuous* grazing management, namely 1800–2200 lb. [*2000–2500* kg.] starch equivalent, is inferior to that obtained from arable crops. On the other hand, my

TABLE 68

The influence of grazing method on yield of herbage in Great Britain

Method of grazing and stock used	Production of nutrients			
	Digestible crude protein		Starch equivalent	
	cwt./acre	[kg./ha.]	cwt./acre	[kg./ha.]
Extensive grazing of young cattle .	2·6	[300]	19·6	[2480]
Extensive grazing of beef cattle on ley	2·5	[318]	20·9	[2650]
Extensive grazing of dairy cattle . .	3·0	[380]	22·6	[2870]
Rotational grazing of young cattle .	3·0	[380]	22·8	[2890]
Rotational grazing of dairy cattle . .	4·5	[570]	31·0	[3940]
Close-folding dairy cattle . . .	6·0	[760]	40·0	[5080]

From Holmes (43).

rationally managed pastures produced a yield superior to that of any arable crop. It should be noted that my arable crops and rationally managed grassland are situated on analogous soils, under similar conditions, namely, flat plateau loam.

TABLE 69

Average production from Arable Land in the Pays de Caux

		lb./S.E./acre			kg./S.E./ha.
Wheat					
27·8 cwt./acre of grain at 80·9 S.E.	=	2249	35 qx./ha. of grain at 72·2 S.E.	=	2527
2·4 tons/acre of straw at 305 S.E.	=	732	6000 kg./ha. of straw at 0·136 S.E.	=	816
		3081			3343
Oats					
20·7 cwt./acre of grain at 70·8 S.E.	=	1465	26 qx./ha. of grain at 63·2 S.E.	=	1643
1·6 tons/acre of straw at 408 S.E.	=	653	4000 kg./ha. at 0·182 S.E.	=	728
		2118			2371
Mangolds					
27·9 tons/acre of roots at 125 S.E.	=	3488	70 tonnes/ha. of roots at 56 S.E.	=	3920
4·8 tons/acre of tops at 173 S.E.	=	830	12 tonnes/ha. of tops at 77 S.E.	=	924
		4318			4844
Red Clover					
(a) Grazing during 1st year:					
89 animal units days at 7·70 S.E.	=	685	220 animal unit days at 3·50 S.E.	=	770
58 lb./acre live-weight gain at 2·50 S.E.	=	145	65 kg./ha. live-weight gain at 2·50 S.E.	=	162
		830			932
(b) 2 cwt. of hay in 2nd year:					
4·0 tons/acre at 784 S.E.	=	3136	10,000 kg./ha. at 0·35 S.E.	=	3350
		3966			4432

N.B. For tethered grazing in the 1st year it has been supposed that 50 animal units graze 17·3 acres [7 ha.] for 30 days. This gives $\frac{50 \times 30}{17·3}$ = 87 animal unit days per acre [approximately *220* per hectare], which is equivalent to a daily ration of 56 sq. yd. [*47* m.²].

Comparative costs

Absolute costs, in agriculture, have a very limited value; comparative costs are more interesting and of more value.

It was stated above that in 1954 my rational grazing represented per acre [per ha.]:

549 lb. [*616* kg.] meat

and   6420 lb. [*5180* kg.] milk,

the monetary value of which is approximately:

$$549 \times \frac{1}{20}^{1} = \pounds 28$$

$$4622 \times \frac{0.125}{10.32}^{2} = \pounds 56$$

$$\pounds 84$$

Sugar beet will be taken as the basis of comparison, since it is the crop with the highest *gross* cash production. On the assumption that the yield per acre [per ha.] is:

> 14 tons [*32 tons*] of roots (1 ton/acre = 2·275 kg./ha.)
> 9 tons [*20 tons*] of tops
> 7 tons [*17 tons*] of pulp

In Britain, where roots are sold outright to refineries, the equivalent value would be, per acre:

> 11½ tons of roots £70 [3]
> 7 tons of tops £ 7
> £77

The *gross* cash production of a pasture under rational grazing management is therefore superior to that of sugar beet, which, moreover, has probably received more than 56 lb./acre [*63 kg./ha.*] nitrogen. And one has only to think of all the various operations and expenses involved in the growing of sugar beet to realise that the *net* returns would be even more in favour of rational grazing.

Statistics true but at the same time false

Statistics always show that the yield of pastures is less than that of arable crops. Klapp has the following to say on the subject (70, p. 7).

> "According to the statistics, German pastures produce only 80–90% of the production of cereal crops on arable land. This is not too bad if one bears in mind that a large proportion of the pasture is on very poor soil where any other type of cultivation would be impossible. Moreover, the attention given to pasture is slight by comparison with the care required by arable crops.
> "*Statistical surveys reveal in particular the extent to which the question of rational pasture management has been neglected in all the countries of the world.*"

My figures confirm Klapp's conclusion. *When I utilised my grassland for continuous grazing, I was producing 1800–2000 lb.* [*2000–2500 kg.*] *starch equi-*

---

[1] 1s. per lb.   [2] 2s. 6d. per gallon.   [3] 123s. per ton for 15½% sugar content.

valent per acre [per ha.] per annum, which is less than my arable crops (vide Table 69). **But when I applied rational management to my pastures, their yield was tripled and the starch equivalent production exceeded that of any arable crop** (on equivalent soils under similar conditions).

Statistics comparing the yields of pasture and arable land present a true picture in the *present* circumstances of pasture management; but they become quite false (at least for many regions) when pasture is *rationally managed*. To make use of an analogy that has frequently been called to mind in this work: if one grazes continuously, the grass is sheared by the animal's teeth *twenty* times per annum as against *six or seven* times under rational grazing. If lucerne was cut *ten* times instead of the normal thrice per annum, the forage yields would be low indeed. *Continuous grazing is analogous to ten cuts of lucerne annually.*

If one wanted to compare lucerne with wheat, one would not take a lucerne crop cut ten times, but a crop cut three times per annum as the basis of comparison. But when the production of continuously grazed pasture is compared with that of wheat, what is being compared (per acre or per ha.) is the yield of pasture sheared *twenty* times a year by the animal's teeth, whereas, in fact, a well-managed sward would be sheared only *six* times.

Is this comparison valid other than in North-West Europe?

It is quite clear that in my region, Normandy, rationally managed pastures are capable of producing much more starch equivalent, protein and cash return, gross or net, than arable crops on the same soil. Is this conclusion valid for other regions? To know this one would have to apply rational management to pastures in these regions, something which can be done only when rational methods have been fully explained and made widely known. Meanwhile, the question cannot be answered.

What is certain, however, is that the improvements achieved in Normandy have been particularly obvious in dry weather. It has also been seen (Table 52, p. 247) that reduced yields of grass, due to an excessively long period of occupation and too short a period of rest, are particularly marked under unfavourable environmental conditions. The principles of rational grazing will no doubt give rise to the greatest improvements in pasture yield in regions with dry summers.

Results from the Grassland Research Station at Cleves (Germany)

My results are confirmed by those of that outstanding scientist Dr. Schütz-hold (Photo 3), who has published an excellent book (96) on methods of calculating pasture yield. Dr. Schützhold is the Director of the North Rhine–Westphalia Grassland Research Station at Cleves.

The figures he quotes for the yields of arable crops per acre [per ha.] are *maxima* rather than averages:

Wheat: 32 cwt. [*4000* kg.] of grain
57 cwt. [*7200* kg.] of straw
Beet: 17 tons [*40* tons] of roots
14 tons [*32* tons] of tops

Dr. Schutzhold uses the cereal unit (Getreidewert) (the details of which cannot be discussed here) and concludes that the production per acre [per ha.] of the maximum figures quoted above is:

20·3 [*51·7*] cereal units for wheat
48·1 [*120·3*] cereal units for beet

(*an average of 34·2 [86·0] cereal units for arable crops*).
He found the following yields per acre [per ha.] for pasture:

Continuous grazing: 8·12 [*20·3*] cereal units (or 1350 lb. [*1500* kg.] starch equivalent).

Semi-rational grazing: 16·2 [*40·5*] cereal units (or 2700 lb. [*3000* kg.] starch equivalent).

Improved rational grazing: 35·1 [*87·7*] cereal units (or 5840 lb. [*6500* kg.] starch equivalent).

Dr. Schützhold concludes that the **production allowed by rational grazing is at least equal to that of arable crops.**
It should be noted that the improved rational grazing included in these measurements is still far from having achieved the maximum possible yields, although 144 lb./acre [*160* kg./ha.] of nitrogen is applied: and Dr. Schützhold has improvements in mind to increase the yields still further.

At all events, this result confirms my own estimates: *permanent pasture under continuous grazing management produces less than arable land, but if it is managed rationally it is capable of producing far more.*

PART TWELVE
DIFFICULTIES OF YESTERDAY AND TO-MORROW

Chapter 1

THE PRINCIPLES HAVE ALWAYS BEEN KNOWN

NEW things always, and with good reason, frighten farmers. But rational grazing, apart from the use of nitrogen, **is not a new idea: its principles were known but have been forgotten**.

For thousands of years shepherds have been conversant with that extraordinary, living electric fence, the dog. An able shepherd could successively graze small areas which he limited with the aid of his dog; in this way he practised the equivalent of strip grazing. But the conscientious shepherd respected the rest periods necessary for the plants. He did not know, it is true, that a grass must reach a certain stage of development before it has accumulated the necessary reserves for normal, vigorous growth; but he did know that if he grazed red clover for a second time when it was too young or too mature, that is, if he gave it either too *short* or too *long* a rest period, the yields he obtained were very mediocre. Unfortunately the principles by which such shepherds worked have been forgotten.

The *time* factor has been forgotten

It could be said that the time factor has been neglected, but this word is not strong enough: it has been *forgotten*, and more and more so with the years. We saw above in studying some of the common mistakes made in rational grazing (pp. 189–192) that James Anderson (Photos 15 and 16) in 1777 was teaching that a paddock should be grazed again 14–19 days after the preceding grazing. Here, however, he was making the grave mistake of not varying the rest period of the grass *according to the season*. Documents published after his work, however, make even less reference to the time factor than he does. The pioneers of the Hohenheim System (pp. 193–201) concentrated their attention on the question of stocking density and its variation according to seasonal fluctuation in grass growth and almost completely neglected the factor "time". The same is true of rationed grazing.

The great protein illusion

Scientific study of animal nutrition led, at the beginning of the twentieth century, to the conclusion (Kellner, Lehmann, Armsby, et. al.), with good

319

reason, that the element lacking in rations, and therefore the limiting factor was protein. Unfortunately, however, *nitrogen and protein were, and still are, confused* (pp. 60–62). With the justifiable aim of providing a ration with a relatively higher protein content, grazing systems were evolved furnishing a herbage extremely rich in nitrogen. The disastrous consequences of this very unbalanced feeding have been outlined (pp. 117–127): a Swiss advocate of rotational grazing going so far as to admit, without a qualm, that his system provided the ruminant with a herbage that made rumination impossible (p. 198).

The combined effects of rationed grazing (badly rationed) and ploughing up of permanent pastures in producing bloat and tetany give rise, it was seen, to much anxiety.

There is no practical dissertation on grazing management

Text-books on animal nutrition devote only a few words, out of politeness, to grazing, although this feeds the animals for eight months out of twelve. They are almost solely concerned with stall feeding.

Books on grassland give detailed botanical descriptions of all herbage plants, good and bad. A little is said about fencing, ordinary and electric, and the remainder of the book is devoted to re-seeding, hay-making and ensilage. Very little is written about the practical management of grazing or its principles.

International Grassland Congresses ignored grazing methods

The *Proceedings of the 6th International Congress* occupy some 1800 pages in two large volumes; not one of these is concerned with grazing management. The *Report of the European Grassland Congress*, 1954, extends to 420 large pages of small print. The only paper dealing with pasture management is the one I myself delivered.

In the course of the discussion at the end of the latter Conference I pointed out that of the eight study sections not one had been devoted exclusively to methods of pasture management, although two had been concerned with sowing down pastures and two with forage harvesting and conservation (which provides food for only four months of the year).

My remarks gave rise to the following, discreet, official conclusion (p. 18 of the *Proceedings*):

> "Intensive grassland farming was presented by a farmer who supported his remarks with lavish documentary evidence and examples from his personal farming experience.
>
> "It would be useful if more papers could be presented on this subject, since the publicity drive now under way in all countries to popularise this method might lead to a valuable exchange of information."

It can only be hoped that this "valuable exchange of information" may one day constitute a *basic* subject of discussion at Grassland Congresses.

Chapter 2

TWO EDUCATIONAL DIFFICULTIES OF THE FUTURE

Course on grassland management

OUR Schools of Agriculture provide courses in Botany and courses in Animal Husbandry. There are also lectures on hay-making and ensiling, that is to say, on *Man's* harvesting of the grass the cows are to eat for four months of the year. But no courses are offered on methods of pasture management or the *cow's* harvesting of her own grass diet for eight months of the year. The difficulty here lies not only in our lack of knowledge of the subject but, it must be admitted, in a certain amount of scorn for a subject considered fit only for shepherds or graziers.

To-day, however, this subject of pasture management appears as a great science with many absorbing aspects worthy of many hours of instruction. Unfortunately, although such courses will be a valuable aid in popularising rational grazing, they will neverthelsss be insufficient.

Difficulties in training grassland advisory officers

Even assuming that agricultural advisers, thanks to these courses, are made completely conversant with the principles of rational grazing, this does not solve the problem of publicising and applying the latter. Advisers will have to gain experience in practical pasture management, which requires not only an understanding of geometry, but above all, a certain skill which cannot be acquired by reading books or listening to lectures. The fine points of the art of grazing management can be inculcated into agricultural advisers only if they are made to follow the course of a rational grazing system day after day for many months.

Need for many visits to the farmer

If the adviser has to follow rational grazing day by day for two seasons in order to learn the system, equally he will have to visit week by week for the same period of time the farmer who is going to try to apply it. Obviously, it is easier for an adviser to recommend ploughing up and re-seeding, in which case all he needs to do is to indicate what mixture should be sown. This will

321

require only one visit compared with the many visits required to guide a farmer along the path of rational grazing.

To use again the illustration of the boy with the untidy hair (p. 271): one visit to the barber is enough to have the boy's hair cut. Repeated advice and much patience is required to teach him to use a brush and comb.

The farmer must first be taught good pasture management

A pasture deteriorates because it is badly managed (p. 269). Management, however, is a tricky problem which no one wanted to face and which was thought to have been successfully avoided in the simple solution of ploughing out and re-seeding. In any circumstances, however, *whether it is permanent or temporary*, pasture must be suitably managed. Before any decision is taken, the method of management must be carefully examined.

The adviser must first look for faults in the management and rectify them

Reference has already been made (pp. 232 and 271) to a system of rationed grazing that was considered perfect because the electric fence was moved forward twice daily. Serious deterioration of the flora, however, was evident on the sections near the watering point, but the adviser accompanying me said that this was of no importance, since the pasture was about to be ploughed out in accordance with normal, four-yearly practice. The watering point will not change its position, so it will always be the same sections of the pasture that are subjected to ever-increasing damage.

If the adviser had known the principles of rational grazing, and particularly the importance of the time factor, he would have been able to diagnose the causes of the deterioration of certain parts of the pasture. Having cured these "disorders" of the pasture, he would have seen the extent of the improvement effected on the sick flora. A doctor tries to treat his patient so as to avoid the amputation of a limb by surgical interference. The agricultural adviser must try to restore the flora before having recourse to the surgical operation of ploughing which destroys the valuable soil structure and kills the Lilliputian ploughmen, *vide* Voisin (122).

Chapter 3

DIFFICULTIES OF RESEARCH INTO PASTURE MANAGEMENT

Hopes which remain

AT the end of a paper read to the French Animal Husbandry Association in 1951, I said:

> "Research and study up till now have dealt with the plant or with the stall-fed animal. The knowledge acquired in the fields of Botany and Animal Husbandry is certainly valuable: but its full fruits cannot possibly be borne until the consequences of the *meeting of plant and animal* have been studied and perfectly understood.
>
> "We must not forget that in many regions cows are out at grass for eight months in the year. Any advance in our knowledge of the grazing animal will bring in a relatively much greater profit than the improvement of stall rations. . . .
>
> "I hope that our young research workers will turn their attentions to the grazing animal although investigations in this field require infinitely greater material and financial resources than small plot experiments with plants or stall-feeding trials. By such investigation alone can the progress be made which is essential if we are to obtain maximum yields from our green pastures. . . ."

These words are as true and as pressing to-day as they were in 1951: their application, however, remains as difficult.

Extensive resources required to study the cow-pasture complex

Study of the cow-pasture complex embraces an enormous number of different factors, the effects of some of which, moreover, can be judged only after at least ten years or so have elapsed. Not a few animals are required, as for experiments in the stall or with the respiration apparatus, but whole herds.

With the limited resources at their disposal, it is understandable that research workers could not do more than they have done. The solution proposed by an eminent Dutch agriculturalist therefore seems worthy of attention.

323

European Research

The problem of grassland research is only one among the many difficulties facing agricultural research in all its aspects to-day. As a science develops, the expenditure incurred in its investigational work becomes ever greater until it can hardly be borne by one country alone. For this reason the suggestion put forward by Mr. Frankena, Director of the Netherlands Ministry of Agriculture, seems to me to be very sound. On May 25, 1955, he wrote me as follows:

> "Propagation of results obtained with an improved technique of pasture management is essential in view of the many farmers one sees adopting bad methods. . . .
>
> "In the light of your work I am convinced that we are ignorant of many fundamental facts concerning herbage growth, fluctuations in which affect the consumption and utilisation of the grass by the animal and consequently influence milk production. What are the elements governing these fluctuations? Many factors enter into play. . . .
>
> "In conclusion may I say that the account you have given is of no mean importance and your conclusions call, on an international basis, for a series of demonstrations of modern grazing techniques and a series of scientific investigations designed to throw light on problems which remain in the air. . . ."

Like Mr. Frankena, I believe that it is urgently necessary for all the problems concerned with the meeting of grass and animal to be studied on an international, or at least on a European, scale. This is the first essential in the improvement of methods of management, methods which have remained more or less the same for centuries. Indeed, they have even deteriorated, as is evident from comparison with the techniques used 6000 years ago by shepherds grazing their flocks on the plateaux of Iran or Palestine.

Chapter 4

RATIONAL GRAZING AND THE GENERAL ECONOMY OF THE FARM

A particular benefit of rational grazing is a higher stocking rate

IT has been seen that rational management allows the efficiency of utilisation of grass by the animal to be improved.

Let us be cautious and assume that the individual yield of the animals can be increased by 15%. It was seen that it is often possible to double and probably triple grass yield as compared with that of continuous grazing.

In consequence, rational grazing allows the stocking of a pasture to be considerably increased by comparison with ordinary continuous grazing.

The success of rational management makes it difficult to remain master of the grass

The great worry of graziers contemplating rational grazing is the cost of erecting fences, installing watering points, etc. Even where fixed fences are used and pipes installed to carry water, the cost, for paddocks little less than 2½ acres [1 ha.] in area, does not exceed £20 per acre. In the case of electric fencing it is rare to spend £5 per acre.

If rational grazing is well managed and the necessary fertilisers applied, three times the stocking possible with continuous grazing and low rates of fertiliser application can be anticipated after a few years. One and a half extra beasts per acre [4 per hectare] represents not much less than £135, which is more than five times the cost of fixed installations and twenty times the cost of movable fencing.

When a grazier knows how to manage rational grazing, therefore, it is difficult for him to graze his grass sufficiently due to shortage of stock. One is always hearing "rotationers" say, "My capital does not allow me to enlarge my herd, and I am snowed under with grass." When contemplating rational grazing, therefore, the farmer must always give a thought to his material or financial resources so that he is in a position to considerably increase the extent of his herd in the years to come.

From the point of view of the whole country, this increase in the numbers of stock carried presents some delicate problems touching both on economics and stock rearing.

Different solutions to a delicate problem

In the case of a grazier who puts beasts out to graze at the beginning of the season and sells them at the end, the problem is solely financial. But where a farmer keeps his animals throughout the year and rears them himself, there is the problem of feeding, bedding and housing this greatly increased herd in winter. If the same area of grassland is to be kept, the whole economy of the farm is necessarily upset.

Of course, cattle can be taken in for feeding, with all the advantages and disadvantages of such a system. Equally the relative balance of permanent pasture and arable land can be altered, care always being taken not to reduce the quantities of farmyard manure applied to the arable land below a dangerous minimum.

Whatever the animal enterprise (fattening, rearing) or the nature of the pasture (temporary, permanent), etc., rational grazing, by multiplying pasture output, profoundly alters the whole farm management. This is one of the important aspects of its introduction into the farming system.

Chapter 5

"GRASS PRODUCTIVITY", A STATE OF MIND ESSENTIAL IN THE FUTURE

The idea of productivity dominates modern civilisation

THIS book has been entitled "Grass Productivity". Productivity is a word one hears reiterated again and again in all the many fields of activity of modern man. It is applied not only to factory products but in all economic spheres. Whether *Homo Productivus* of to-day is happier than *Homo sinanthropus* of the past, I do not know. What I do know, however, is that a strong nation to-day is a nation, all the branches of whose activity are highly productive.

In applying the word "productivity" to agriculture one thinks first and foremost of the productivity of manual labour and machinery, which is both normal and necessary. But agriculture presents other special problems, productivity per acre, for example, a factor obviously meriting no (or very little) attention in factories producing cars or sewing-machines.

In the case of pastures I have evolved a special conception of productivity (Figs. 3 and 4, pp. 15 and 16). This problem peculiar to grass is nevertheless bound up with the great problem of productivity as a whole.

Scientific management and grassland management

Without carrying imagination and fantasy too far, I should like to try to compare the problems of *grass* productivity with those of productivity in *industry*, that is, with the classical questions of "Scientific Management".

A well-known principle of scientific management is the fact that it is not the highest output *per minute* which produces the highest output *per day*, for the resulting fatigue of the worker (or machine) reduces daily production in the long run. The same is true of grassland management: maximum annual production is not obtained by forcing the grass to provide three or four shearings per month (as in continuous grazing). The rhythm of shearings *per month* should be reduced in order that the *annual* production of grass may be increased.

The first study made by Taylor, the founder of "Scientific Management", which is often also called Taylorism, allows a telling comparison to be made between productivity of grass and productivity of the worker.

327

Taylor's study of the handling of pig-iron

Taylor's first study was undertaken at the Bethlehem Steel Company in 1897, and was concerned with the handling of pig-iron. The pigs in this factory were handled by a team of seventy-five men of average quality under the supervision of a good foreman who took *special care to see that the men did not loaf.* The work, which was very simple and involved only the hands and arms of the workmen, no special tools being required, was carried out as quickly and as economically as anywhere else at that time.

The time study in itself presented no difficulties. As this was heavy, physical work, it was less a case of determining the time taken to go loaded and return empty than knowing the number of complete journeys there and back a man could undertake in the course of his day. This was a different problem.

Using various tests, they tried to determine the degree of fatigue of the workers. To their surprise they found that fatigue did not depend so much on the weight of pig carried by a man as on the rate at which he carried it. The least-tired worker was the one who carried his pigs *quickest*, so that he could come back slowly and *loaf* without attracting the notice of the foreman.

Barth, the man engaged by Taylor to study these operations, therefore came to the conclusion that *to get maximum productivity from these workmen they would have to be allowed to relax their muscles sufficiently, that is, to take a sufficiently long rest period.* Calculations showed that if a man had sufficient rest in the course of the day he could handle 47 tons of pig against his previous 13 tons. In other words, *judicious resting allowed the output to be at least trebled (in this particular case).*

Worker Smith was then called in and given three surprises. He was told that:

1. His salary would be increased by 60%.
2. He would have to shift 47 tons of pig per day instead of 13 tons.
3. He was to rest when the timekeeper told him instead of constantly shifting pig as he had done previously.

On the first day, without increased fatigue, Smith shifted 47 tons of pig and made his name famous in the history of Taylorism and Scientific Management.

Grass, like workers, needs rest

In the course of a grazing season, grass needs rest to renew its strength, just as the worker carrying pig-iron has to rest to relax his muscles. Given this condition, it will treble its productivity, like worker Smith.

Motion study and rational grazing

Taylor and Barth's study was only the first of a long series of analogous studies grouped together under the heading "motion studies"; without any new machinery being introduced, at the most after slight modification of an implement, the worker is enabled to increase his productivity merely by having his movements altered.

The case of rational grazing is analogous. No new machines are introduced at most, one simple instrument is used, the electric fence. *But a fundamental "motion" on the part of the grass has been altered, namely, the time during which it is at rest.*

In agriculture, as in industry, **it is often possible to achieve enormous increases in production without introducing costly machinery.** Too often, in both cases, engineers can visualise increased production only after numerous and expensive pieces of machinery and equipment have been installed. In the case of pastures it was frequently believed that productivity could be increased only by using special machinery to plough up and then sowing special, very expensive seeds mixtures.

Grass "motion studies" avoid such expenditure while at the same time much increasing the output of pastures, just as Barth tripled the productivity of the pig handlers without introducing heavy and complicated machinery.

Productivity, a state of mind

Among the many definitions of productivity I tried to find one applicable to the special case of grass. In the course of a conversation with a high official of the "Commissariat Général au Plan", I asked him point blank if he could give me a *universal* definition of the idea of productivity. The answer came: "The best and most general definition is that productivity is a state of mind".

I remarked that this was a truly remarkable definition, but I could not see how I could produce such a state of mind into a grass sward. However, on reflection, I saw that this definition applies exactly to the case of grass. The productivity of grass will be increased, that is, its yield doubled or trebled, only if this state of mind is inculcated into all those interested in the problem.

Once we are all convinced of the immense possibilities of rationally managed pastures, the problem of pasture productivity will cease to exist and means for developing methods of practical application will be found.

CONCLUSIONS
GREEN PASTURES

GRASS LYRICS

AT the last annual reunion of the French *Académie d'Agriculture* at his Château at Harcourt, our president, M. Jean Lefèvre, read a few verses from Professor Blay whose poems sing of sylvan beauty. The setting was indeed fitting: a historic dwelling (Photo 41, facing p. 291) surrounded by one of the most beautiful forests in France. In the distance I thought I heard the murmurs of the verses of Ronsard and Wordsworth, but in my heart I felt a certain sadness. Why, I wondered, have the poets who so ably sing the praises of trees forgotten the beauty of grass?

Green Symphony

The poetry of the pasture is no less than that of the forest. What loveliness! What shades of colour all blending to form an even more magnificent picture where rational grazing is applied. The different paddocks, at different stages of re-growth, are not all of the same hue. Moreover, in a well-managed system the paddocks are not grazed in the same order as they stand, and so the colour tones, like reflections on the sea, do not gradually and uniformly diminish in intensity. Between two dark greens one glimpses a paddock lighter in colour, like the depth of a wave. A part where the grass has already begun to flower takes on an undulating, wavy aspect. What enchantment a pasture grazed in this way offers to the eye!

Grass must be loved

From one point on my farm (Photo 33) eleven grass paddocks are visible, none with the same appearance or of the same shade as the other. I remarked one day on the beauty of this green symphony to some English friends, and one of them, Mr. Currie from Dartington Hall Study Centre in Devon (England), replied: "To manage grass well, one must love it with a love that is truly profound. What could be more magnificent? The loveliest hour of the day for me is when I rise and see the dew drops sparkling on the blades of grass. . . ." And he quoted the Bible:

"The King's wrath is as the roaring of a lion; but his favour is as dew upon the grass" (Prov. 19, 12).

Let us respect grass

I felt obliged to quote in return from the Apocalypse (9, 4):

"And it was commanded that they should not hurt the grass of the earth, neither any green thing. . . ."

To which Mr. Currie replied dreamily: "Yes, let us learn to love the grass. . . . May Man be able to respect what the Creator has given him!"

The pastures of Prometheus

I left Mr. Currie for a moment to his reverie. For fear of hurting his feelings I did not dare recall a memory I had of his own native country. I decided, however, to do so and said very gently:

"I am not quite certain that the Creator gave us pastures as they are to-day. . . . Do you mind if I tell you what a Derbyshire farmer gleefully told me?

"I was congratulating Mr. Brocklehurst on the wonderful pastures he had succeeded in establishing and maintaining on unproductive granitic soil (Photo 21), to which he replied: "I have had a lot of trouble! Nobody understands the effort and years it takes to get pastures like those. The other day the vicar called to see me and we took a walk through the fields. When he saw my lovely grass, the reverend gentleman said: 'How you must thank God for giving you such good pasture!' I could not help answering: 'I wish you had seen the sorry state they were in when God had them!' "

We were sitting down at the foot of a Celtic cairn (Photo 22). A storm was building up on the heights of the Pennine Chain. I felt that amidst the rolling of the thunder, the granite hills were re-echoing, through the voice of this peasant Prometheus, the defiant challenge of the legendary Titan:

"Zeus, who was it who helped me?
. . . If not myself, myself alone."

Mr. Currie was perhaps somewhat shocked by these slightly impious words. Very smartly he said: "I do not know whether God kept his earthly pastures in a bad state before He gave them to Man, but I know for certain that the pastures of Heaven are perfectly green and lovely."

A bell rang: my wife was summoning us to dinner.

I said to Mr. Currie: "Before going in to enjoy the very worldly pleasures of Norman cooking, and as I have been perhaps a little blasphemous, will you allow me to say that you are right without however saying that Mr. Brocklehurst was wrong?"

Symbols of Serenity

I will remind you of the words of an American director in North Carolina, Voisin (117), vol. II, p. 379:

"Few things are able to move us as much as a luxuriant meadow standing out against a background of dark trees under blue sky. From time immemorial this sight has inspired musicians to translate its beauty into pastoral symphonies. Bathing their eyes in its wonder, painters have presented us with their most beautiful landscape pictures.

"Green pastures have become the symbol of serenity, of stability of peace and of plenty. Man's feeling of respect for grass pastures is so profound that he has associated them in his mind with eternal rest, dreaming nostalgically of the 'green pastures and still waters' that await him in the world to come."

Is it not a pleasant thought that rational grazing helps to realise this dream on earth?

BIBLIOGRAPHY

1. ALBRECHT, W. A. "More and better proteins make better food and feed", *Better Crops with Plant Food Magazine* (1952).

2. ALLCROFT, Ruth. "Hypomagnesæmia: A summary of evidence", *Agricultural Review*, **1**, 47–50 (1955).

3. ALLCROFT, Ruth. "Hypomagnesæmia of cattle and sheep in Britain", *Journal of the British Grassland Society*, **11**, 119–120 (1956).

4. ANDERSON, James. *Essays Relating to Agriculture and Rural Affairs.* (*a*) 2nd edition. Edinburgh (1777); (*b*) 4th edition. London (1797).

5. ANONYMOUS. *Le Premier pas: L'exploitation rationnelle des Herbages* (*The First Step: Rational Management of Grassland*). Brochure de Propagande du Ministère de l'Agriculture (Paris, 1951).

6. ANONYMOUS. "Urea as a protein replacement for ruminants", *Nutrition Reviews*, **12**, 43–45 (1954).

7. ANONYMOUS. "Bigger toll from grass sickness: severe outbreaks in Northumberland," *Farmers Weekly* (July 16, 1954).

8. BALFOUR, Eve. "9600 miles in a station wagon. IV. Some findings by agricultural scientists", *Journal of the Soil Association*, **8**, 48–64 (1951).

9. BONNER, J., and GALSTON, J. W. *Principles of Plant Physiology*, San Francisco (1952).

10. BROWN, D. "Methods of surveying and measuring vegetation", *Commonwealth Agricultural Bureaux Bulletin No. 42* (1954).

11. BRUNDAGE, L., and PETERSEN, W. E. "A comparison between daily rotational grazing and continuous grazing", *Journal of Dairy Science*, **35**, 623–630 (1952).

12. CAPUTA, J. *Organisation et exploitation des pâturages* (*Organisation and Management of Pasture*), Association pour le Développement de la Culture Fourragère. Berne (1952).

13. CASTLE, M. E., FOOT and HALLEY. "Some observations on the behaviour of dairy cattle with particular reference to grazing", *Journal of Dairy Research*, **17**, 215–230 (1950).

14. CHALMERS, M. I., CUTHBERTSON, D. P., and SYNGE, R. L. M. "Ruminal ammonia formation in relation to the protein requirement of sheep. I. Duodenal administration and heat processing as factors influencing fate of casein supplements", *Journal of Agricultural Science*, **44**, 254–262 (1954).

15. CZERWINKA, W. "Bäuerlicher Futterbau: mehr und besseres Futter von kleinerer Fläche" ("Forage cultivation on the small farm: more and better forage in a smaller area"), *Graz* (1952).

16. DANCEY, R. J. "The influence of management on the subsequent productivity of pasture". B.Sc. (with Honours) thesis at the University of Nottingham (1955).

17. ELLISON, W. "The animal is the best judge", *Farmers Weekly* (Oct. 22, 1948).

18. ETTER, A. G. "Animal behaviour: study of their habits has much to teach us", *Journal of the Soil Association*, **8**, 73-78 (1954).

19. EVANS, A. C. "The importance of earthworms", *Farming* (February 1948).

20. FALKE. " Die Dauerweiden" (*Permanent Pastures*). Hanover (1907).

21. FINCK, A. "Okologische und bodenkundliche Studien über die Leistungen der Regenwürmer für die Bodenfruchtbarkeit" ("Ecological and pedological studies on the influence of earthworms on soil fertility"), *Zeitschrift für Pflanzenzernährung, Düngung, Bodenkunde*, **58**, 120-145 (1952).

22. FISSMER. "Beiträge zur Frage der Sättigung bei der Milchkuh" ("How to satisfy a dairy cow's appetite"), *Zeitschrift für Tierernährung und Futtermittelkunde*. Band. 8 (1941).

23. FLANDIN, P. "Dans une ferme du Berry, les pâturages tournants améliorent la production" ("On a Berry farm rotational grazing improves production"), *France Agricole* (August 30, 1954).

24. FRANZ H. *Bodenzoologie als Grundlage der Bodenpflege* (*Zoology of the Soil as a Basis for Methods of Cultivation*). Berlin (1950).

25. FRASER, Allan. "But is it science?" *Farmers Weekly* (November 9, 1948).

26. GEERING, J. "Uber den Einfluss der Häufigkeit des Wiesenschnittes auf Pflanzenbestand, Nährstoffgehalt und Nährstoffertrag" ("Influence of frequency of cuts in mown fields on the flora and content and yield of nutrients"), *Landwirtschaftliche Jahdbücher der Schweiz*, **55**, 579-595 (1941).

27. GEITH, R. "Der Nährstoffgehalt des Weidegrases und sein Einfluss auf die Bewirtschaftung der Deutschen Weiden" ("Nutrient content of pastures and their effects on systems of management of German grassland"), *Landwirtschaftliche Jahrbücher*, **82**, 187-196 (1936).

28. GEITH, R. "Die Verbesserung der Normen zur Ermittlung des tierischen Nutzertrags einer Weide" ("Improvement of norms used to determine animal production from a pasture"), *Fourth International Grassland Congress*. Aberystwyth, 434-440 (1937).

29. GEITH, R. *Neuzeitliche Weidewirtschaft* (*Modern Methods of Grassland Management*). Berlin (1943).

30. GEITH, R., and FUCHS, K. *Grünlanfibel* (*Grassland Manual*). (Berlin 1943).

31. GILBERT, M. "Recherches sur les espèces de prairies artificielles qu'on peut cultiver avec le plus d'avantage dans la généralité de Paris" ("Enquiry into the types of temporary pastures, which can be most profitably developed in the region of Paris"), *Mémoires d'Agriculture, d'Economie Rurale et Domestique*. La Société Royale d'Agriculture (Paris, 1788).

32. GRAFF, O. "Bodenzoologische Untersushungen mit besonderer Berücksichtigung der terrikolen Oligochæten" ("Enquiry into zoology of soils and especially the Oligochæta), *Zeitschrift für Pflanzenernährung, Düngung und Bodenkunde*, **61**, 72–77 (1953).

33. GREGOR, J. W. "The use of complementary grassland", *Scottish Agriculture*, **26**, 104–109 (1946).

34. GRÜNINGEN, F. Von. "Die Bedeutung des Unkrautes für die Ernährung des Rindviehs" ("The importance of weeds in the feeding of beef cattle"), *Fifth International Grassland Congress, Holland* (1949).

35. HANCOCK, John. "Grazing habits of dairy cows in New Zealand", *Empire Journal of Experimental Agriculture*, **18**, 249–263 (1950).

36. HANCOCK, John. "Studies in monozygotic cattle twins. IV. Uniformity trials: grazing behaviour", *New Zealand Journal of Science and Technology*, **2** (32), 22–59 (1950).

37. HANCOCK, John. "Grazing behaviour of identical twins in relation to pasture type, intake, and production of dairy cattle", *Proceedings of the Sixth International Grassland Congress*, 1399–1407 (1952).

38. HEIM, G. "Au sujet du pâturage intensif" ("About intensive grazing"), *Revue Romande d'Agriculture, de Viticulture et d'Arboriculture*, No. 11 (1949).

39. HEINE, G. O. "Nutzbarmachung des Elektrozaunes in der Weidetechnik" ("Use of the electric fence in the management of pastures"), *Das Grünland* (April and May 1954).

40. HODGSON, R. E. "Influence of pasture management upon the grazing habits of dairy cattle", *Journal of Agricultural Research*, **47**, 417–424 (1933).

41. HOLMES, W., WAITE, R., FERGUSSON, D. L., and CAMPBELL, Jean I. "Studies in grazing management. I. A comparison of the production obtained from close-folding and rotational grazing of dairy cows", *Journal of Agricultural Science*, **40**, 381–391 (1950).

42. HOLMES, W., WAITE, R., FERGUSSON, D. L., and CAMPBELL, Jean I. "Studies in grazing management. IV. A comparison of close-folding and rotational grazing of dairy cows on intensively fertilized pasture", *Journal of Agricultural Science*, **42**, 304–313 (1952).

43. HOLMES, W. "The feeding value of grass and grassland products", *Proceedings of the British Society of Animal Production*, 90–112 (1952).

44. HOLMES, W. "High milk yields per acre from grassland", *Journal of the British Grassland Society*, **9**, 17–27 (1954).

45. HOLMES, W. "Modern methods of grassland management in feeding the dairy herd", *Proceedings of the Nutrition Society*, **13**, 19–22 (1954).

46. HOLMES, W. "The intensive management of pasture for dairy cows", *Sixth International Grassland Congress*, **1**, 377–383. Pennsylvania State College (1952).

47. IVINS, J. D. "The relative palatability of herbage plants", *Journal of the British Grassland Society*, **7**, 43–54 (1952).

48. Johns, A. T. "Ruminant metabolism", *Proceedings of the New Zealand Society of Animal Production*, **13**, 106–111 (1953).

49. Johnstone-Wallace, D. B. "The influence of wild white clover on the seasonal production and chemical composition of pasture herbages, and upon soil temperature, soil moisture, and erosion control", *Fourth International Grassland Congress*, 188–196. Aberystwyth (1937).

50. Johnstone-Wallace, D. B. "Pasture improvement and management", *Cornell Extension Bulletin No. 393* (October 1938).

51. Johnstone-Wallace, D. B., and Kennedy, K. "Grazing management practices and their relationship to the behaviour and grazing habits of cattle", *Journal of Agricultural Science*, **34**, 190–197 (1944).

52. Johnstone-Wallace, D. B. "The principles of pasture management", *Proceedings of the New York Farmers*, 1944–45.

53. Johnstone-Wallace, D. B. "Grass and the grazing animal", *Farmers Weekly* (November 17, 1950; November 24, 1950; December 1, 1950; December 8, 1950).

54. Johnstone-Wallace, D. B. "Animal behaviour and grazing management", *Journal of the Royal Agricultural Society of England*, **114** (1953).

55. Jones, Iorwerth. "The effect of varying the period of rest in rotational grazing", *Welsh Journal of Agriculture*, **9**, 159–170 (1933).

56. Jones, Martin. "Grassland management and its influence on the sward. Part I. Factors influencing the growth of pasture plants", *Empire Journal of Experimental Agriculture*, **1**, 43–57 (1933).

57. Jones, Martin. "Grassland management and its influence on the sward. Part II. The management of a clovery sward and its effects", *Empire Journal of Experimental Agriculture*, **1**, 122–128 (1933).

58. Jones, Martin. "Grassland management and its influence on the sward. Part III. The management of a 'grassy' sward and its effects", *Empire Journal of Experimental Agriculture*, **1**, 223–234 (1933).

59. Jones, Martin. "Grassland management and its influence on the sward. Part IV. The management of poor pastures. Part V. Edaphic and biotic influences on pastures", *Empire Journal of Experimental Agriculture*, **1**, 361–368 (1933).

60. Jones, Martin. "Grassland management and its influence on the sward", *Journal of the Royal Agricultural Society of England*, **94**, 21–41 (1933).

61. Jones, Martin. "The improvement of grassland by its proper management", *Fourth International Grassland Congress*, 470–473. Aberystwyth (1937).

62. Kauter, A. "Weiderwirtschaft und Weidetechnik" ("Management and technique of grazing"), *Arbeitsgemeinschaft zur Förderung des Futterbaues*. Wildegg (1950).

63. Khatchadourian, Der L. *L'exploitation intensive des prairies* (*Intensive Grassland Management*). Paris (1955).

64. Klapp, E. "Uber einige Wachstumsregeln mehrjähriger Pflanzen unter der Nachwirkung verschiedener Nutzungsweise" ("Some rules of growth for perennial plants according to different forms of usage"), *Pflanzenbau*, **14**, 209–224 (1937).

65. KLAPP, E. "Entwicklung, Wurzelbildung und Stoffspeicherung von Futter-pflanzen" ("Development, root formation and accumulation of substance in forage plants"), *Pflanzenbau*, **18** (1941–1942).

66. KLAPP, E. "Borstgrasheiden der Mittelgebirge" ("Mat Grass Pastures (Mit-telgebirge)"), *Zeitschrift für Acker = und Pflanzenbau*, **93**, 400–444 (1951).

67. KLAPP, E. "Leistung, Bewurzelung und Nachwuchs einer Grasnarbe unter verschieden häufiger Mahd und Beweidung" ("Production, root develop-ment and regrowth of a sward according to frequency of cut and grazing"), *Zeitschrift für Acker = und Pflanzenbau*, **93**, 269–286 (1951).

68. KLAPP, E. "Zum zeitlichen Verlauf des Graszuwachses auf Weiden und seiner Beeinflussung durch Stickstoffgaben" ("Seasonal fluctuations in grass growth and the effect of nitrogen application on growth"), *Zeitschrift für Acker = und Pflanzenbau*, **95**, 69–72 (1952).

69. KLAPP, E. "Should fertilizers for grassland be worked into the soil?" *Sixth International Grassland Congress*, **1**, 821–826. Pennsylvania State College (1952).

70. KLAPP, E. *Wiesen und Weiden* (*Meadows and Pastures*). Berlin (1954).

71. LATTEUR, J. P. "La tétanie d'herbage" ("Grass tetany"), *Revue de l'Agriculture* (*Bruxelles*), No. 7.(July 1953).

72. LAVOINNE, André. *La Race Bovine Normande: Cinquante ans d'élevage au Bosc-aux-Moines* (*The Norman Cattle Breed: Fifty years of raising at Bosc-aux-Mines*). Yvetot (1948).

73. LINEHAN, P. "Output of pasture", *Farming*, **1**, 173–176 (1947).

73a. McINTYRE, G. A., and DAVIES, J. G. "Small plot studies in the evaluation of pasture intended for grazing", *Sixth International Grassland Congress*, **2**, 1361–1366. Pennsylvania State College (1952).

74. MEHNER, A., and GRABISCH, W. "Verzehr, Leistung und Futterverwertung bei der Nutzung von Dauergrünland durch Abweiden und Abmähen" ("Con-sumption, yield and conversion efficiency of animals on permanent and mown pastures"), *Züchtungskunde*, **28**, 80–88 (February 1956).

75. MELVILLE, J. "Pasture quality and animal production", *Proceedings of the New Zealand Society of Animal Production*, **13**, 65–73 (1953).

76. MOORE, H. I. *Grassland Husbandry*. London, 1943.

77. MOTT, Barbara. "Ein Beitrag zur Feststellung des Geschmackswertes der Grünlandpflanzen" ("A contribution to the determination of the palatability of herbage plants"), *Das Grünland* (April and May 1955).

78. MUIR, W. R. "Pasture herbage as a causal factor in animal disease", *Journal of the British Grassland Society* (December 1948).

79. MULDER, E. G. "Fertilizer versus legume nitrogen for grasslands", *Sixth Inter-national Grassland Congress*, **1**, 740–748. Pennsylvania State College (1952).

80. MÜNZINGER and BABO. "Das Hohenheimer Weidesystem" ("The Hohenheim system of grazing"), *Landwirtschaftliche Jahrbücher*, **73**, 139–168 (1931).

81. OHMS, E. "Vorbereitung auf die Weidezeit" ("Preparation of animals for grazing"), *Tierzucht*, **8**, 117–119 (1954).

82. Osieczanski, E. *Biologie und Nützung des Grünlandes* (*Biology and Utilisation of Grassland*). (Translation from the Polish.) Berlin (1954).

83. Pérignon, K. Th. "Die Unterteilung des Weideviehs in Leistungsgruppen" ("Division of the grazing herds into groups according to production"), *Das Grünland* (November 1954).

84. Potasses d'Alsace, Sté des. *La Fumure des prairies* (*The Manuring of Fields*). Mulhouse (1950).

85. Rozier. *Cours complet d'Agriculture* (en 10 volumes) (*Complete Course in Agriculture*) (in 10 volumes). Paris. (The first volume published in 1785, the last in 1800.)

86. Ruhr-Stickstoff Aktiengesellschaft. *Versuchs-Erfahrungen im Gebiet Weser-Ems, 1942–1952* (*Education with Trials in the District between the Weser and the Ems, 1942–1952*). Oldenburg (1953).

87. Scharrer, K. *Die biochemischen Grundlagen der Tierernährungslehre* (*The Biochemical Bases of Animal Feeding*). Stuttgart (1950).

88. Schlipf. *Praktisches Handbuch der Landwirtschaft* (*Practical Manual of Agriculture*). 27th edition. Berlin (1941).

89. Schmidt, J., Mehner, A., and Schelper, E. "Beobachtungen über Leistung und Futterausnutzung bei drei Höhenrassen" ("Observations on the utilisation of grass by three milk breeds with their yields"), *Zeitschrift für Tierzüchtung und Züchtungsbiologie*, **50**, 298–338 (1950).

90. Schmidt, J., Mehner, A., and Piel, H. "Untersuchungen über Weideerträge zweier Betriebe, sowie über Verzehr und Leistung auf der Weide bei Milchkühen dreier Rassen" ("Enquiry into the yield of pastures of two crops and the quantities of grass harvested by in-milk cows of three breeds with their yields"), *Züchtungskunde*, **23**, 110–122 (1951).

91. Schulze, E. "Ueber Oberflächen =, Tiefendüngung und Düngereinarbeitung auf Dauergrünland" ("Application of fertilisers on the surface, in the soil, and by working the soil in the case of permanent grassland"), *Das Grünland*, 12–14 (February 5, 1953).

92. Schulze, E. "Zusammenhänge zwischen Düngungserfolg, Pflanzenbestand und Schnitthäufigkeit auf Dauerwiesen" ("Relation between the efficacy of fertilisers, composition of the flora and frequency of cuts of permanent grassland"), *Das Grünland*, **3**, 73–75 (1954).

93. Schulze, E. "Aufgeteilte Stickstoffgaben" ("Distribution of nitrogen applications"), *Deutsche Landwirtschaftiche Presse* (June 16, 1956).

94. Schuppli, P. "Die Eintagsweide, ihre Einrichtung und Vorteile" ("Daily grazing area: its establishment and advantages"), *Deutsche Landwirtschaftliche Presse*, **63**, 185 (1936).

95. Schützhold, C. "Leistungsvermögen der Dauerweiden und Kleegrasweiden" ("Productive capacity of permanent and temporary pastures of clover and gramineae"), *Zeitnahe Fragen der Grünland = und Futterwirtschaft*, 22–46 (Hanover, 1951).

96. Schützhold. *Was leistet meine Weide?* (*What Does my Pasture Produce?*). Völkenrode (1952).

97. SINCLAIR, John. *Code of Agriculture*. (3rd Edition. London, 1821).

98. SÖDING, H. *Die Wuchsstofflehre* (*Theory of Growth Factors*). Stuttgart (1952).

99. STAEHLER, H. *Die neue Mähweide-Uhr* (*The New Clock to Regulate the Combination of Grazing and Cutting*). München (1951).

99a. STAEHLER, H. "Portionsweide besser als Umtriebsweide?" ("Is rationed better than rotational grazing?"), *Deutsche Landwirtschaftliche Presse* (January 10, 1953).

100. STAPLEDON, R. G. *The Plough-up Policy and Ley Farming*. London (1947).

101. STAPLEDON, R. G. "Pastures old and new: the animal's point of view", *Journal of the Ministry of Agriculture*, **55**, 6 (September 1948).

102. SULLIVAN, J. T., and SPRAGUE, V. G. "Composition of the roots and stubble of perennial rye-grass following partial defoliation", *Plant Physiology*, **18**, 656–670 (1943).

103. SULLIVAN, J. T., and GERBER, R. J. "Chemical composition of pasture plants", *Pennsylvania State College Bulletin No. 489* (November 1947).

104. SYNGE, R. L. M. "The utilization of herbage protein by animals", *The British Journal of Nutrition*, **6** (1), 100–104 (1952).

105. TACKE, B. "Ueber die Wirkung von Kali auf schweren Alluvialböden" ("Effect of potassium on heavy alluvial soils"), *Die Ernährung der Pflanzen*, 433 (1931).

105a. TAYLER, J. C. "The grazing behaviour of bullocks under two methods of management", *British Journal of Animal Behaviour*, **1**, 72–77 (1953).

106. TERROINE, Thérèse. "Valeur alimentaire de l'azote non protéique: urée, amides, ammoniaque" ("Nutritive value of non-protein nitrogen: urea, amides, ammonia"), *Annales de la Nutrition et de l'Alimentation*, **3**, 49–73 (1949).

107. TESSIER, THOUIN, BONDAROY, BOSC, etc. *Encyclopédie méthodique de l'Agriculture* (*Methodological Encyclopædia of Agriculture*). Paris and Liège. Vol. 1 (1787); Vol. II (1791); Vol. III (1793); Vol. IV (1796); Vol. V (1813); Vol. VI (1816).

108. THOMAS, M. T., DAVIES, A. G., and DAVIES, W. "Field trials with pedigree (stationbred) and indigenous strains of grasses", *Welsh Journal of Agriculture*, **15**, 202–(1939).

109. TRAPPMANN. *Die Enchytreen* (*The Enchytræid Worms*). Landbau Forschung (Völkenrode) Heft. 2 (1953).

110. TRIBE, D. E. "The importance of animal behaviour studies in animal production", *Journal of the Australian Institute of Agricultural Science*, **19**, 28–33 (1953).

111. TRIBE, D. E. "The behaviour of the grazing animal", *Journal of the British Grassland Society* (September 1950).

112. VERDEYEN, J. "La relation entre la plante et l'animal" ("The relation between plant and animal"), *Comptes rendus des Recherches Herbagères et Fourragères*, No. 9 (January 1953).

113. VOISIN, André. "La rotation des herbages" ("Grassland rotation"), *Revue de l'Élevage*, Special Number on Grassland (1949).

114. VOISIN, André. "Rotation des herbages" ("Rotational grazing"), *Bulletin du Herd-Book normand* (August 1950).

115. VOISIN, André. "Pourquoi la vache préfère les herbages en rotation" ("Why cows prefer grassland rotation"), *Le Figaro Agricole*, Special Number, (June 1951).

116. VOISIN, André. "Comportement de la vache au pâturage" ("Behaviour of the grazing animal"). (Conference en 1951 aux Journées d'Études sur l'Alimentation à la Prairie.) Published in *L'Alimentation a la Prairie*. L'Association Française de Zootechnie, 16, rue Claude-Bernard, Paris 5ᵉ (1952).

117. VOISIN, André. "Journal de voyage aux U.S.A. de la mission: 'Production Fourragère'" ("Travel diary, U.S.A. mission: 'Forage Production'"). Two volumes published in 1952 by the *Ministère de l'Agriculture*, 72, rue de Varenne, Paris.

118. VOISIN, André. "Le problème du rassasiement de la vache laitière" ("The problem of the saturation of the dairy cow"), *Annales de l'Institut national de Recherches Agromiques*, Series D, 1–49 (1952).

119. VOISIN, André. "Rotation des herbages" ("Rotational grazing"), *Revue de l'Élevage* (February 1953).

120. VOISIN, André. "Comment la vache s'alimente-t-elle au pâturage?" ("How does the grazing cow feed?") *Revue de l'Élevage* (March 1953).

121. VOISIN, André. "L'application pratique de la rotation des herbages" ("Practical application of rotational grazing"), *Bulletin du Herd-Book normand* (July 1953).

122. VOISIN, André. "Grandeurs et faiblesses du Ley-Farming" ("Strength and weakness of Ley farming"), *Bulletin Technique d'Information des Ingénieurs des Services Agricoles, No. 82* (1953).

123. VOISIN, André. "Production du fumier et productivité agricole" ("Manure production and agricultural productivity"), *Bulletin de la Société Française d'Économie Rurale* (September 1953).

124. VOISIN, André. "Devons-nous retourner nos herbages pour les améliorer?" ("Should we plough-up to improve grassland?"), *Bulletin de la Direction des Services Agricoles du Nord* (March–April 1954).

125. VOISIN, André. "Aspects biochimiques de l'ensilage" ("Biochemical aspects of ensilage"). Published in *La Conservation des Fourrages*, Paris (1954).

126. VOISIN, André. "Le principe fondamental de la rotation des herbages" ("The fundamental principle of rotational grazing") *European Grassland Conference*. Paris (June 1954).

127. VOISIN, André. "Verhalten und Sättigung der Kuh auf der Weide und im Stall" ("Behaviour and saturation of the grazing and stall-fed cow"), *Vorträge der 8 Hochschultagung* (Bonn, 1954).

128. VOISIN, André. "Die vier Grundgesetze einer rationnellen Weidewirtschaft" ("The four fundamental laws for rational grazing"), *Mitteilungen der Deutschen Landwirtschafts-Gesellschaft* (January 6, January 13, January 20, 1955).

129. VOISIN, André. "Théorie de l'exploitation intensive des herbages" ("Theory of intensive grassland management"). Published in *La Prairie Charolaise et son exploitation*. Ministère de l'Agriculture (Direction des Services de Agricoles de Saône-et-Loire) (1955).

130. VOISIN, André. "Le principe de base du pâturage rationnel" ("The basic principle of rational grazing"), *Comptes rendus de l'Académie d'Agriculture.* Session of February 29, 1956.

131. VOISIN, André. "Comment un impôt mal appliqué décourage l'effort du paysan" ("How a badly applied tax discourages agricultural effort"), *Comptes rendus de l'Académie d'Agriculture.* Session of March 21, 1956.

132. VOISIN, André. "L'exploitation rationnelle des pâturages très dégradés les transforme en pâturages de haute qualité" ("The rational management of broken-down pastures transforms them to high quality pastures"), *Comptes rendus de l'Académie d'Agriculture.* Session of April 11, 1956.

133. VOISIN, André. "Le caractère héréditaire de la vache détermine la quantité d'herbe qu'elle récolte" ("The hereditary characteristics of the cow governs the quantity of grass she harvests"), *Comptes rendus de l'Académie d'Agriculture.* Session of May 23, 1956.

134. VOISIN, André. "Faisons pâturer en appliquant un principe connu, mais oublié" ("Application of a known, but forgotten principle to grazing"), *Bulletin des Centres d'Études Techniques Agricoles, No. 29* (May 1956).

135. WAGNER, R. E., and WILKINS, H. L. "The effect of legumes on the percentage of crude protein in orchard grass and bromegrass", *Journal of the American Society of Agronomy*, **39**, 141–145 (1947).

136. WAITE, R., HOLMES, W., CAMPBELL, Jean I., and FERGUSSON, D. L. "Studies in grazing management. II. The amount and chemical composition of herbage eaten by dairy cattle under close-folding and rotational methods of grazing", *Journal of Agricultural Science*, **40**, 392–402 (1950).

137. WAITE, R., MACDONALD, W. B. and HOLMES, W. "Studies in grazing management. III. The behaviour of dairy cows grazed under the close-folding and rotational system of management", *Journal of Agricultural Science*, **41**, 163–173 (1951).

138. WATSON, S. J., PROCTER, J. and FERGUSON, W. S. "The effect of nitrogen on the yield, composition and digestibility of grassland herbage", *Journal of Agricultural Science*, **22**, 251–290 (1932).

139. WATSON, S. J. *Grassland and Grassland Products.* London (1951).

140. WEINMANN, Hans. "Carbohydrate reserves in grasses", *Sixth International Grassland Conference*, **1**, 655–660. Pennsylvania State College (1952).

141. WEISE, F. "Der Einfluss von Düngung und Beweidung auf die Veränderung von Pflanzenbeständen des Dauergrünlandes" ("Influence of fertiliser application and of grazing on modifications of the flora of permanent pastures"), *Schriftenreihe. AID, No. 50* (1954).

142. WELSH PLANT BREEDING STATION. "Yield and productivity trials with individual species and strains", *An Account of the Organisation and Work of the Station from its Foundation in April 1919 to July 1933*, 72–86. Aberystwyth (1933).

143. WELSH PLANT BREEDING STATION. "Rotational grazing", *An Account of the Organisation and Work of the Station from its Foundation in April 1919 to July 1933*, 140–143.

144. WHYTE, R. O. "The physiological nature of a herbage plant", *Seventh International Botanical Conference*, 162–165. Stockholm (1950).

145. WOODMAN, H. E., and UNDERWOOD, E. J. "Nutritive value of pasture. VIII. The influence of intensive fertilising on the yield and composition of good permanent pasture", *Journal of Agriculture Science*, **22**, 26–71 (1932).

146. WOODMAN, H. E., and EVANS, R. E. "Nutritive value of pasture. XII. The influence of cutting at monthly intervals over nine seasons on the quality and productivity of a heavy-land pasture", *Journal of Agricultural Science*, **28**, 581–591 (1938).

147. WOODWARD, T. E., SHEPHERD, J. B., and HEIN, M. A. "The Hohenheim system in the management of permanent pastures for dairy cattle", *U.S. Department of Agriculture Technical Bulletin, No. 660* (October 1938).

148. ZÜRN, F. "Neuere Forschungsergebnisse über Grünlandwirtschaft" ("Results of recent research in the management of grazing"), *Veröffentlichungen der Bundesanstalt für alpine Landwirtschaft in Admont*, Heft. 8. Vienna (1953).

149. ZÜRN, F. "Der Futterwuchs auf den Weiden" ("The growth of grass on pastures"), *Das Grünland*, **3**, 89–91 (1954).

INDEX OF NAMES

347

INDEX

(The principal subject headings are printed in capitals)

Also Available from Island Press

Land and Resource Planning in the National Forests
By Charles F. Wilkinson and H. Michael Anderson
Foreword by Arnold W. Bolle

This comprehensive, in-depth review and analysis of planning, policy, and law in the National Forest System is the standard reference source on the National Forest Management Act of 1976 (NFMA). This clearly written, non-technical book offers an insightful analysis of the Fifty Year Plans and how to participate in and influence them.

1987. xii, 396 pp., index.
Paper ISBN 0-933280-38-6. **$19.95**

Reforming the Forest Service
By Randal O'Toole

Reforming the Forest Service contributes a completely new view to the current debate on the management of our national forests. O'Toole argues that poor management is an institutional problem; he shows that economic inefficiencies and environmental degradation are the inevitable result of the well-intentioned but poorly designed laws that govern the Forest Service. This book proposes sweeping reforms in the structure of the agency and new budgetary incentives as the best way to improve management.

1988. xii, 256 pp., graphs, tables, notes.
Cloth, ISBN 0-933280-49-1. **$34.95**
Paper, ISBN 0-933280-45-9. **$19.95**

Last Stand of the Red Spruce
By Robert A. Mello
Published in cooperation with Natural Resources Defense Council

Acid rain—the debates rage between those who believe that the cause of the problem is clear and identifiable and those who believe that the evidence is inconclusive. In *Last Stand of the Red Spruce*, Robert A. Mello has written an ecological detective story that unravels this confusion and explains how air pollution is killing our nation's forests. Writing for a lay audience, the author traces the efforts of scientists trying to solve the mystery of the dying red spruce trees on Camels Hump in Vermont. Mello clearly and succinctly presents both sides of an issue on which even the scientific community is split and concludes that the scientific evidence uncovered on Camels Hump elevates the issues of air pollution and acid rain to new levels of national significance.

1987. xx, 156 pp., illus., references, bibliography.
Paper, ISBN 0-933280-37-8. **$14.95**

Western Water Made Simple, by the editors of High Country News
Edited by Ed Marston

Winner of the 1986 George Polk Award for environmental reporting, these four special issues of *High Country News* are here available for the first time in book form. Much has been written about the water crisis in the West, yet the issue remains confusing and difficult to understand. *Western Water Made Simple,* by the editors of *High Country News,* lays out in clear language the complex issues of Western water. This survey of the West's three great rivers — the Colorado, the Columbia, and the Missouri — includes material that reaches to the heart of the West — its ways of life, its politics, and its aspirations. *Western Water Made Simple* approaches these three river basins in terms of overarching themes combined with case studies — the Columbia in an age of reform, the Colorado in the midst of a fight for control, and the Missouri in search of its destiny.

1987. 224 pp., maps, photographs, bibliography, index.
Paper, ISBN 0-933280-39-4. **$15.95**

The Report of the President's Commission on Americans Outdoors: The Legacy, The Challenge
With Case Studies
Preface by William K. Reilly

"If there is an example of pulling victory from the jaws of disaster, this report is it. The Commission did more than anyone expected, especially the administration. It gave Americans something serious to think about if we are to begin saving our natural resources."
—Paul C. Pritchard, President,
National Parks and Conservation Association

This report is the first comprehensive attempt to examine the impact of a changing American society and its recreation habits since the work of the Outdoor Recreation Resource Review Commission, chaired by Laurance Rockefeller in 1962. The President's Commission took more than two years to complete its study; the Report contains over sixty recommendations, such as the preservation of a nationwide network of "greenways" for recreational purposes and the establishment of an annual $1 billion trust fund to finance the protection and preservation of our recreational resources. The Island Press edition provides the full text of the report, much of the additional material compiled by the Commission, and twelve selected case studies.

1987. xvi, 426 pp., illus., appendixes, case studies.
Paper, ISBN 0-933280-36-X. **$24.95**

Public Opinion Polling: A Handbook for Public Interest and Citizen Advocacy Groups
By Celinda C. Lake, with Pat Callbeck Harper

"Lake has taken the complex science of polling and written a very usable 'how-to' book. I would recommend this book to both candidates and organizations interested in professional, low-budget, in-house polling."—Stephanie Solien, Executive Director, Women's Campaign Fund.

Public Opinion Polling is the first book to provide practical information on planning, conducting, and analyzing public opinion polls as well as guidelines for interpreting polls conducted by others. It is a book for anyone—candidates, state and local officials, community organizations, church groups, labor organizations, public policy research centers, and coalitions focusing on specific economic issues—interested in measuring public opinion.

1987. x, 166 pp., bibliography, appendix, index.
Paper, ISBN 0-933280-32-7. **$19.95**
Companion software now available.

Green Fields Forever: The Conservation Tillage Revolution in America
By Charles E. Little

"*Green Fields Forever* is a fascinating and lively account of one of the most important technological developments in American agriculture. . . . Be prepared to enjoy an exceptionally well-told tale, full of stubborn inventors, forgotten pioneers, enterprising farmers—and no small amount of controversy."—Ken Cook, World Wildlife Fund and The Conservation Foundation.

Here is the book that will change the way Americans think about agriculture. It is the story of "conservation tillage"—a new way to grow food that, for the first time, works *with,* rather than against, the soil. Farmers who are revolutionizing the course of American agriculture explain here how conservation tillage works. Some environmentalists think there are problems with the methods, however; author Charles E. Little demonstrates that on this issue both sides have a case, and the jury is still out.

1987. 189 pp., illus., appendixes, index, bibliography.
Cloth, ISBN 0-933280-35-1. **$24.95**
Paper, ISBN 0-933280-34-3. **$14.95**

Federal Lands: A Guide to Planning, Management, and State Revenues
By Sally K. Fairfax and Carolyn E. Yale

"An invaluable tool for state land managers. Here, in summary, is everything that one needs to know about federal resource management policies." — Rowena Rogers, President, Colorado State Board of Land Commissioners.

Federal Lands is the first book to introduce and analyze in one accessible volume the diverse programs for developing resources on federal lands. Offshore and onshore oil and gas leasing, coal and geothermal leasing, timber sales, grazing permits, and all other programs that share receipts and revenues with states and localities are considered in the context of their common historical evolution as well as in the specific context of current issues and policy debates.

1987. xx, 252 pp., charts, maps, bibliography, index.
Paper, ISBN 0-933280-33-5. **$24.95**

Hazardous Waste Management: Reducing the Risk
By Benjamin A. Goldman, James A. Hulme, and Cameron Johnson for the Council on Economic Priorities

Hazardous Waste Management: Reducing the Risk is a comprehensive sourcebook of facts and strategies that provides the analytic tools needed by policy makers, regulating agencies, hazardous waste generators, and host communities to compare facilities on the basis of site, management, and technology. The Council on Economic Priorities' innovative ranking system applies to real-world, site-specific evaluations, establishes a consistent protocol for multiple applications, assesses relative benefits and risks, and evaluates and ranks ten active facilities and eight leading commercial management corporations.

1986. xx, 316 pp., notes, tables, glossary, index.
Cloth, ISBN 0-933280-30-0. **$64.95**
Paper, ISBN 0-933280-31-9. **$34.95**

An Environmental Agenda for the Future
By Leaders of America's Foremost Environmental Organizations

". . . a substantive book addressing the most serious questions about the future of our resources." — John Chafee, U.S. Senator, Environmental and Public Works Committee. "While I am not in agreement with many of the positions the authors take, I believe this book can be the basis for constructive dialogue with industry representatives seeking solutions to environmental problems." — Louis Fernandez, Chairman of the Board, Monsanto Corporation.

The chief executive officers of ten major environmental and conservation organizations launched a joint venture to examine goals that the environmental movement should pursue now and into the twenty-first century. This book presents policy recommendations for implementing the changes needed to bring about a healthier, safer world. Topics discussed include nuclear issues, human population growth, energy strategies, toxic waste and pollution control, and urban environments.

1985. viii, 155 pp., bibliography.
Paper, ISBN 0-933280-29-7. **$9.95**

Water in the West
By Western Network

Water in the West is an essential reference tool for water managers, public officials, farmers, attorneys, industry officials, and students and professors attempting to understand the competing pressures on our most important natural resource: water. Here is an in-depth analysis of the effects of energy development, Indian rights, and urban growth on other water users.

1985. *Vol. III: Western Water Flows to the Cities*
v, 217 pp., maps, table of cases, documents, bibliography, index.
Paper, ISBN 0-933280-28-9. **$25.00**

These titles are available directly from Island Press, Box 7, Covelo, CA 95428. Please enclose $2.75 shipping and handling for the first book and $1.25 for each additional book. California and Washington, DC residents add 6% sales tax. A catalog of current and forthcoming titles is available free of charge. Prices subject to change without notice.